Photographic Atlas of Botany
and
Guide To Plant Identification

By
James L. Castner

Department of Biology
Pittsburg State University
Pittsburg, Kansas   66762

Published by

Feline Press
P.O. Box 357219
Gainesville, FL 32635   USA
e-mail: jlcastner@aol.com
www.felinepress.com

Photographic Atlas of Botany
and
Guide To Plant Identification

By
James L. Castner

Published by:
Feline Press
P.O. Box 357219
Gainesville, FL
32635   USA

© 2004 by Feline Press, Inc.
Fourth printing 2010
Printed in China
ISBN 0-9625150-0-0

Library of Congress
Control Number:
2004099232

Preface

I created this photo atlas and identification guide for botany at the urging of professors and students who used my entomology photo atlas. The widespread acceptance and success of that work convinced me that a similar publication treating botany and plant taxonomy would be equally valuable. Although several photo-oriented botany aids already existed, their coverage lacked thoroughness and the quality of the photographs used was greatly inconsistent. In most cases, it appeared as if the authors had gone through their photo collections and then constructed a study aid around them. This guide was created with one purpose in mind. To be used as a tool by students enrolled in General Botany, Plant Taxonomy, and related botanical courses. Almost every photograph included in this book was taken to serve as an illustration for this guide and nothing else. In this way I have endeavored to eliminate photos of subjects casually arranged on colored paper and with glaring shadows that make it nearly impossible to see all but the largest of anatomical features. On the contrary, whenever possible the photos included here were taken under studio conditions with consistent lighting and close-up lenses that permit a clear representation of difficult to see features. To further clarify the identification and labeling of structures, the background was often 'dropped out' and replaced with a neutral gray or dark blue.

I would have enjoyed traveling all over the country for years throughout the seasons to obtain the needed photos for this book, but expediency demanded otherwise. Therefore, approximately two years were spent, primarily in the southeastern United States, which is reflected in the species used in the illustrations. However, the familial characters used in identification do not change with locality, and this guide will help students regardless of their geographical location.

Some botany courses are taught without many living plant specimens or cuttings available for study. Temperate institutions without an abundance of greenhouse space must often make do with herbarium specimens and/or limited local specimens. This guide will be a boon to students in such classes. Even when large numbers of living plants are available for laboratory sessions, the use of this book will afford students greater independence and hopefully result in more efficient use of their time. It will certainly aid them in remembering physical characteristics and anatomical details by providing a photographic reference to consult even when they are out of the laboratory.

The use of this guide also provides the potential for a greater degree of consistency in how a general botany laboratory or plant taxonomy course is taught. It is common practice in many universities for the laboratory sections of courses to be taught by graduate students or teaching assistants. One lab section of the same course may therefore vary greatly from the next, resulting in a difference in the quality and exposure students receive to the subject matter. Use of this guide will assure that certain information will be made available to each and every student.

The greatest difficulty I faced in preparing this publication was in how to create a book that both botanists following a traditional classification philosophy and those using a cladistic or phylogenetic approach could use. I have usually split families into more narrowly construed taxa. When there is disagreement over a family's recognition as distinct, I have tried to note that. I shall leave the individual professors and instructors to follow the method they feel is best. Whichever they choose, it is hoped that they shall find the thousands of illustrations contained in this book as a useful instructional aid.

Sincerely,   *James L. Castner*   jlcastner@aol.com   www.felinepress.com

Acknowledgments

Many people and institutions contributed in a variety of ways to the production of this book. Academic colleagues provided their expertise, while botanical gardens, nurseries, growers, farmers, and plant enthusiasts in general afforded me access to their plants. I can honestly say that I have never met so many helpful and generous people as while working on this project. Although it is impossible to list every single person with whom I have had contact, certain individulas need to be recognized for their contributions.

I am extremely fortunate in having two botanist friends that represent both the traditional train of taxonomic thought, as well as the more recent cladistic philosophy based on phylogenetics as determined by DNA studies. Stephen Timme, who has taught botany for over 15 years, patiently answered my questions at all hours of the day and night. (I'm sure he now has caller-ID.) He might have easily acted in a 'territorial' manner, but instead chose to serve as a mentor and gave his sincere help in guiding me to make this book a valuable teaching tool.

Richard Abbott, currently a doctoral student of botany, maintains an encyclopedic knowledge of plant taxonomy along with the diagnostic skills of a field taxonomist extraordinaire. His passion for plants and seemingly limitless knowledge of all subjects botanical reminds me of Al Gentry. It was my good fortune that my work on this project coincided with Richard conducting floristic surveys throughout the state of Florida. As a result, I was the constant recipient of plants, cuttings, or exact locality data leading me to species on my photographic 'hit list'. In addition, Mr. Abbott provided positive identifications for my own collections and specimens, as well as authoritative answers to the endless e-mail queries I would pose. His review of the manuscript greatly enhanced its accuracy and overall usefulness to students.

I would also like to offer individual thanks to Don Goodman, Director and moving force behind Kanapaha Botanical Gardens in Gainesville, Florida. For as long as I have known him, Don has happily provided answers to any questions on plants or natural history. As director, he allowed me to not only go off trail to photograph while roaming the garden grounds, but also to take cuttings or even entire plants back to my home studio for close-up work. The most pleasant aspect of this project was walking through Kanapaha Botanical Gardens on a regular basis. It is the prettiest and most tranquil place associated with Gainesville, and the cheerful supervision of Don Goodman is reflected in the quality of his employees and staff of dedicated volunteers.

I give my appreciation to Walter Judd, who also kindly answered many botanical questions. By permitting me to sit in on his Plant Taxonomy classes, I was able to envision exactly what form I wanted this photographic atlas to take. I often consulted the book *Plant Systematics* by he and his colleagues.

I would like to thank Barbara Carlsward for providing expertise on the identification of tissues pictured in photomicrographs throughout this book. Her corrections and attention to detail allowed me to include such illustrations with a high degree of confidence in their accuracy.

My thanks to Barry Davis who helped in plant identifications based on his floristic survey of Kanapaha Botanical Gardens.

I would like to offer my thanks and appreciation to the following growers, nurseries, botanical gardens, and individuals: Russell Adams of Gainesville Tree Farm, Ann Bigelow of the Department of Recreation and Parks of the City of Gainesville, Eric Bjerregaard of Boondox Tropicals, Dave Chiapinni of Chiapinni Nursery, Patty Clendenin, Ron Dieterman of the Atlanta Botanical Garden, David Dilcher, Marjorie DeMartino, Nancy Edmondson of the Marie Selby Botanical Garden, Hardy Eshbaugh, Don Evans and Fairchild Tropical Garden, Garden Gate Nursery, David Hall, Harmony Gardens, Cathleen Kabat, Paul Lyrene, the Montgomery Botanical Center, Richard Moyroud of Mesozoic Landscapes, Al Muzzell of New World Bromeliads, Larry and Angela Neal of Rainbow Nursery and Ponds, Jim Notestein, Peter Raven, Joyce Raye, Rogers Farm, April Rosnik of the Fruit and Spice Park, Larry Schokman of the Kampong, Sherry Seabrook of Seabrook Nursery, Alan Shapiro of Grandiflora Nursery, Sheffield's Nursery, Ralph Thomson, Claire Tingling, Reggie Whitehead, Mark Whitten, and Norris Williams.

I am greatly indebted to Carolina Biological Supply for generously providing microscope slides of botanical tissues and structures. Access to this material was of incalculable value.

Photo Credits

One of the reasons I decided to create the *Photographic Atlas of Botany and Guide to Plant Identification* was because similar attempts by others had resulted in publications with extremely inconsistent quality of photographs. I have provided the majority of the illustrations in an attempt to maintain high and consistent quality, contributing over 2,000 of the total photos that appear in this book. However, with realistic constraints to time and budget, I also had to rely on friends and colleagues to fill photographic 'holes' that existed. Several people generously provided slides or photos of various subject matter that allowed me to present a much more thorough coverage for this work. I have placed the photographer's name next to the photograph in a credit line, but would also like to acknowledge them here as well. My thanks and appreciation to: Richard Abbott, Robin Giblin-Davis, Bill Howard, Steven Madigosky, Stephen Timme.

I would also like to thank Ralph Thomson who took time from his busy schedule to collect plant material for me from Kentucky and send it to me, and to Bill Howard and Steve Madigosky who photographed particular plants for me. The absence of any of these images is not a reflection of quality, but of space availability in this book.

Photography accounted for only a portion of preparing the illustrations in this book. Due to the digital nature of the majority of the photos, color control, background manipulation, and labeling were also of major importance. I would like to thank Pat Payne whose knowledge of Adobe Photoshop far exceeds my own, and who willingly offered his expertise and suggestions on many aspects of image preparation. I would also like to thank Steve Orf of All Systems Colour who exercised remarkable patience in guiding me through various steps of digital photo treatment. The presentation of accurate color in the photos is largely responsible due to the color correction curve that he created for this publication.

Other Titles By The Author

A Field Guide To Medicinal And Useful Plants Of The Upper Amazon. 1998. James L. Castner, Stephen L. Timme, and James A. Duke. Feline Press. Gainesville, FL.

The Amazon Rainforest - An Exploration Of Countries, Cultures, And Creatures - A Learning Center For Secondary School Students. 1999. James L. Castner. Feline Press. Gainesville, FL.

Explorama's Amazon - A Journey Through The Rainforest Of Peru. 2000. James L. Castner. Feline Press. Gainesville, FL.

Photographic Atlas Of Entomology And Guide To Insect Identification. 2000. James L. Castner. Feline Press. Gainesville, FL.

Amazon Insects - A Photo Guide. 2000. James L. Castner. Feline Press. Gainesville, FL.

Forensic Insect Identification Cards. 2000. James L. Castner and Jason H. Byrd. Feline Press. Gainesville, FL.

Shrunken Heads - Tsantsa Trophies And Human Exotica. 2002. James L. Castner. Feline Press. Gainesville, FL.

Forensic Entomology - The Utility Of Arthropods In Legal Investigations. 2000. Jason H. Byrd and James L. Castner (Editors). CRC Press. Boca Raton, FL.

Deep In The Amazon - Surviving In The Rainforest
Deep In The Amazon - Partners And Rivals
Deep In The Amazon - Layers Of Life
Deep In The Amazon - River Life
Deep In The Amazon - Native Peoples
Deep In The Amazon - Rainforest Researchers
2002. James L. Castner. Benchmark Books, Marshall Cavendish Corp. Tarrytown, N.Y.

Rainforests - A Guide To Research And Tourist Facilities At Selected Tropical Forest Sites In Central And South America. 1990. James L. Castner. Feline Press. Gainesville, FL. (Out of Print)

For more information on any of the above titles, contact:

Dr. James L. Castner
c/o Feline Press
P.O. Box 357219
Gainesville, FL 32635 USA
E-mail: jlcastner@aol.com
www.felinepress.com

Table of Contents

Table of Contents

TAXONOMY

Table of Contents

TAXONOMY

Table of Contents

TAXONOMY

Table of Contents

TAXONOMY

Table of Contents

TAXONOMY

Flowering Plants - *Angiosperms*
Core Eudicots

Table of Contents

TAXONOMY

Flowering Plants - *Angiosperms*
Monocots

Table of Contents

TAXONOMY

Flowering Plants - *Angiosperms*
Monocots
Order Zingiberales

Introduction

This book is divided into two primary sections. The first covers Plant Anatomy and will familiarize you with both the basic and specialized structures found on most plants. It will also introduce you to the terminology necessary for describing and discussing plants. Initial contact with the language of botany is like beginning to study any other foreign language. You start with learning the basic vocabulary. In this case, we will start with the descriptive terms that pertain to the vegetative, floral, and fruiting structures of the flowering plants, also know as the angiosperms. The remaining groups of vascular plants and their specific terminology will be discussed in separate sections.

If we look at a flowering plant we will almost always find three specific vegetative features. These are the **roots**, **stems**, and **leaves**. Remember, there are always exceptions, and sometimes one or more of these features may be reduced, absent, or not readily visible. This is especially true in species modified for a specialized environment, such as aquatics, epiphytes, and drought-adapted species. However, in the majority of angiosperm species these characters are large and obvious. Fertile plant material will also show either **flowers** or **fruits** which may be essential in the identification process. The anatomy section covers these five categories in detail, providing both detailed photographs of whole specimens as well as photomicrographs of important microscopic structures and tissues.

The second section which makes up the majority of this book is devoted to Plant Taxonomy. It provides a description and photographs of 153 of the most commonly studied families of plants including Seedless Vascular Plants, Non-Flowering Seed Plants, and Flowering Plants. For all angiosperm families the following characters are listed and described: **Habit, Leaves, Inflorescences, Flowers, Calyx, Corolla, Stamens, Carpels, Ovary, Fruit,** and **Genera**. In some cases a note about the taxonomic status of the family or other significant information is provided. The last page of the book provides a quick-reference listing of all the families covered and the pages on which they can be found.

Standard abbreviations have been used in most cases in the figure captions in order to save space. Therefore LS = Longitudinal Section, XS = Cross Section, and CU = Close Up. Captions typically appear below the image they describe.

This publication is primarily concerned with identification to the family level. In the Plant Taxonomy section, the name of the family being described appears on each page. The name of the order in which that family is placed appears in the upper left corner of the first page of the description. In the upper right corner of any page is the broad group of plants to which the family under consideration belongs.

In some introductory sections cladistic terminology has been used in an effort to clarify plant relationships. However, it is not the purpose of this book to define cladistic philosophy and how it relates to plant taxonomy. For such information and a discussion of cladistic theory I suggest the reader consult a text such as *Plant Systematics - A Phylogenetic Approach* by Judd et al.

Plant Anatomy

ROOTS

Functions

Anchorage - Roots anchor plants firmly in the ground and provide support for vertical growth.
Absorption - Water and nutrients such as minerals are absorbed by roots and transported through the plant to the locations where they are needed.
Storage - Many plant species use roots as a storage organ, primarily for carbohydrates.
Growth - Dividing cells in the root cap are responsible for the elongation of the root (primary growth).

Terminology

adventitious root - Root or root system that does not grow from other roots or root tissue.
aerial root - Root or root system growing above ground or water.
branch root - Smaller, laterally-growing root that branches off to the side of the main taproot.
buttress root - Large, flattened, fin-like extension of the trunk base of some trees.
contractile root - Anchoring roots that pull the plant down deeply into the soil.
fibrous root - Root or root system characterized by many branching roots that are all of approximately the same size.
flank root - Large, flattened fin-like extension of the trunk base of some trees.
haustorial root - The invasive root of some parasitic plants that penetrates the vascular system of their hosts.
lateral root - Smaller, laterally-growing root that branches off to the side of a main taproot.
mycorrhizal root - Root in or near which symbiotic fungi occur.
pneumatophore - Specialized aerial root involved in gas exchange that is characteristic of certain species of mangrove.
primary root - The large main downward-growing root(s) in a root system.
prop root - Adventitious root growing from the base of the stem or trunk that functions to support the plant.
root hair - Single-cell extension of the growing root tip that increases the surface area in contact with the soil.
root nodule - Rounded swelling or tubercle found on the roots of some legumes and other plants in which nitrogen-fixing bateria occur.
secondary root - The smaller lateral-growing roots that branch off the primary root(s).
storage root - Large, swollen root that serves as a storehouse for plant carbohydrates.
taproot - Root or root system characterized by one large, downward-growing main root that is significantly larger than the lateral roots growing in association with it.

Root Types

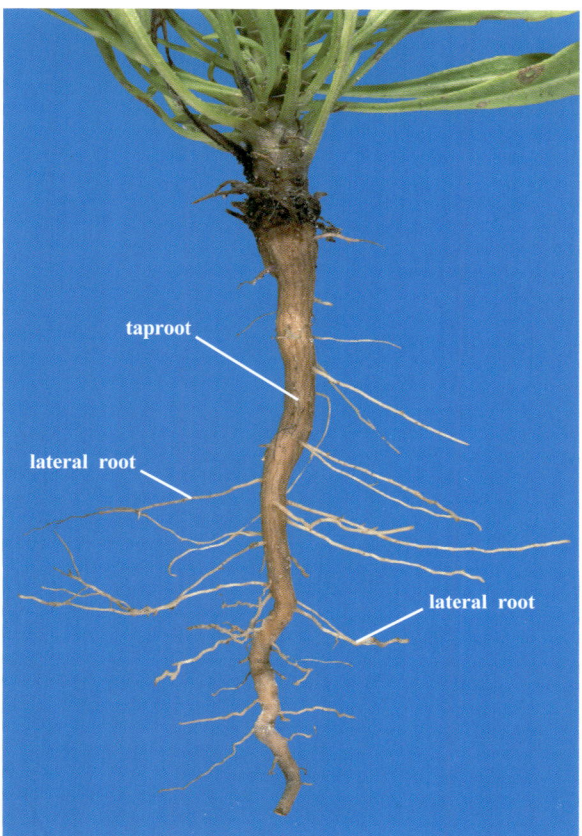

Fig. 1 Horseweed is an aster with a taproot system. There is one primary root that grows downward and smaller lateral or secondary roots branching from it. *Conyza canadensis*   Family Asteraceae

Fig. 2 The carrot is a taproot that has been modified for storage and is thus much bigger than the lateral roots growing from it. *Daucus carota*   Family Apiaceae

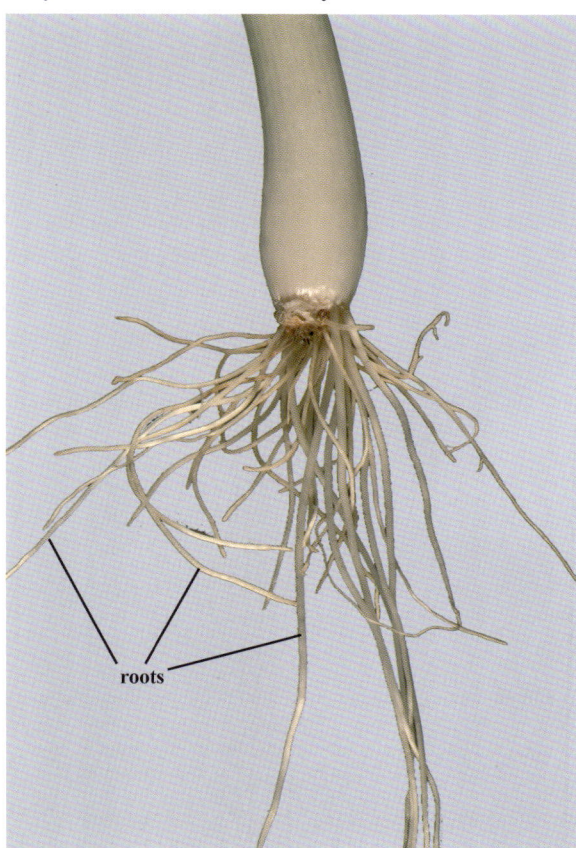

Fig. 3 Monocots like onions have a fibrous root system with many roots approximately the same size. *Allium cepa*   Family Alliaceae

Fig. 4 Cattails are also monocots and have fibrous root systems as seen here. *Typha domingensis*   Family Typhaceae

Root Types

Fig. 5   Adventitious roots on the stem of a trumpet creeper. *Campsis radicans*   Bignoniaceae

Fig. 6   Prop roots at the base of a corn plant. *Zea mays*   Poaceae

Fig. 7   Prop roots at the base of a screw pine tree. *Pandanus* sp.   Pandanaceae

Fig. 8   Prop roots of the red mangrove tree. *Rhizophora mangle*

Fig. 9   Pneumatophores growing in wet area at base of mangroves. *Rhizophora mangle*

Fig. 10   Huge buttress roots of tropical silk cotton tree. *Ceiba pentandra*

Fig. 11   Nodules of nitrogen-fixing bacteria growing along the roots of a leguminous plant.   Fabaceae

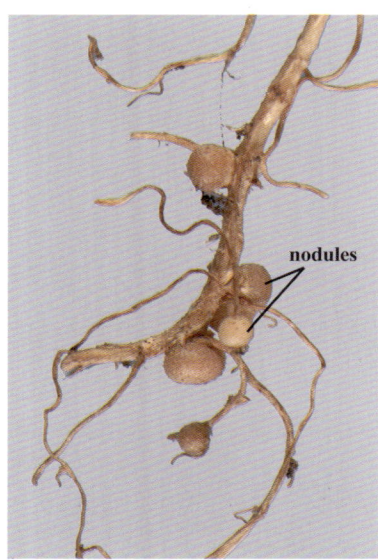

Fig. 12   Close-up of root nodules with nitrogen-fixing bacteria. Fabaceae

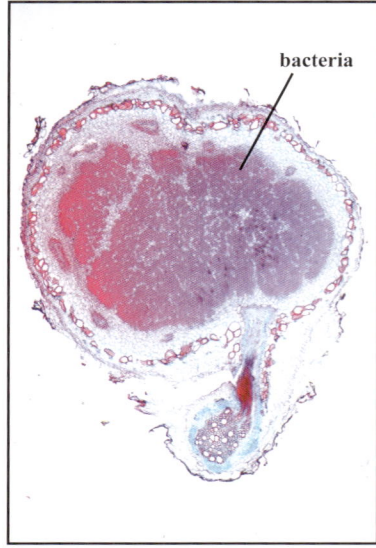

Fig. 13   LS of root nodule with nitrogen-fixing bacteria. Fabaceae

Root Microscopic Anatomy

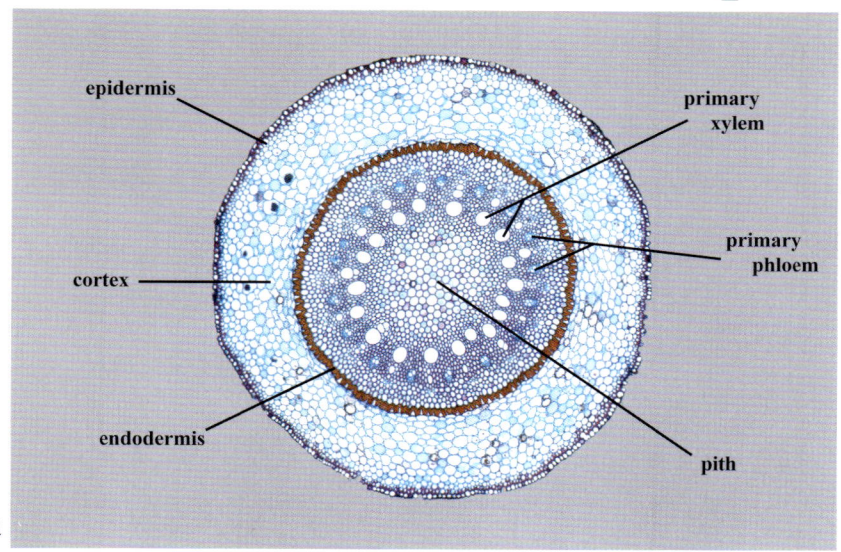

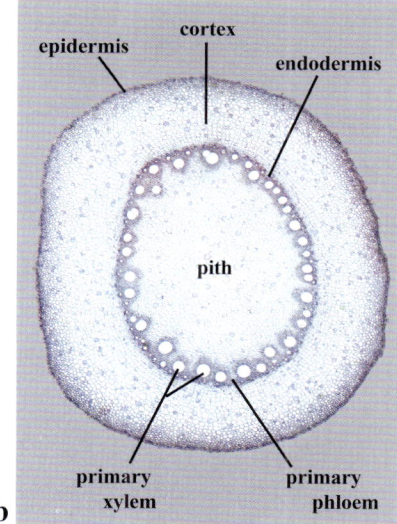

Fig. 14   Monocot roots in XS with major tissues identified.
a) *Smilax* sp.   b) *Zea mays*

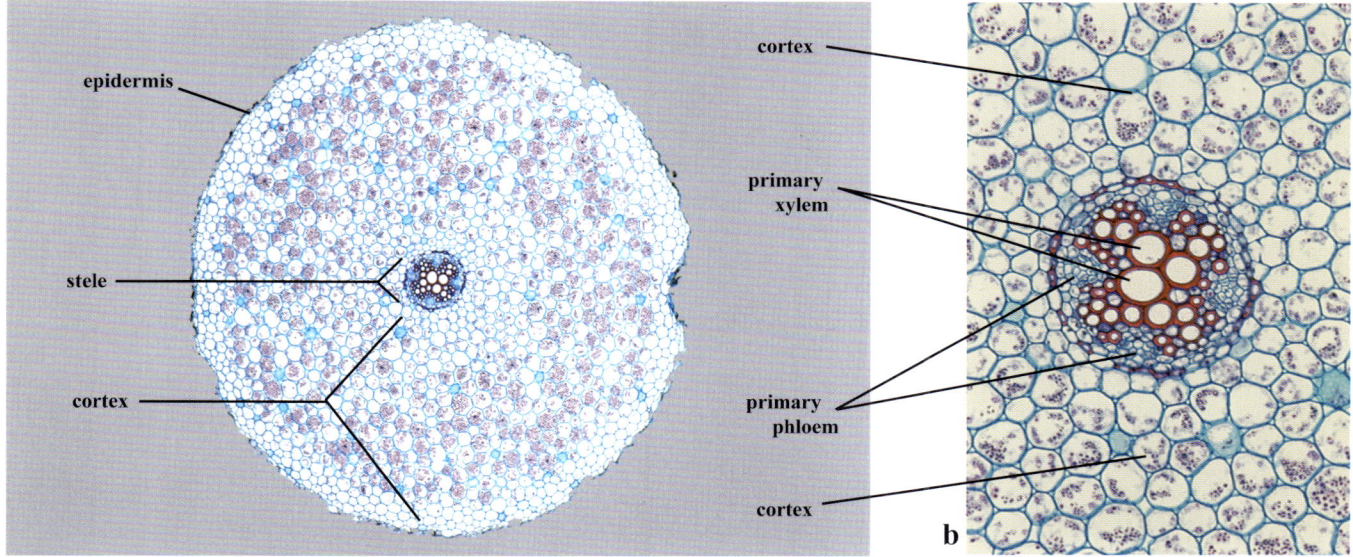

Fig. 15   Eudicot root in XS with major tissues identified.
a) *Ranunculus* sp. in overall view.   b) *Ranunculus* sp. with stele area magnified.

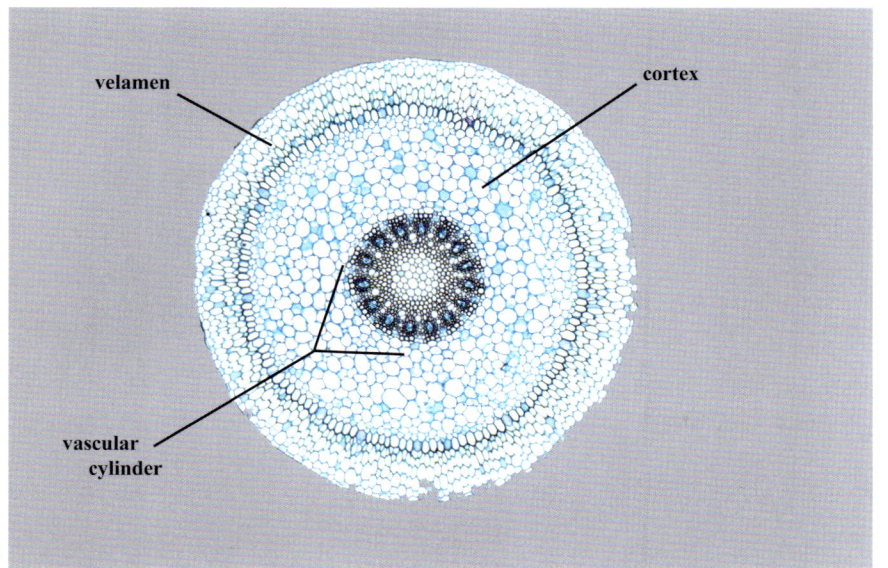

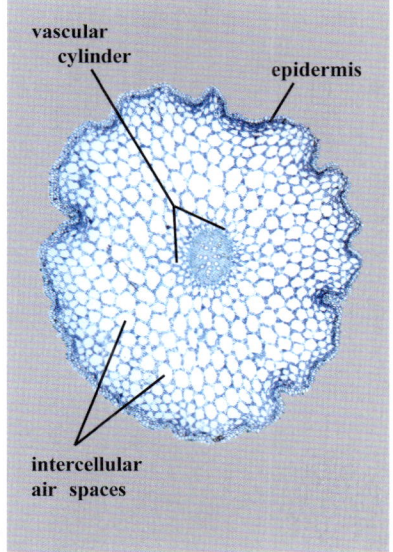

Fig. 16   Aerial and aquatic roots in XS. a) Aerial root of orchid with major tissues identified.
b) Aquatic root of water lily.   *Nymphaea* sp.

STEMS

Functions

Support - Stems provide support and a point of attachment for leaves, flowers and fruit, optimizing their position for exposure to sunlight, and pollination and dispersal, respectively.

Transport - The presence of a vascular system within the stem provides for the conduction of water and dissolved minerals from the roots upward, and transport of carbohydrates produced in the leaves downward and throughout the plant. Via the stem, all parts of the plant are connected to one another.

Growth and Tissue Production - Via cell division in specialized areas, the stem grows and produces essential tissues and structures such as buds, leaves, flowers, and fruits.

Storage - Specialized stems, usually underground, serve as storage reservoirs for carbohydrates and other nutrients.

Photosynthesis - Many stems, especially when young, have tissues with chloroplasts that contribute to the plant's total carbon balance. Other specialized stems may be leaf-like in shape or carry out photosynthesis in plants where leaves are greatly reduced or absent.

Terminology

acaulescent - When a stem is not present or apparent.

aerial stem - Stem that grows above ground.

bud - Embryonic stem, often covered by some protective structure for protection.

bulb - Short, flattened underground stem surrounded by layers of thick fleshy modified leaves (scales).

cladophyll - Photosynthetic stem that is flattened and resembles a leaf.

corm - Short, vertical underground stem surrounded by thin papery leaves.

herbaceous - Stem or plant that is soft and without woody tissue.

internode - Area of a stem between two consecutive nodes.

leaf - Typically flat photosynthetic structure produced from the nodes of stems and branches of plants.

liana - Woody vine.

long shoot - Stem or branch characterized by long internodal areas, when both long and short shoots are present on the same plant.

node - Area of a stem from which leaves or buds are produced.

prickle - Sharp superficial outgrowth of epidermal plant tissue that may be found anywhere on the plant.

rhizome - Horizontal underground stem with scale-like leaves.

runner - Elongated horizontal stem that grows on the surface of the ground and that roots and produces new plants at the nodes and at the tip.

shoot - Young stem or branch.

short shoot - Stem or branch characterized by short internodal areas, when both long and short shoots are present on the same plant.

spine - Sharp, modified leaf (or portion of a leaf) or stipule that arises at the node.

stolon - Elongated horizontal stem that grows on the surface of the ground and that roots and produces new plants at the nodes and at the tip.

subterranean stem - Underground stem.

tendril - Modified stem or leaf that is used for attachment and climbing.

thorn - Sharp, modified branch that arises at the node from the leaf axil or at the branch tip.

trunk - Main stem of a tree.

tuber - Short, fleshy underground stem or thickened portion of a rhizome.

twig - Small branch or shoot from a tree.

vine - Herbaceous plant with a stem that trails on the ground or climbs.

woody - Hard stem or plant formed by the presence of secondary xylem.

Fig. 17 Stem features of a peperomia plant.
Peperomia obtusifolia　　　Piperaceae

Fig. 18 Stem features of a camellia.
Camellia japonica　　　Theaceae

Fig. 19 Twig features of boxelder.
Acer negundo　　　Aceraceae

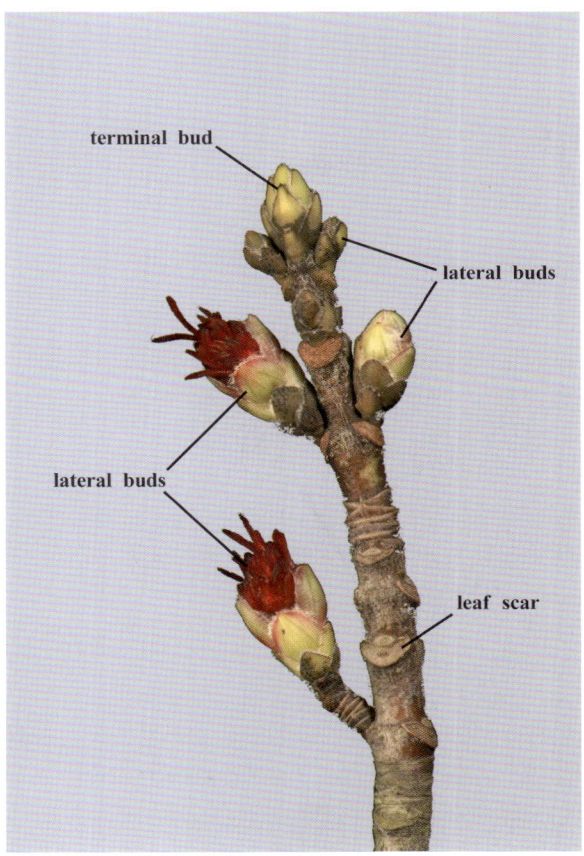

Fig. 20 Twig features of red maple.
Acer rubrum　　　Aceraceae

Buds - Leaf Scars

Imbricate Bud

scales

Fig. 21 Bud with imbricate scales.
Quercus sp.　　Fagaceae

Imbricate Bud

scales

Fig. 22 Bud with imbricate scales.
Camellia japonica　　Theaceae

Imbricate Bud

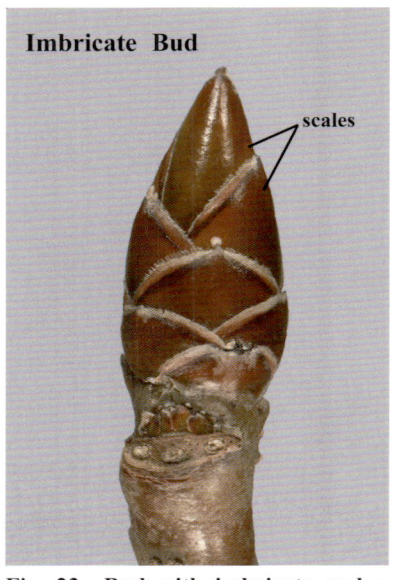

scales

Fig. 23 Bud with imbricate scales.
Liquidambar styraciflua
Altingiaceae

Valvate Bud

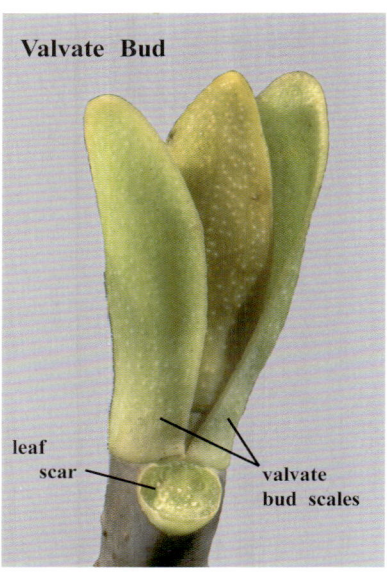

leaf
scar

valvate
bud scales

Fig. 24 Bud with valvate scales.
Liriodendron tulipifera
Magnoliaceae

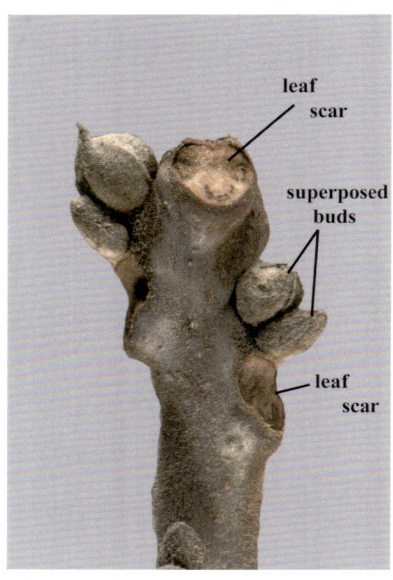

leaf
scar

superposed
buds

leaf
scar

Fig. 25 Buds of black walnut.
Juglans nigra　　Juglandaceae

**Appressed
Bud**

Fig. 26 Bud of river birch.
Betula nigra　　Betulaceae

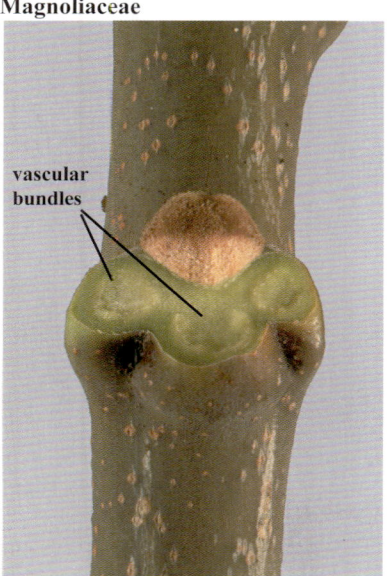

vascular
bundles

Fig. 27 Leaf scar of chinaberry.
Melia azedarach　　Meliaceae

vasc.
bundle

Fig. 28 Leaf scar of black walnut.
Juglans nigra　　Juglandaceae

Fig. 29 Leaf scar of ash.
Fraxinus sp.　　Oleaceae

Fig. 30 THORNS are modified branches and therefore occur at nodes and arise from an axillary position. There is usually a leaf or leaf scar beneath a thorn. a) *Citrus aurantifolium* Rutaceae b) *Citrus aurantifolium* Rutaceae

Fig. 31 SPINES are modified leaves or stipules and therefore occur at nodes. They may have a bud or modified portion of a leaf above them. a) *Opuntia* sp. Cactaceae b) *Berberis julianae* Berberidaceae

Fig. 32 PRICKLES are modified hairs and may occur anywhere on a plant, not just at the nodes.
a) *Rosa* sp. Rosaceae b) *Smilax bona-nox* Smilacaceae c) *Smilax bona-nox* Smilacaceae

Rhizomes - Bulbs - Corms

Fig. 33 RHIZOMES are enlarged horizontal stems lying on or below the ground that can be adapted for vegetative reproduction. a) Intact rhizome of iris showing young vegetatively-produced plant and root. *Iris* sp. Iridaceae b) Iris rhizome in LS (top) and intact (bottom). *Iris* sp. Iridaceae

Fig. 34 BULBS are short, disk-like underground stems surrounded by fleshy, enlarged leaf bases (scale leaves) w. stored food. a) Intact sprouting onion bulb. b) LS of onion bulb. *Allium cepa* Alliaceae c) LS of amaryllis bulb. *Amaryllis* sp. Amaryllidaceae

Fig. 35 CORMS are swollen underground stems with the material surrounding the stem solid and not layered as in bulbs. a) Intact daffodil corm. b) LS of daffodil corm. *Narcissus* sp. Amaryllidaceae c) Top and bottom views of gladiolus corm. *Gladiolus* sp. Iridaceae

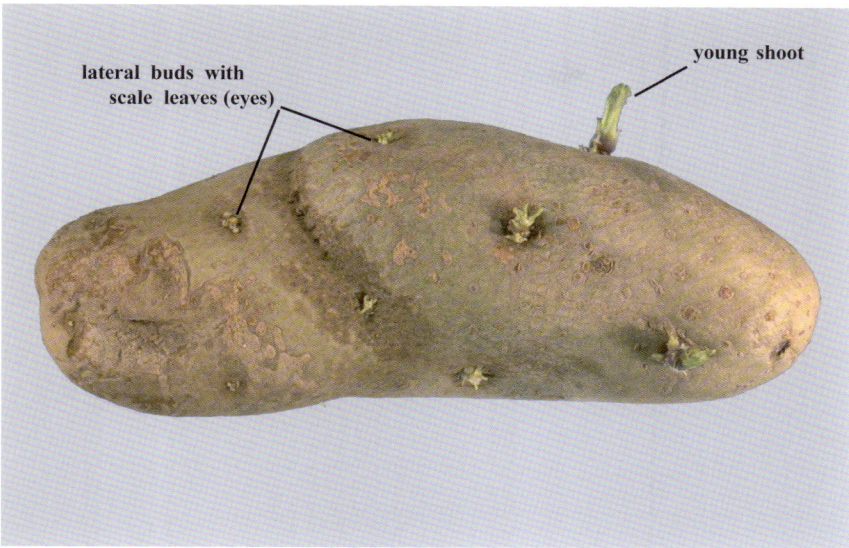

Fig. 36 Tubers are enlarged, short underground stems that store reserve food and are adapted for vegetative reproduction. a) Intact potato tuber with 'eyes' just starting to give rise to new plants. b) Potato tuber that has produced several new upright stems with roots that will form new plants. *Solanum tuberosum* Solanaceae

Fig. 37 Aerial stems have evolved in some plants to function in vegetative reproduction. The air potato plant, a yam relative, produces such structures that separate from the parent plant and are called bulbils.
a) & b) Bulbils or vegetatively-reproducing aerial stems of the air potato. *Dioscorea bulbifera* Dioscoreaceae

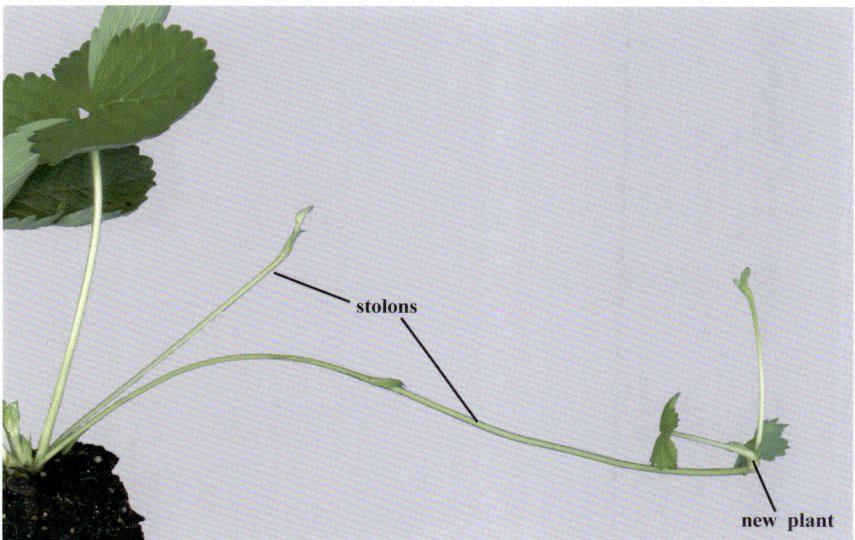

Fig. 38 Stolons (runners) are horizontal stems on or under the ground that produce new plants at their tips as in this strawberry plant. *Fragaria x ananassa* Rosaceae

Fig. 39 Leaf-like photosynthetic stem of the paddle cactus. *Opuntia* sp. Cactaceae

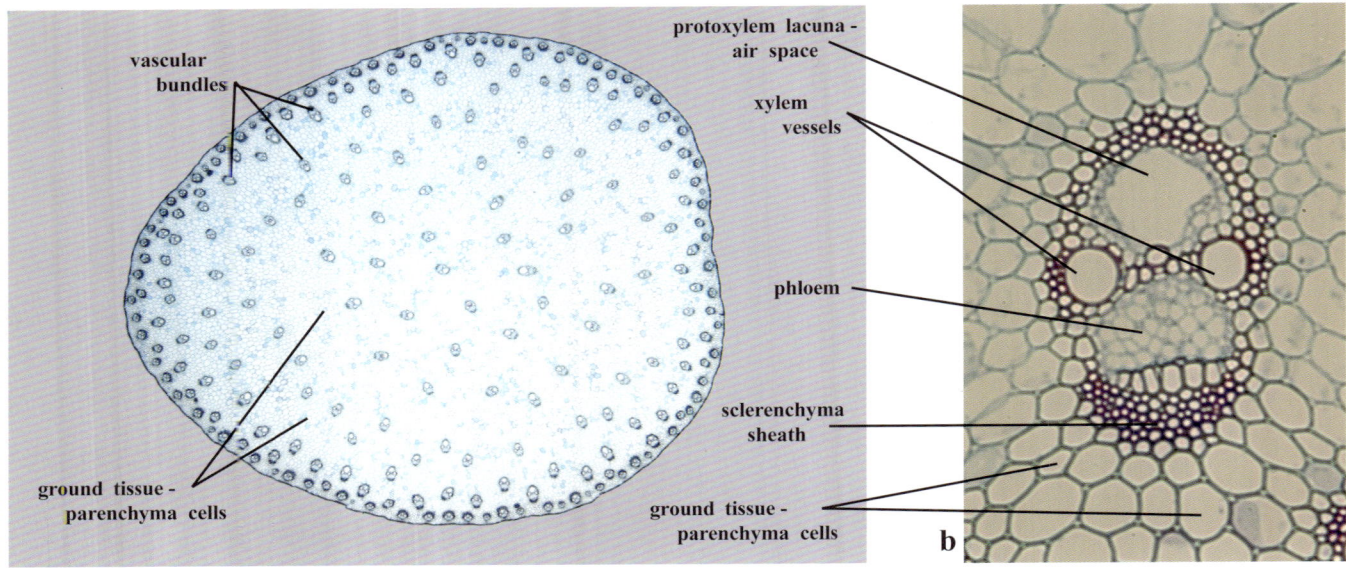

Fig. 40   a) XS of a corn (monocot) stem showing the scattered arrangement of the vascular bundles.
b) CU of a corn (monocot) vascular bundle.   *Zea mays*   Poaceae

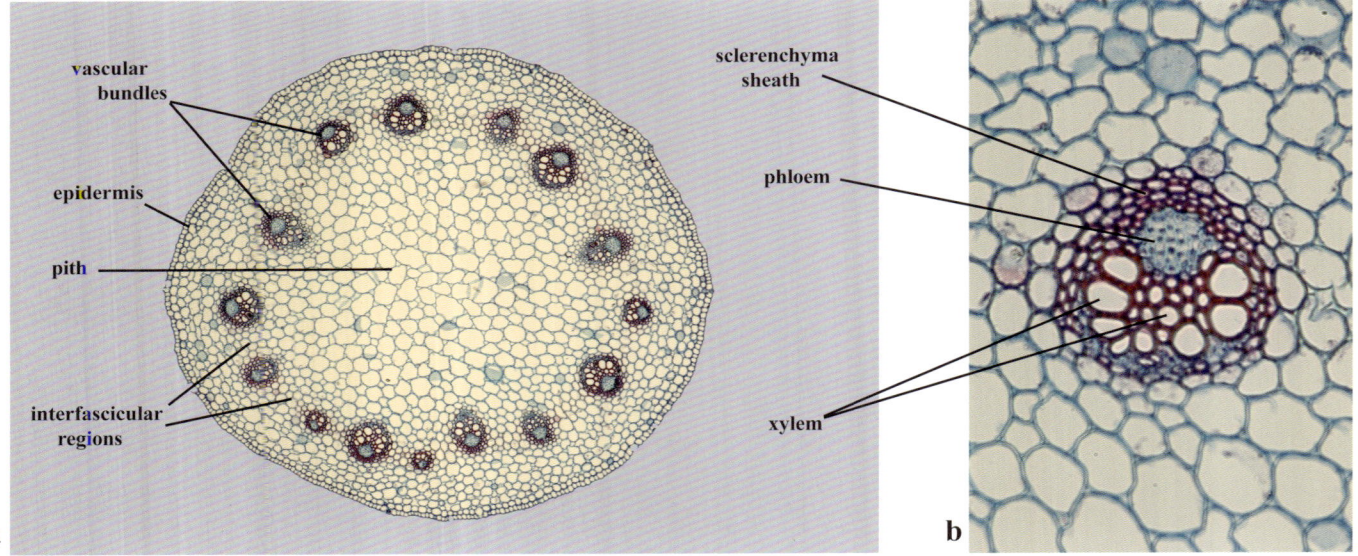

Fig. 41   a) XS of a buttercup (eudicot) stem showing the regular ring-like arrangement of the vascular bundles.
b) CU of a buttercup (eudicot) vascular bundle.   *Ranunculus* sp.   Ranunculaceae

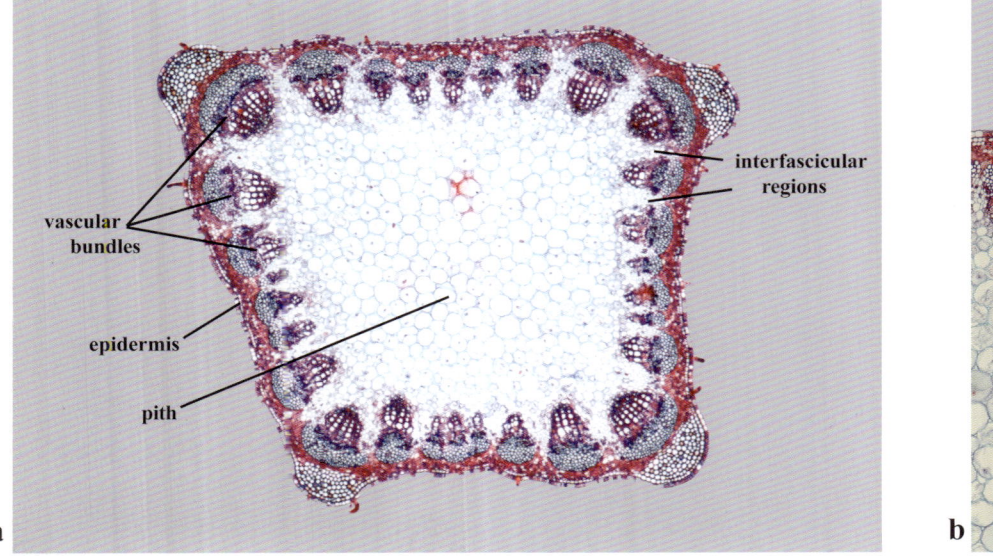

Fig. 42   a) XS of an alfalfa (eudicot) stem showing the distinctive arrangement of the vascular bundles.
b) CU of the vascular bundles of alfalfa (eudicot).   *Medicago sativa*   Fabaceae

Stem Microscopic Anatomy

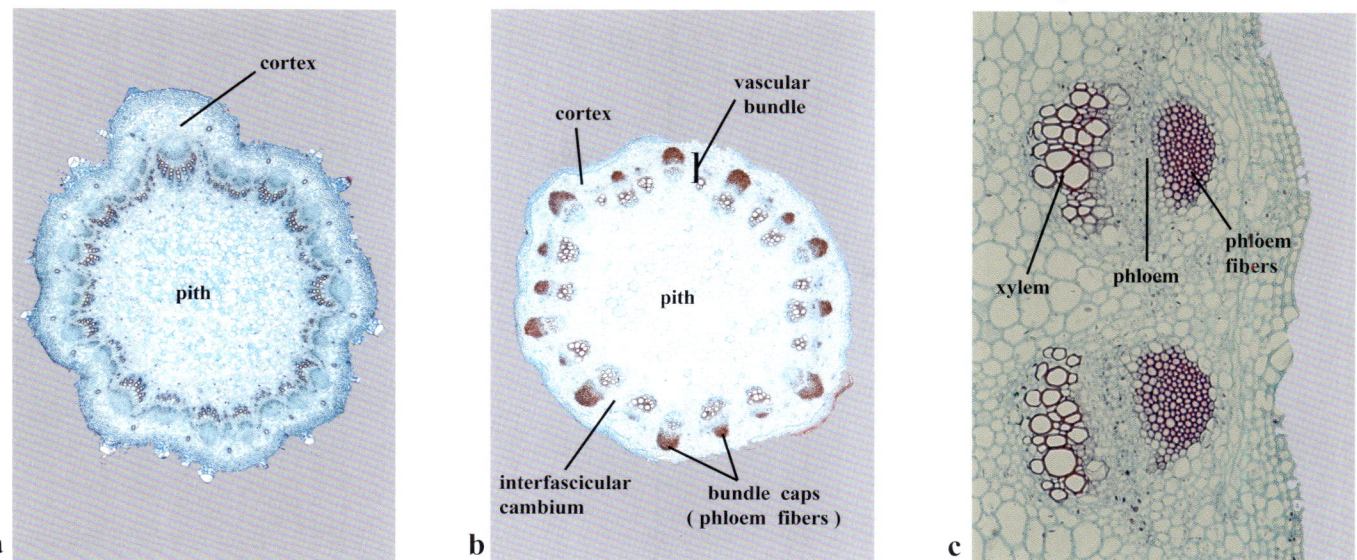

Fig. 43   Tissue organization in the stem of a sunflower (eudicot).   a) XS of young plant.   b) XS of mature plant.
c) CU of vascular bundle.   *Helianthus* sp.   Asteraceae

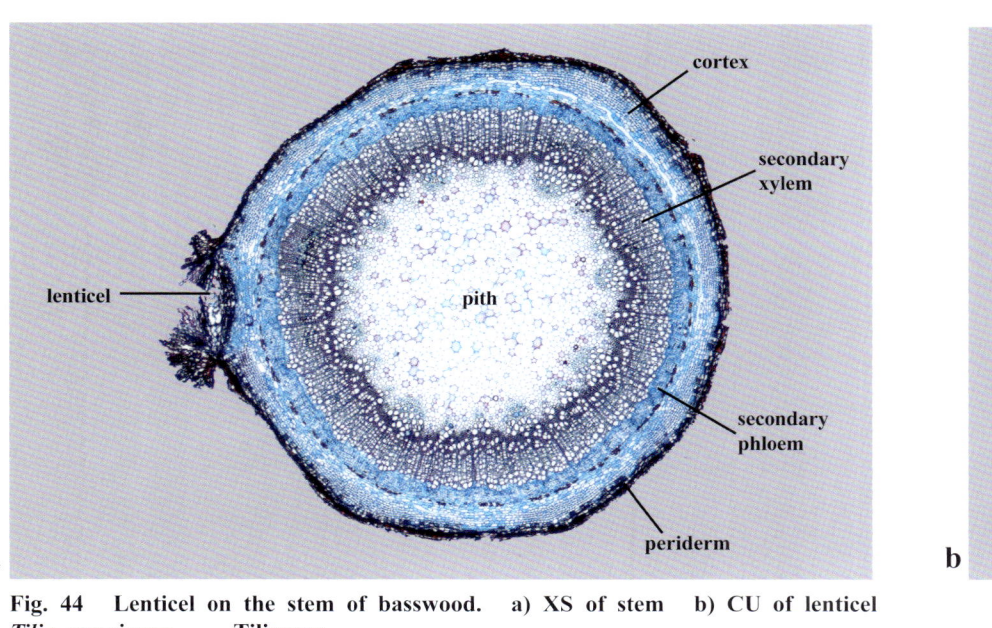

Fig. 44   Lenticel on the stem of basswood.   a) XS of stem   b) CU of lenticel
Tilia americana   Tiliaceae

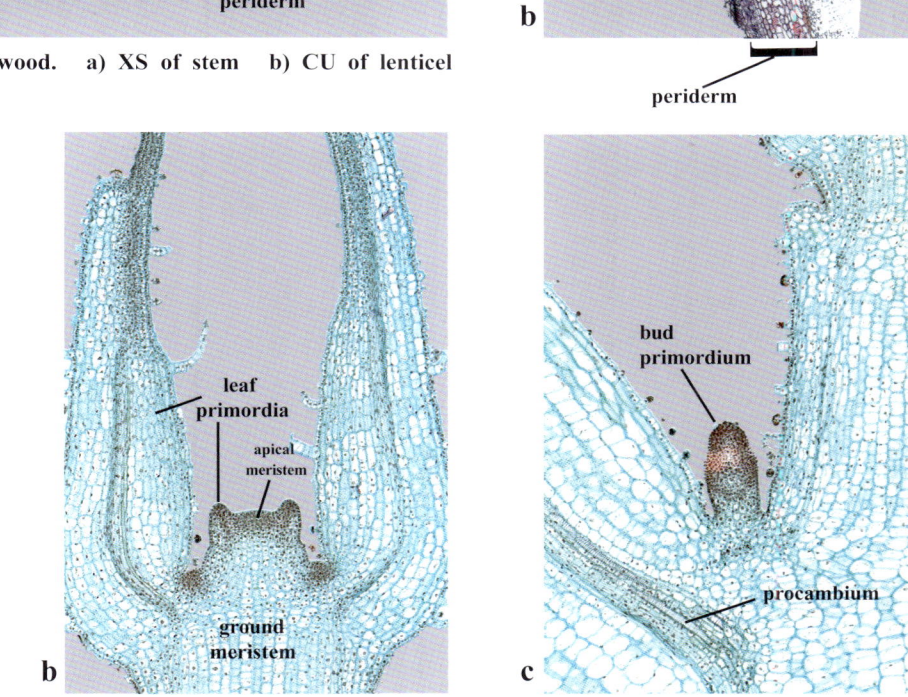

Fig. 45   a) LS of apical meristem of coleus.   b) CU of leaf primordia.   c) CU of bud primordium.
Coleus sp.   Lamiaceae

LEAVES

Functions

Photosynthesis - Light absorption is facilitated by increasing the surface area available.
Maximize CO_2 Uptake & Retention - Stomata allow diffusion of atmospheric CO_2 into plant tissues.
Minimize Water-Gas Loss - Stomata in epidermis regulate exposure of inner tissue to the environment.
Prey Capture - Plants in nutrient-poor soils have developed modified leaves to trap and digest insects.
Vegetative Propagation - Tiny plants or plantlets grow from the leaves of certain plants.
Support - Leaves modified into spines or tendrils may function to anchor stems in climbing plants.

Terminology

abaxial surface - Facing away from the stem axis, such as the underside of a leaf.
acute - Terminating in a point of less than 90 degrees.
adaxial surface - Facing towards the stem axis, such as the upper side of a leaf.
alternate - Leaf arrangement with one leaf per node, alternating from one side of the stem to the other.
blade - Flat expanded photosynthetic portion of a leaf. (Also known as the **lamina**.)
compound - Leaf divided into two or more distinct segments or **leaflets**.
crenate - Leaf margin characterized by rounded teeth or appearing scalloped.
dentate - Leaf margin that is toothed, with the teeth at right angles to the line of the margin.
elliptic - Leaf shape where the middle is wider than the base and apex. Oval in shape.
entire - Leaf margin that is continuous and lacks teeth or serrations or divisions of any kind.
heterophylly - Situation where leaves with distinctive or different shapes occur on the same plant.
leaf - Lateral outgrowth at nodes of stem, usually with a stalk and expanded flat part for photosynthesis.
lobed - Leaf margin characterized by two to many large, rounded or sharp projections.
midrib - The central and principal vein of a leaf, usually seen as a continuation of the petiole.
oblong - Leaf shape where the edges are more or less parallel-sided.
obovate - Leaf shape where the apex of the blade is broader than the base.
obtuse - Blunt. Terminating in an angle that is greater than 90 degrees.
opposite - Leaf arrangement where two leaves arise from the stem at opposite sides of the same node.
ovate - Leaf shape where the base of the blade is broader than the apex.
palmate venation - Arrangement of three or more major veins from or near the base of the blade.
palmately compound - Compound leaf with more than three leaflets extending out from a common point like the fingers on the palm of a hand.
parallel venation - Arrangement where major veins are more or less parallel to the midrib or leaf edge.
peltate - Leaf with the petiole attached at the center rather than at the edge.
petiole - Stalk of a leaf. (Some leaves are sessile, arising directly from the stem without petioles.)
petiolule - Stalk of a leaflet.
pinnate venation - Arrangement of veins from either side of the midrib like the pinnae of a feather.
pinnately compound - Compound leaf with more than three leaflets arranged in two rows from a central axis or rachis like the pinnae of a feather.
plinervy venation - Palmate venation with 3 primary veins diverging from near the base.
pulvinus - Enlarged portion of the petiole base or apex (or both), that functions in leaf movement.
reticulate venation - An arrangement of veins forming a net-like pattern. (Also called **netted venation**.)
rotate venation - Arrangement where major veins project out from center like the spokes of a wheel.
serrate - Leaf margin that is saw-toothed with the teeth pointing forward towards the apex.
simple - Leaf with a single blade, not divided into leaflets.
stipule - Typically leafy appendage occurring on the stem at the base of a petiole, often in pairs.
trifoliate - Compound leaf with three leaflets.
unifoliate - Compound leaf with one leaflet, usually with a joint between the blade and petiole.
whorled - Leaf arrangement where three or more leaves arise from the same node circling the stem.

Fig. 46 Typical leaves have two main anatomical structures: a stalk-like petiole and a flattened blade. *Pyrus calleryana*   Rosaceae

Fig. 47 Alternate leaf arrangement has one leaf per node. This branch is 2-ranked with leaves arising in alternate pattern from two sides of the stem. *Camellia japonica*   Theaceae

Fig. 48 Opposite leaf arrangement occurs when two leaves grow from opposite sides of the same node. *Ligustrum sinensis*   Oleaceae

Fig. 49 Whorled leaf arrangement occurs when three or more leaves grow from the same node and extend in a circle around the stem. *Allamanda cathartica*   Apocynaceae

Simple Leaves

Fig. 50 OVATE leaves are broader at the base than the apex.
Ligustrum japonicum
Oleaceae

Fig. 51 OBOVATE leaves are broader at the apex than the base.
Clusia rosea
Clusiaceae

Fig. 52 ELLIPTIC leaves are wider at the middle than the base or the apex.
Cinnamomum camphora
Lauraceae

Fig. 53 OBLONG leaves have the margins more or less parallel-sided.
Rhamnus caroliniana
Rhamnaceae

Leaf Shapes

Fig. 54 PELTATE
Hydrocotyle umbellata
Apiaceae

Fig. 55 SPATULATE
Quercus nigra
Fagaceae

Fig. 56 CORDATE
Cercis canadensis
Fabaceae

Fig. 57 SAGITTATE
Syngonium podophyllum
Araceae

Fig. 58 LOBED
Sassafras albidum
Lauraceae

Fig. 59 DELTOID
Populus deltoides
Salicaceae

Fig. 60 LANCEOLATE
Sagittaria lancifolia
Alismataceae

Fig. 61 LINEAR
Aechmea cylindrata
Bromeliaceae

Compound Leaves

UNIFOLIATE

Fig. 62 *Citrus aurantifolium*
Rutaceae   Inset: CU of petiole.

BIFOLIATE

Fig. 63 *Bignonia capreolata*
Bignoniaceae

TRIFOLIATE

Fig. 64 *Toxicodendron radicans*
Anacardiaceae

PALMATE

Fig. 65 *Aesculus pavia*
Hippocastanaceae

EVEN PINNATE

Fig. 66 *Cassia alata*
Fabaceae

EVEN PINNATE

Fig. 67 *Sesbania punicea*
Fabaceae

ODD PINNATE

Fig. 68 *Tecoma capensis*
Bignoniaceae

ODD PINNATE

Fig. 69 *Carya glabra*
Juglandaceae

EVEN BIPINNATE

Fig. 70 *Leucaena leucocephala*
Fabaceae

Leaf Apices

MUCRONATE

Fig. 71 *Buxus microphylla*
Buxaceae

ACUMINATE

Fig. 72 *Photinia* sp.
Rosaceae

ATTENUATE

Fig. 73 *Cinnamomum camphora*
Lauraceae

ROUNDED

Fig. 74 *Crassula* sp.
Crassulaceae

EMARGINATE

Fig. 75 *Chionanthus retusa*
Oleaceae

TRUNCATE

Fig. 76 *Passiflora murucuja*
Passifloraceae

Leaf Bases

CORDATE

Fig. 77 *Cercis canadensis*
Fabaceae

AURICULATE

Fig. 78 *Dioscorea bulbifera*
Dioscoreaceae

SAGITTATE

Fig. 79 *Syngonium podophyllum*
Araceae

CUNEATE

Fig. 80 *Eriobotrya japonica*
Rosaceae

TRUNCATE

Fig. 81 *Populus deltoides*
Salicaceae

ASYMMETRICAL

Fig. 82 *Tilia americana*
Tiliaceae

ENTIRE

Fig. 83   An entire margin continues unbroken and lacks lobes, teeth, serrations, or divisions of any kind.
Illicium floridanum
Illiciaceae

LOBED

Fig. 84   A lobed margin is broken by angular or rounded projections.
Acer rubrum
Aceraceae

DISSECTED

Fig. 85   A dissected margin is characterized by obvious sinuses or divisions into the blade.
Merremia dissecta
Convolvulaceae

SERRATE

a

b

Fig. 86   A serrate margin has a sawtoothed edge with the points of the teeth directed forward.
Ulmus parvifolia   Ulmaceae

DOUBLY SERRATE

a

b

Fig. 87   A doubly serrate or compoundly serrate margin has sawteeth of different sizes on the edge.
Betula nigra   Betulaceae

CRENATE

a

b

Fig. 88   A crenate margin has a scalloped edge of rounded teeth.
Ardisia crenata   Myrsinaceae

DENTATE

a

b

Fig. 89   A dentate margin has an edge with teeth whose points project directly outward.
Vitis rotundifolia   Vitaceae

Leaf Venation

Fig. 90   a) & b) Venation of an elm leaf illustrating the classes and some of the patterns of veins used in describing a leaf. Here secondary veins diverge from the midvein pinnately and reach the point of the serrate teeth. *Ulmus parvifolia*   Ulmaceae

Fig. 91   *Rhamnus caroliniana*
Rhamnaceae

Fig. 92   *Cocculus laurifolius*
Menispermaceae

Fig. 93   *Tibouchina* sp.
Melastomataceae

Fig. 94   *Aleurites fordii*
Euphorbiaceae

Fig. 95   Grass
Poaceae

Fig. 96   *Hydrocotyle umbellata*
Apiaceae

Fig. 97   a) - c)   Variation in leaf shape from the same sassafras bush.
Sassafras albidum   Lauraceae

Fig. 98   a) Juvenile (bottom) and adult (top) leaves of the juniper. b) Pointed needle-like juvenile leaves.
c) Flattened scale-like adult leaves.   *Juniperus chinensis*   Cupressaceae

Fig. 99   a) - b)   Variation in leaf shape from the same greenbrier plant.
Smilax auriculata   Smilacaceae

Fig. 100   Light triggers new leaf
shape in tropical aroid on tree trunk.

Leaf Microscopic Anatomy

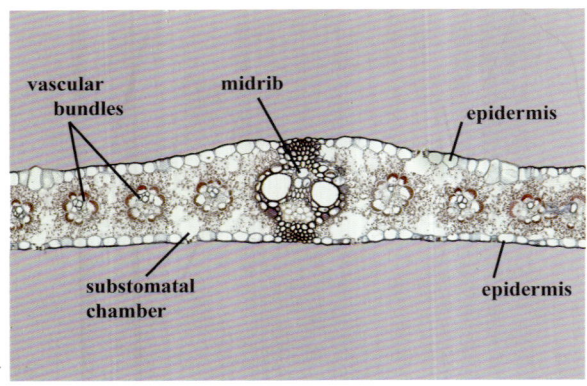
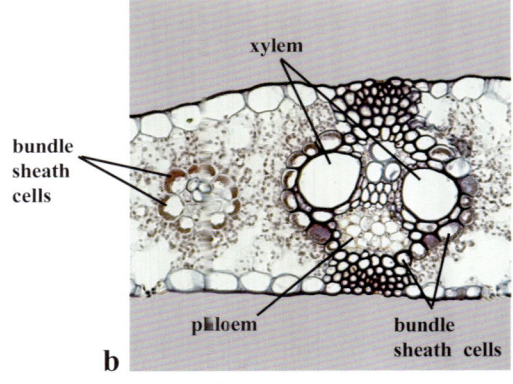

Fig. 101 a) & b) XS and CU of a monocot leaf showing the arrangement of the vascular bundles and leaf tissues. *Zea mays*   Poaceae

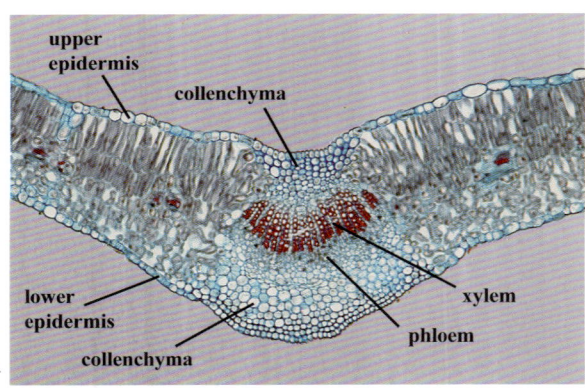
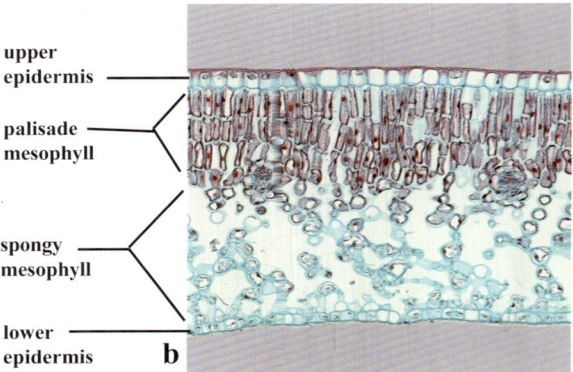

Fig. 102 a) & b) XS and CU of a eudicot leaf from a mesophyte, a plant that has evolved for growth in a mesic habitat. *Syringa vulgaris*   Oleaceae

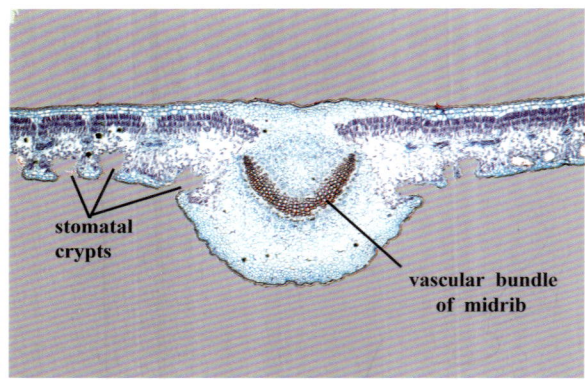
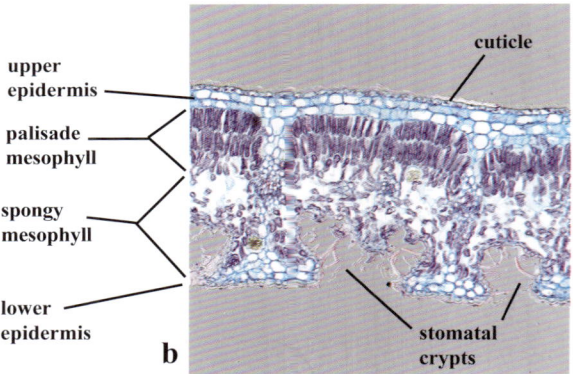

Fig. 103 a) & b) XS and CU of the leaf of a xerophyte or plant adapted to arid conditions and habitats. The cuticle is thick and the epidermis is several layers thick. Stomata and trichomes are restricted to stomatal crypts or invaginations of the lower epidermis. *Nerium oleander*   Apocynaceae

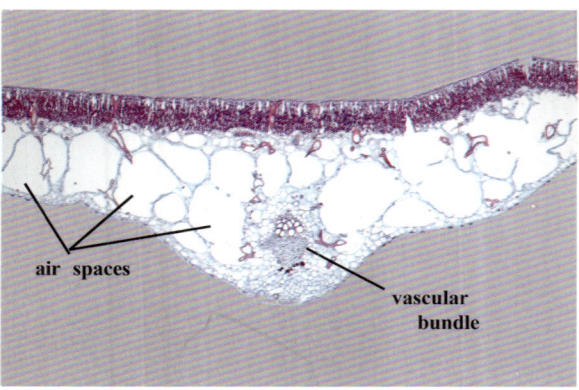
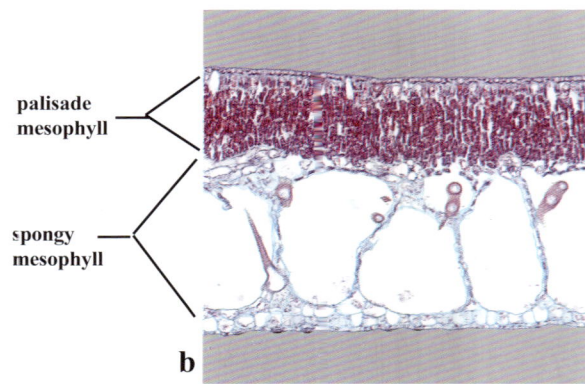

Fig. 104 a) & b) XS and CU of the leaf of a hydrophyte or plant that grows entirely or partially submerged in water. Stomata are restricted to the upper epidermis and large air spaces increase buoyancy. Vascular tissue is greatly reduced. *Nymphaea odorata*   Nymphaeaceae

Leaf Microscopic Anatomy

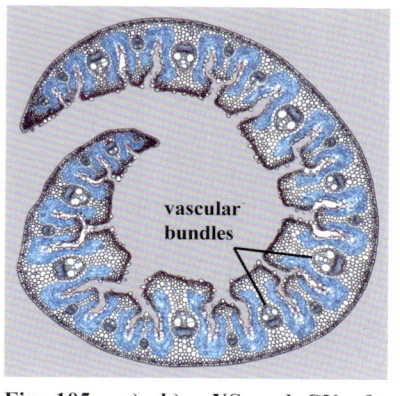

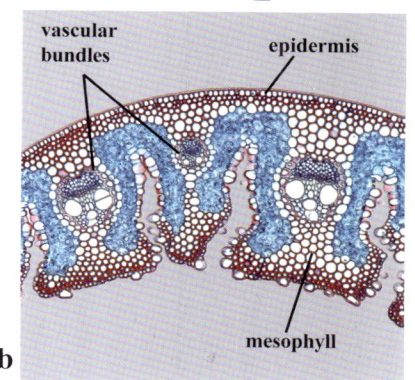

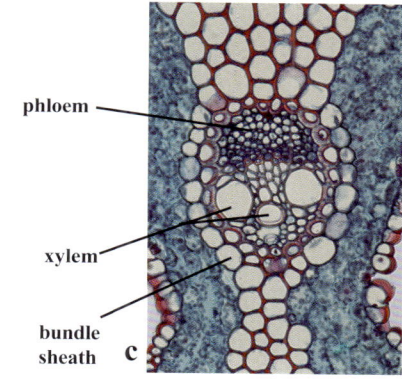

Fig. 105 a) - b) XS and CU of a beach grass leaf. c) Vascular bundle.
Ammophila sp. Poaceae

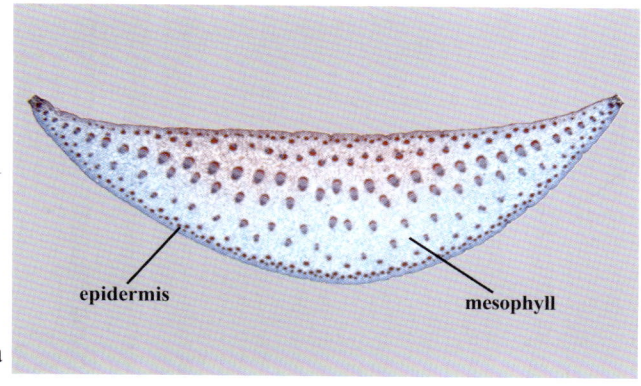

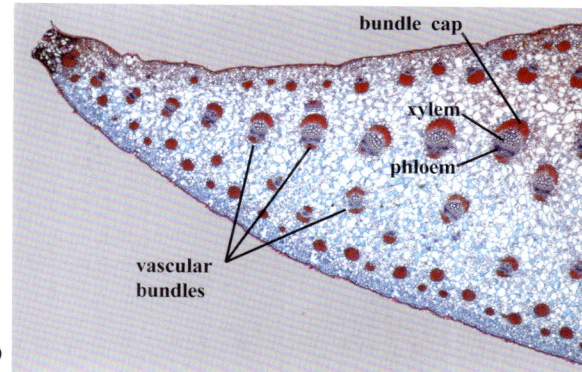

Fig. 106 a) & b) XS and CU of a yucca leaf that has evolved to minimize water loss. The cuticle is thicker than usual, and the leaf itself is succulent in nature.
Yucca sp. Agavaceae

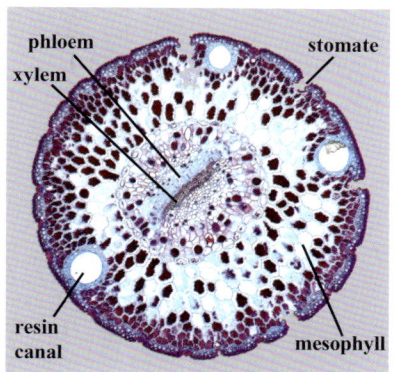

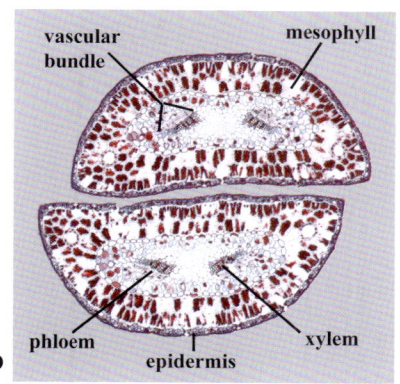

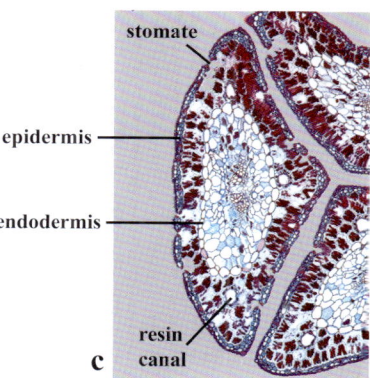

Fig. 107 XS of pine needles or leaves. a) One needle per fascicle species. b) Two needles per fascicle.
c) Three needles per fascicle.
Pinus spp. Pinaceae

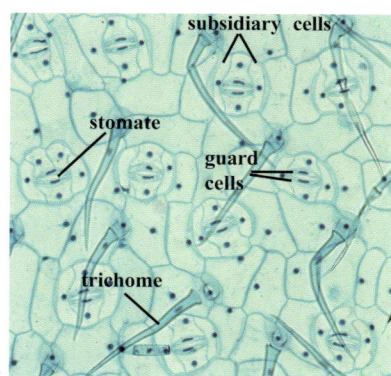

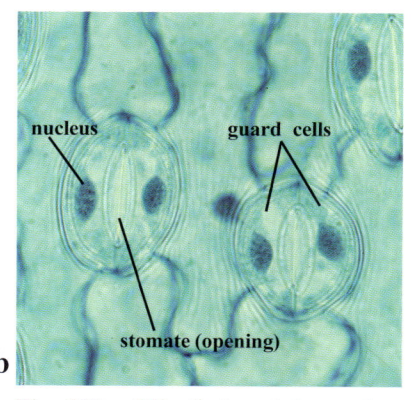

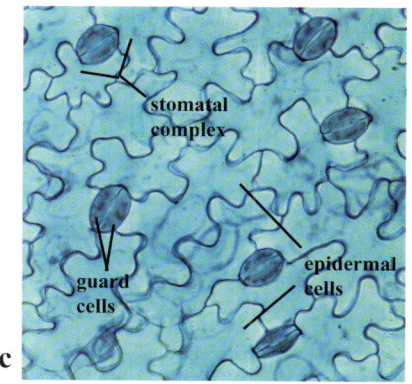

Fig. 108 Stomatal complexes and trichomes in the epidermis of a monocot.

Fig. 109 CU of stomatal complexes in the epidermis of a monocot.
Lilium sp. Liliaceae

Fig. 110 Stomatal complexes in the epidermis of a eudicot.
Pisum sp. Fabaceae

FLOWERS

Functions

Reproduction - Flowers contain the reproductive organs of angiosperm plants and are responsible for sexual reproduction.

Attract Pollinators - Although some flowers are wind- or self-pollinated, most rely on other living organisms to transfer pollen from one flower or plant to another. The color, scent, shape, and physical properties of flowers make it possible for pollination to take place.

Terminology

actinomorphic - Flower with radial symmetry.

adnate - Fusion of dissimilar organs or parts to one another (ex. stamens to petals).

androecium - Collective term for all the stamens of a flower. All male reproductive structures.

anther - Apical portion of the stamen that contains the pollen sacs and produces the pollen.

asymmetrical - Not symmetrical. Flower that lacks a plane of symmetry.

bilateral symmetry - Flower that is divisible into equal halves by only a single plane.

bisexual - Flower containing both a gynoecium and androecium.

bract - Reduced or modified leaf associated with a flower or inflorescence.

bracteole - Small bract.

calyx - Collective term for all the sepals of a flower.

carolla - Collective term for all the petals of a flower.

carpel - Individual female reproductive structure composed of a stigma, style, and ovary. Equivalent to the pistil if there is only one carpel, or if all carpels are separate and unfused.

carpellate - Flower with a functional gynoecium but without a functional androecium. Female flower.

complete - Flower containing a gynoecium, androecium, calyx, and corolla.

connate - Fusion of similar organs or parts to one another (ex. petals to petals).

connective - Structure or tissue connecting two pollen sacs of an anther.

dioecious - Describes when plants bear either staminate flowers or carpellate flowers, but not both.

distinct - Clear separation of similar organs or parts from one another.

epigynous - Flower with an inferior ovary and the perianth and stamens arising from the floral cup or hypanthium that is adnate to the ovary.

extrafloral nectary - Nectar producing gland that is **not** physically associated with a flower.

fertilization - Fusion of reproductive material (sperm) from a pollen grain with the reproductive material (egg) in an ovule. Accomplished only after pollen reaches a stigma (pollination) and a pollen tube grows down through the style and into an ovule in a carpel.

filament - Basal stalk of a stamen atop which sits the anther.

flower - Reproductive structure of angiosperms composed of a modified shoot and modified leaves.

free - Clear separation of dissimilar organs or parts from one another.

funiculus - Stalk of an ovule where it attaches to the placenta.

gynoecium - Collective term for all the pistils or all the carpels of a flower. All female reproductive structures.

gynophore - Stalk that bears the gynoecium.

hilum - Scar on a seed caused by the attachment of the funiculus.

hypanthium - Cup-like or tubular structure that bears the perianth and stamens, and is sometimes adnate to the ovary. Equivalent to a floral cup or floral tube.

hypogynous - Flower with a superior ovary and the perianth and stamens arising from below the ovary.

imperfect - Flower missing either a functional gynoecium or functional androecium.

incomplete - Flower missing either a gynoecium, androecium, calyx, or corolla.

Terminology (Continued)

inferior ovary - Ovary located in a position below the point where the perianth parts are attached to the receptacle or flower stalk, or below the basal point of a hypanthium or floral cup.

irregular - Flower with bilateral symmetry.

monoecious - Both staminate and carpellate flowers are found on the same plant.

nectary - Nectar producing gland that may vary in shape or location.

ovary - Portion of the pistil in which the ovules or seeds are produced, and which matures into a fruit.

ovule - Organ that develops into a seed following fertilization, within a locule of the ovary. Attached to the ovarian wall at the placenta by a stalk known as a funiculus.

pedicel - Stalk of an individual flower in an inflorescence.

perfect - Flower containing both a functional gynoecium and a functional androecium.

perianth - Collective term for all the outer floral parts. All of the tepals, or all of the sepals and petals.

perigynous - Flower with a superior ovary that has the perianth parts and stamens attached to the margin of a hypanthium that surrounds but is separate from the ovary.

petal - One of the inner perianth parts that functions in attracting pollinators. Typically showy and colorful.

pistil - Female reproductive structure consisting of the stigma, style, and ovary. If the carpels are **not** fused, then pistil and carpel are equivalent terms and the number of carpels equals the number of pistils. If there are two or more carpels that are fused, they constitute a single pistil.

pollination - Transfer of pollen from an anther to a stigma.

polygamous - Describes when both bisexual and unisexual flowers are found on the same plant.

radial symmetry - Form of symmetry that permits a flower to be divided into equal halves by two or more planes. Equivalent with the terms **actinomorphic** and **regular**.

receptacle - Tip or apex of the flower stalk that bears the flower parts.

regular - Flower with radial symmetry.

sepal - One of the outer perianth parts that protects the inner petals while in the bud stage. Typically green and leaf-like.

stamen - Pollen producing organ consisting of pollen sacs at the end of a stalk. The male reproductive structure.

staminate - Flower with a functional androecium but without a functional gynoecium. Male flower.

stigma - Apical part of the pistil that receives the pollen.

style - Elongated portion of the pistil that connects the stigma with the ovary.

superior ovary - Ovary located in a position above where the perianth parts are attached to the receptacle or flower stalk, or above the basal point of a hypanthium or floral cup.

tepal - One of the outer floral parts when perianth is not divided into distinct sepals and petals.

unisexual - Flower missing either a gynoecium or androecium.

zygomorphic - Flower with bilateral symmetry.

Prefixes

apo-   Distinct. (Ex. apotepalous = With distinct tepals.)

epi-   Arising from or adnate to. (Ex. epipetalous = Adnate to the petals or corolla.)

syn-   Connate. (Ex. synsepalous = With connate sepals.)

General Floral Features

Fig. 111   Large sepals and small notched petals.
Stellaria media   Caryophyllaceae

Fig. 112   Large petals and smaller sepals.
Ludwigia sp.   Onagraceae

Fig. 113   Large thick sepals and large hairy petals.
Rhizophora mangle   Rhizophoraceae

Fig. 114   Perfect flower with both stamens and pistil.
Tulipa sp.   Liliaceae

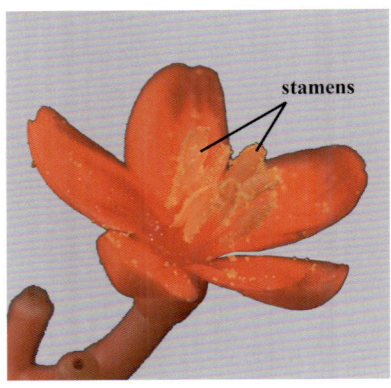

stamens

Fig. 115   Imperfect staminate flower that lacks gynoecium.
Jatropha sp.   Euphorbiaceae

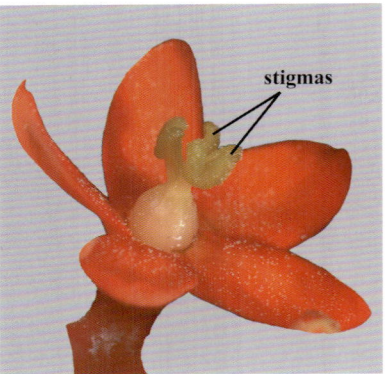

stigmas

Fig. 116   Imperfect carpellate flower that lacks androecium.
Jatropha sp.   Euphorbiaceae

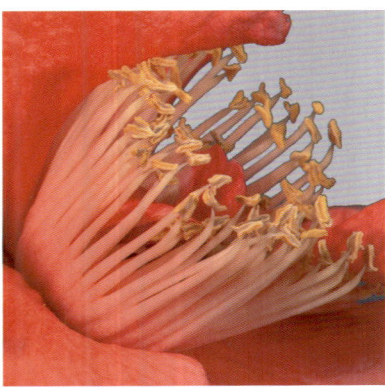

Fig. 117   Connate stamens united to one another at their bases.
Camellia japonica   Theaceae

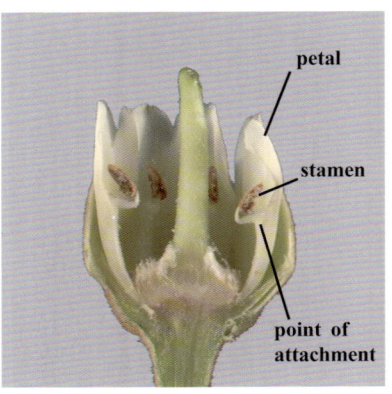

petal

stamen

point of attachment

Fig. 118   Adnate stamens attached to petals.
Manilkara zapota   Sapotaceae

Fig. 119   Apetalous carpellate flowers of live oak.
Quercus virginiana   Fagaceae

Fig. 120 Large bracts of dogwood with small flowers at center of inflorescence. ***Cornus florida*** Cornaceae

Fig. 121 Bougainvillea with colorful bracts enclosing tubular flowers. ***Bougainvillea spectabilis*** Nyctaginaceae

Fig. 122 Large green bracts with nectaries protect the flower bud of this passion vine. ***Passiflora*** sp. Passifloraceae

Extrafloral Nectaries

Fig. 123 Nectariferous stipules on stem of candle tree. *Cassia alata* Fabaceae

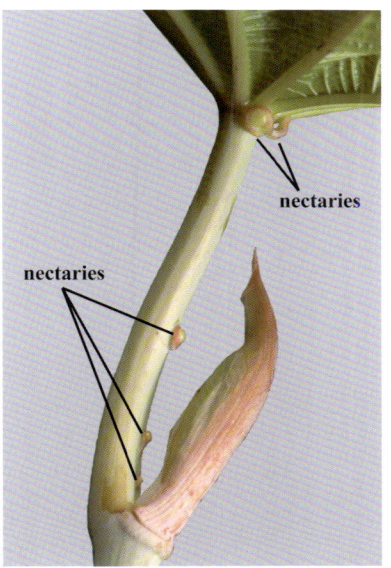

Fig. 124 Nectaries on underside of castor bean leaf and petiole. *Ricinus communis* Euphorbiaceae

Fig. 125 Labiate nectaries along midrib of ice cream bean tree leaf. *Inga* sp. Fabaceae

Floral Nectaries

Fig. 126 Nectary of the delphinium flower is a tubular spur. *Delphinium* sp. Ranunculaceae

Fig. 127 Nectary of the colubrina flower is in the form of a disk. *Colubrina asiatica* Rhamnaceae

Fig. 128   Flower of the spotted lily, a monocot.
Lilium sp.   Liliaceae

Fig. 129   Tepals forming the inner perianth whorl.

Fig. 130   Tepals forming the outer perianth whorl.

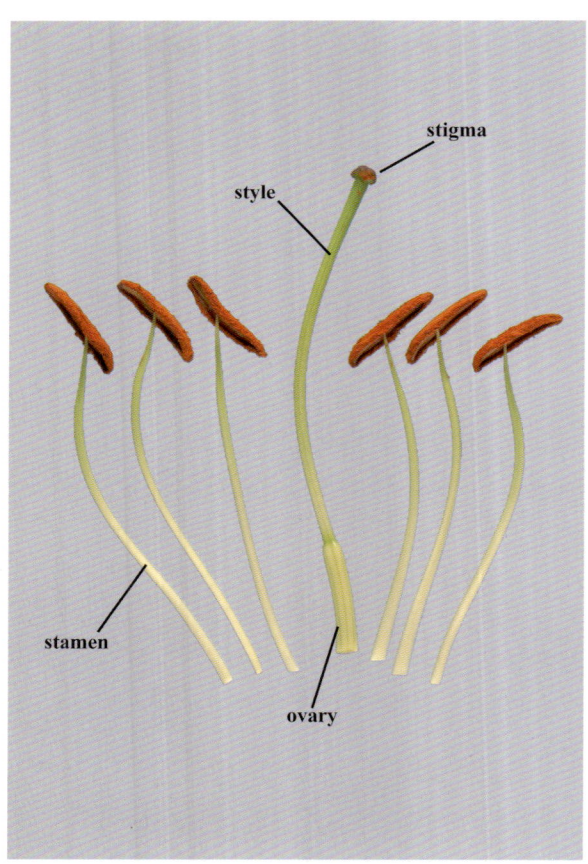

Fig. 131   Androecium (6 stamens) and gynoecium
(pistil) of the spotted lily.

Fig. 132   CU of anther with pollen.

Fig. 133   CU of stigma and apical part of style.

Fig. 134   CU of ovary and basal part of style.

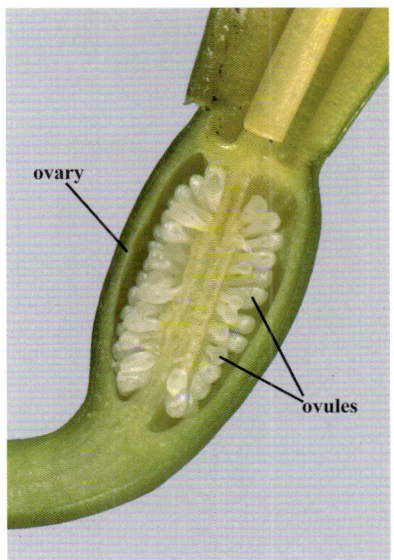

Fig. 135   a) LS of flower of daffodil.   b) CU of LS of inferior ovary showing developing ovules.
Narcissus sp.   Amaryllidaceae

Fig. 136   a) Flower of Mexican prickly poppy with one petal removed. Note multiple stamens and red stigma.
b) CU of LS of superior ovary showing developing ovules and yellow sap.   *Argemone mexicana*   Papaveraceae

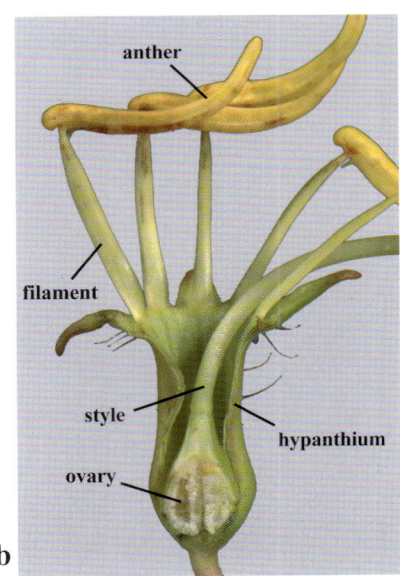

Fig. 137   a) Flower and floral cup (hypanthium) of meadow beauty.   b) CU of LS of hypanthium with petals and some
stamens removed showing developing ovules.   *Rhexia mariana*   Melastomataceae

Flower Symmetry

Radial Symmetry = Actinomorphic or Regular Flowers

Fig. 138   *Tradescantia ohiensis*
Commelinaceae

Fig. 139   *Ludwigia peruviana*
Onagraceae

Fig. 140   *Rosa palustris*
Rosaceae

Bilateral Symmetry = Zygomorphic or Irregular Flowers

Fig. 141   *Digitalis* sp.
Scrophulariaceae

Fig. 142   *Commelina erecta*
Commelinaceae

Fig. 143   *Trichostema dichotomum*
Lamiaceae

Asymmetrical Flowers

Fig. 144   *Canna* sp.
Cannaceae

Fig. 145   *Canna* sp.
Cannaceae

Flower Shapes

TUBULAR

Fig. 146   *Lonicera sempervirens*
Caprifoliaceae

TRUMPET-SHAPED

Fig. 147   *Campsis radicans*
Bignoniaceae

**FUNNEL-
SHAPED**

Fig. 148   *Brugmansia* sp.
Solanaceae

CRUCIFORM

Fig. 149   *Raphanus raphanistrum*
Brassicaceae

CAMPANULATE

Fig. 150   *Platycodon grandiflorus*
Campanulaceae

URCEOLATE

Fig. 151   *Lyonia lucida*
Ericaceae

SALVERFORM

Fig. 152   *Phlox drummondii*
Polemoniaceae

PAPILIONACEOUS

Fig. 153   *Crotalaria spectabilis*
Fabaceae

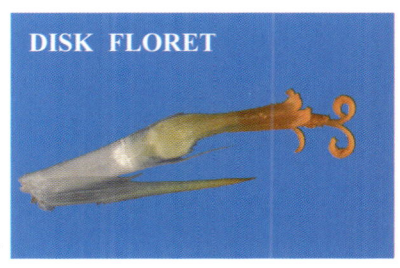

DISK FLORET

a

RAY FLORET

b

Fig. 154   *Tithonia* sp.
Asteraceae

Gynoecium Variation

Fig. 155   Multiple ovaries, styles, and stigmas.   *Magnolia grandiflora* Magnoliaceae

Fig. 156   Multiple ovaries, styles, and stigmas.   *Illicium floridanum* Illiciaceae

Fig. 157   One ovary, three styles, and three stigmas. *Malpighia glabra*   Malpighiaceae

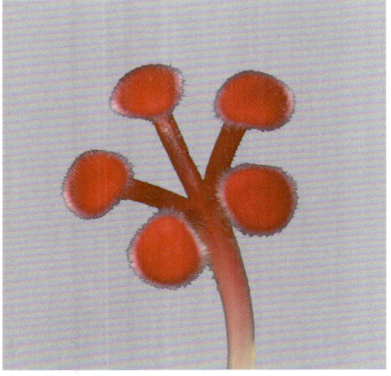

Fig. 158   Five stigmas and one style. *Hibiscus* sp.   Malvaceae

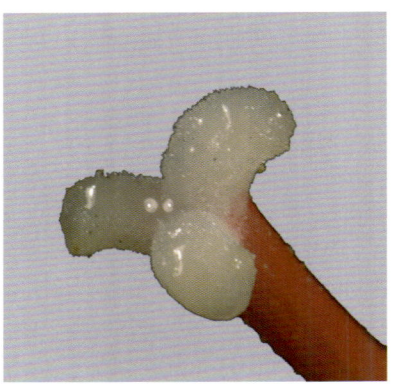

Fig. 159   Three-lobed stigma and one style. *Amaryllis* sp.   Amaryllidaceae

Fig. 160   Three-lobed stigma, one style, and one ovary. *Tulipa* sp.

Placentation Types

placentation - Arrangement and attachment of ovules within and to the ovaries.

locule - One or more chambers into which the ovary is divided, with usually one locule per carpel.

septum - Wall separating adjacent locules.

placenta - Part of ovary to which the ovules are attached.

axile placentation - Ovules are attached to the central axis or inner angle formed by septa of an ovary with two or more locules. The number of locules is typically the same as the number of carpels.

parietal placentation - Ovules are attached to the wall of the ovary, or to extrusions of the wall.

free-central placentation - Ovules are attached to the central axis/column of an ovary with one locule.

basal placentation - Ovule is attached to the base of the ovary.

apical placentation - Ovule is attached at the apex of the ovary.

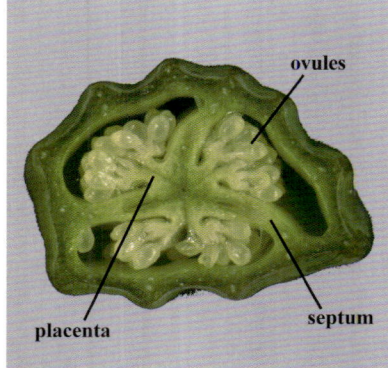

Fig. 161   Axile Placentation Ovary with three locules. *Bilbergia nutans*   Bromeliaceae

Fig. 162   Parietal Placentation *Argemone mexicana* Papaveraceae

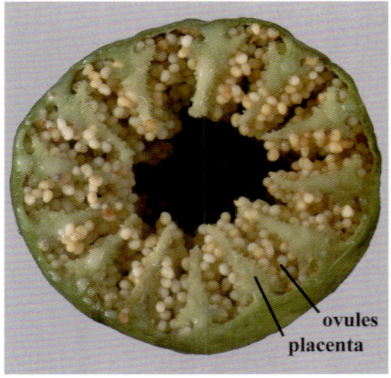

Fig. 163   Parietal Placentation Placentae deeply intruded. *Papaver somniferum*   Papaveraceae

Insertion Types

SUPERIOR OVARIES are located in a position above or *superior* to the point where the perianth parts attach to the flower stalk, or above the basal point of a hypanthium or floral cup.

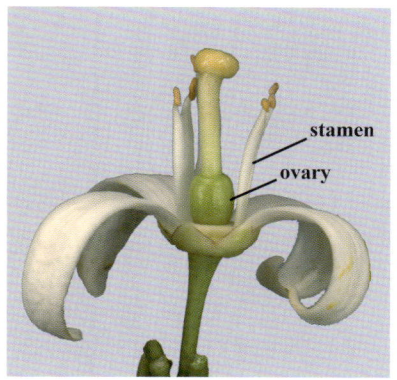

Fig. 164   *Citrus* sp.
Rutaceae

Fig. 165   *Argemone mexicana*
Papaveraceae

Fig. 166   *Ilex* sp.
Aquifoliaceae

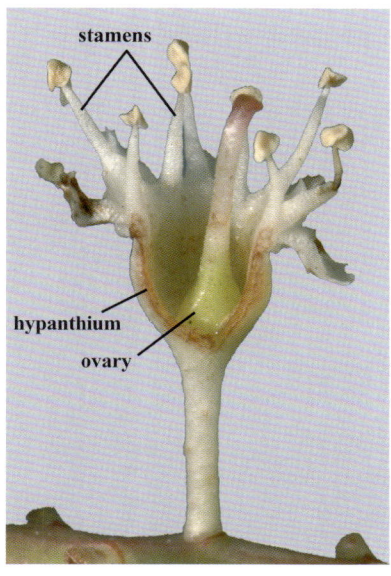

Fig. 167   *Prunus campanulata*
Rosaceae

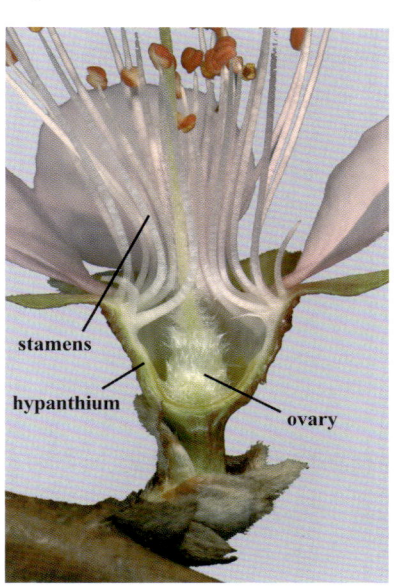

Fig. 168   *Prunus persica*
Rosaceae

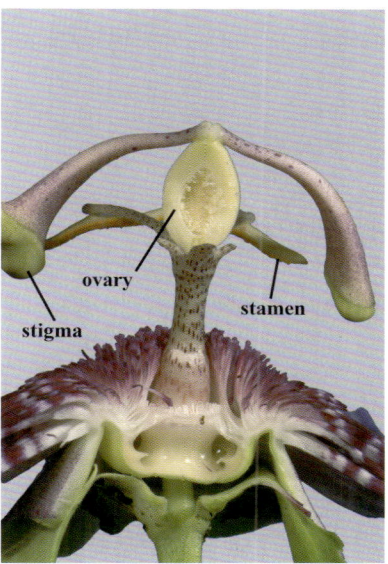

Fig. 169   *Passiflora* sp.
Passifloraceae

INFERIOR OVARIES are located in a position below or *inferior* to the point where the perianth parts attach to the flower stalk, or below the basal point of a hypanthium or floral cup.

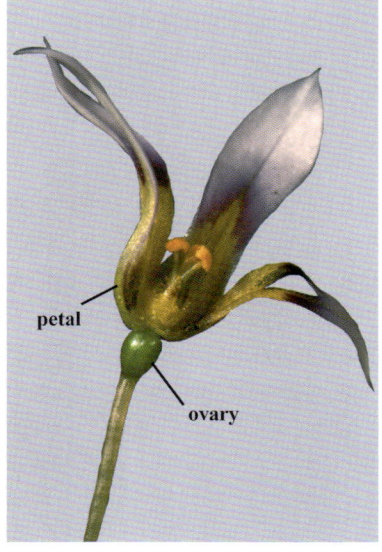

Fig. 170   *Sisyrinchium atlanticum*
Iridaceae

Fig. 171   *Cucurbita* sp.
Cucurbitaceae

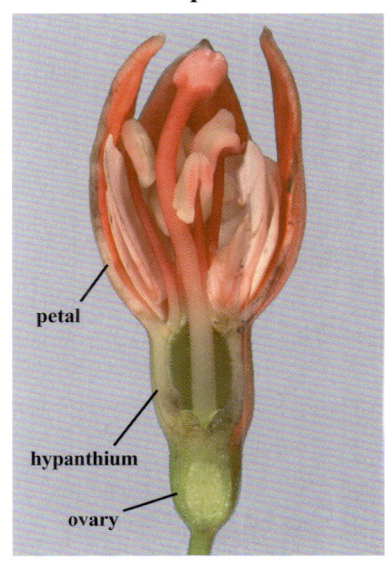

Fig. 172   *Fuchsia* sp.
Onagraceae

Monocot Floral Diversity

The corolla or flower parts of monocots are typically three or multiples of three. The term **-merous** describes the numerical plan of a flower. Most monocots are 3-merous or 6-merous.

Fig. 173 *Sagittaria lancifolia*
Alismataceae

Fig. 174 *Trillium maculatum*
Trilliaceae

Fig. 175 *Tradescantia fluminensis*
Commelinaceae

Fig. 176 *Limnocharis flava*
Limnocharitaceae

Fig. 177 *Iris* sp.
Iridaceae

Fig. 178 *Allium* sp.
Alliaceae

Fig. 179 *Narcissus* sp.
Amaryllidaceae

Fig. 180 *Lilium* sp.
Liliaceae

Fig. 181 *Oncidium maculatum*
Orchidaceae

Eudicot Floral Diversity

The corolla or flower parts of eudicots are typically made up of four or five or multiples of four or five. The term **-merous** describes the numerical plan of a flower. Most eudicots are 4-merous or 5-merous.

Fig. 182 *Oenothera speciosa*
Onagraceae

Fig. 183 *Rhizophora mangle*
Rhizophoraceae

Fig. 184 *Rhexia mariana*
Melastomataceae

Fig. 185 *Aquilegia* sp.
Ranunculaceae

Fig. 186 *Malpighia glabra*
Malpighiaceae

Fig. 187 *Colubrina asiatica*
Rhamnaceae

Fig. 188 *Nandina domestica*
Berberidaceae

Fig. 189 *Nelumbo lutea*
Nelumbonaceae

Fig. 190 *Schlumbergera bridgesii*
Cactaceae

Inflorescences

Fig. 191   SOLITARY
Single flower at the end
of a stalk called a scape.
Tulipa sp.
Liliaceae

Fig. 192   SPIKE
Sessile or stalkless flowers
from a single axis.
Languncularia sp.
Combretaceae

Fig. 193   RACEME
Stalked flowers from a
single axis.
Galphimia sp.
Malpighiaceae

Fig. 194   PANICLE
Raceme w. compound
branching.
Ligustrum sp.
Oleaceae (infructescence)

Fig. 195   UMBEL
Flowers w. stalks of same
length from one point.
Allium sp.
Alliaceae

**Fig. 196   COMPOUND
UMBEL**
An umbel of umbels.
Foeniculum vulgare
Apiaceae

Fig. 197   CYME
Terminal flower w. two
opposite laterals or two
lateral clusters of three.
Rhexia sp. Melastomataceae

Fig. 198   HELICOID CYME
Coiled cyme w. lateral
flowers on same side.
Heliotropium amplexicaule
Boraginaceae

Fig. 199   HEAD
Dense spike of disk flowers
on a short discoid axis.
Emilia fosbergii
Asteraceae

Fig. 200   HEAD
Dense spike of ray and
disk flowers on short axis.
Helianthus sp.
Asteraceae

Fig. 201   CATKIN
Pendent or erect dense
spike of apetalous flowers.
Salix caroliniana
Salicaceae

Fig. 202   SPATHE & SPADIX
Spike w. thickened axis (spadix)
subtended by leafy bract (spathe).
Spathiphyllum sp.
Araceae

Pollination Syndromes

a

b

Fig. 203   a) Stamens of male flowers on catkin of Shumard oak that produce large amounts of pollen. b) Apetalous female flowers receive pollen carried by the wind.   *Quercus shumardii*   Fagaceae

Fig. 204   Red and yellow bracts of heliconia attract hummingbirds. *Heliconia* sp.   Heliconiaceae

Fig. 205   Bee-pollinated flowers often have visual honey guides to direct pollinators. *Digitalis* sp.   Scrophulariaceae

Fig. 206   Brightly colored flowers attract bees and butterflies. *Tithonia* sp.   Asteraceae

Fig. 207   An aquatic orchid that is pollinated by moths. *Habenaria repens*   Orchidaceae

Fig. 208   Carrion flowers have the appearance and smell of rotting meat and attract carrion flies as pollinators.   *Stapelia* sp.   Asclepiadaceae

Fig. 209   Some orchids lure insects into mating with them . *Tolumnia henekenii*   Orchidaceae

Flower Bud Anatomy

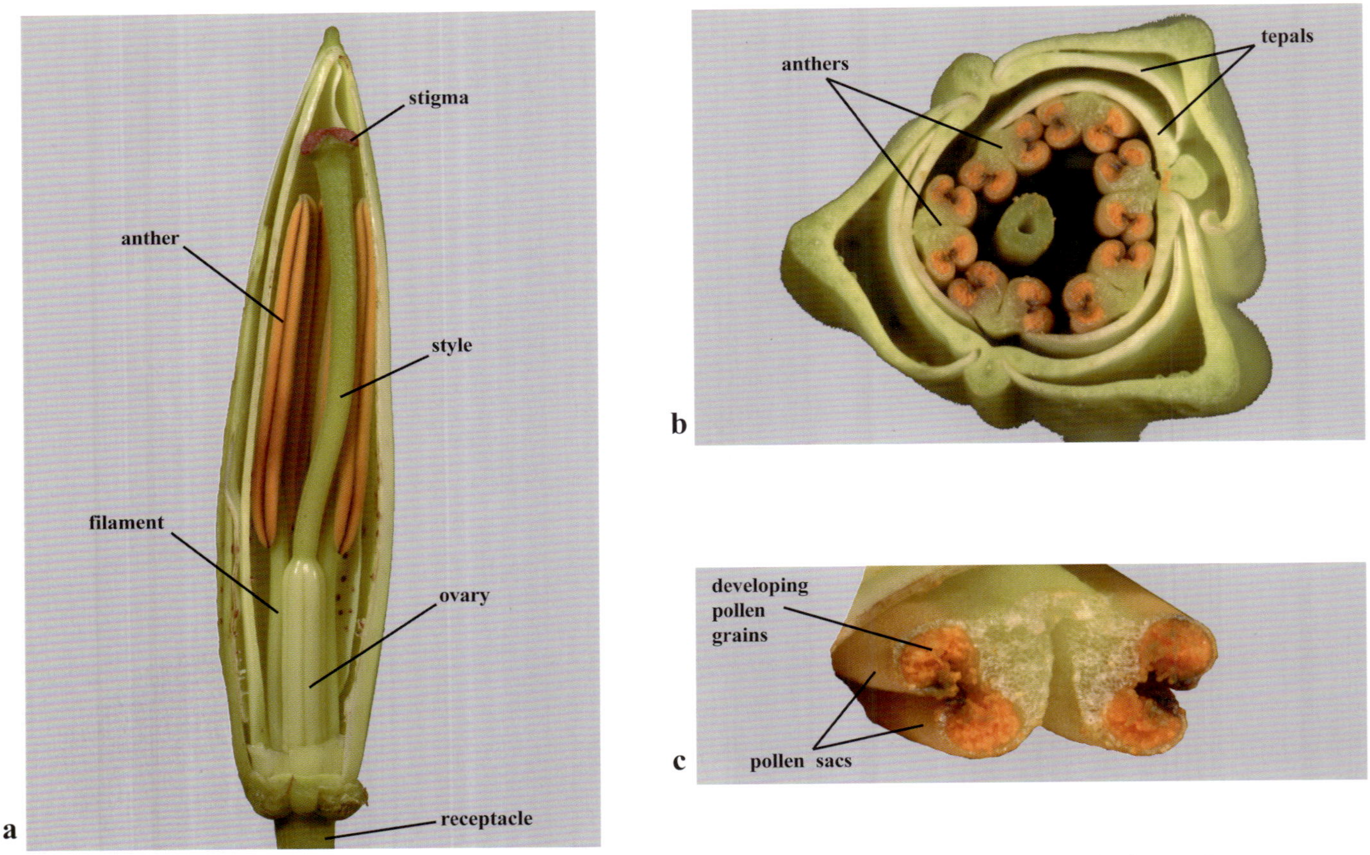

Fig. 210 Views of dissected lily flower bud. a) LS of bud. b) XS of bud. c) CU of XS of developing anther.
Lilium sp. Liliaceae

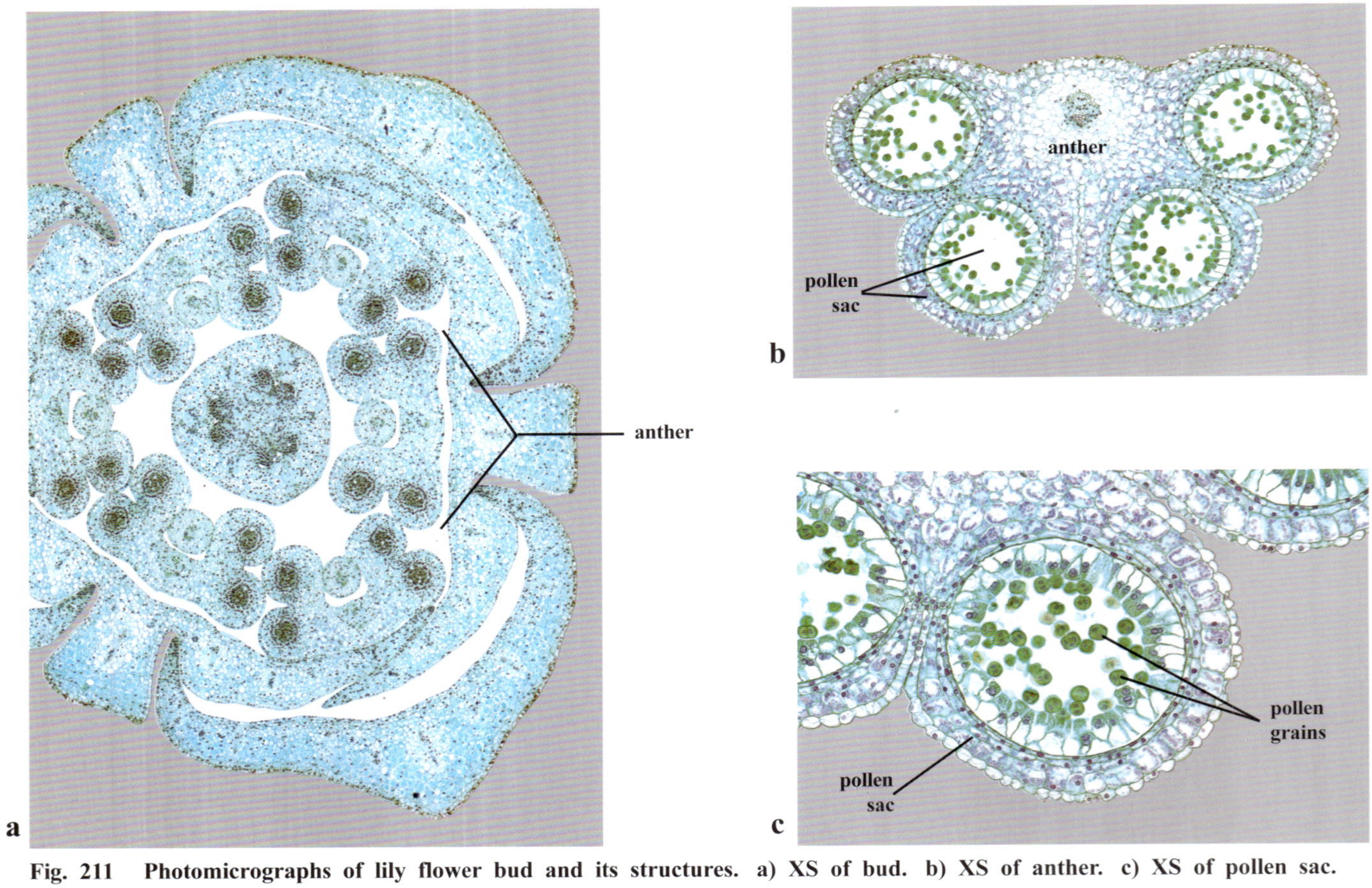

Fig. 211 Photomicrographs of lily flower bud and its structures. a) XS of bud. b) XS of anther. c) XS of pollen sac.
Lilium sp. Liliaceae

accessory fruit - Fruit or cluster of fruits that includes tissue that is not derived from the ovary.

achene - Dry indehiscent fruit with a thin wall surrounding a single seed.

aggregate fruit - Fruit formed from several to many carpels from a single flower.

berry - Indehiscent fruit that is fleshy and typically contains several to many seeds that are not surrounded by pits. The exterior of the berry may be the same consistency as the flesh, or it may vary from being soft to leathery to hard.

capsule - Dry dehiscent fruit produced from two or more carpels, usually categorized based upon how the seeds are released.

circumscissile capsule - Capsule that splits along a horizontal line that circumscribes the entire fruit, resulting in the top portion falling off like a lid.

caryopsis - A grain characteristic of grasses and consisting of a small, dry indehiscent fruit which has a thin coat fused to a single seed.

drupe - Fleshy indehiscent fruit with one or more hard pits called stones or pyrenes in the center. Each pit contains a single seed.

follicle - Dry dehiscent fruit that originates from a single carpel and that splits along a single suture.

hesperidium - Specialized berry with a leathery exterior or rind and a fleshy, often juicy interior that is divided into several to many sections. Only applied to citrus and its relatives.

indehiscent pod - Typically dry fruit that does not split open at maturity. Seeds are sometimes released by the actions of animals that chew the pods open.

legume - Dry fruit that originates from a single carpel and typically dehisces by means of two longitudinal sutures. Some legumes split completely and equally along both the sutures, resulting in a V-shaped fruit after dehiscence. Others split primarily along one suture or only minimally and may resemble a follicle or an indehiscent pod. Legumes are only found in the Family Fabaceae (= Leguminosae).

loculicidal capsule - Capsule that splits or dehisces into the locule, forming segments or valves that have parts of two adjacent locules and that are longitudinally divided by a septum.

loment - Dry fruit produced from a single carpel that breaks along transverse lines into segments of one seed each. Only found in the Family Fabaceae (= Leguminosae).

multiple fruit - Fruit formed from multiple, closely packed flowers where the developing structures coalesce into a single unit.

nut - Indehiscent dry fruit consisting of a single seed surrounded by a hard smooth wall. Larger fruits are termed nuts and smaller fruits nutlets, although the determination is subjective.

nutlet - Indehiscent dry fruit consisting of a single seed surrounded by a hard smooth wall. Smaller fruits are termed nutlets and larger fruits nuts, although the determination is subjective.

pepo - Specialized berry with a leathery to hard exterior or rind and a fleshy interior. These fruits are characteristic of the Family Cucurbitaceae.

pome - Fleshy indehiscent fruit where the outer portion is soft and formed from the expansion of the floral structures that surround the ovary and enlarge as the fruit matures. The central portion contains the seeds which are surrounded by a cartilaginous or papery structure that represents the ovary wall, although it is sometimes indistinct and difficult to see. Pomes are only found in the Subfamily Maloideae of the Family Rosaceae.

poricidal capsule - Capsule that releases its seeds by means of pores such as holes or flaps.

samara - Dry indehiscent winged fruit that contains a single seed.

schizocarp - Typically dry but sometimes fleshy fruit derived from two or more carpels that splits into two or more mericarps, each of which contains one seed.

septicidal capsule - Capsule that splits along the septa to release the seeds, resulting in segments or valves that represent a single entire locule without a septum dividing it.

silicle - A silique that is approximately as long as it is wide. (See silique.)

silique - Fruit produced from a gynoecium with two carpels, with seeds attached to a persistent thickened rim (replum) and membrane. Seeds and membrane are not visible until the two valves of the fruit have split away. Repla sometimes remain attached to the plant. Only found in the Family Brassicaceae.

Berry

An indehiscent fruit that is fleshy and typically contains several to many seeds that are not surrounded by pits. The exterior of the fruit may be the same consistency as the flesh, or it may vary from leathery to hard.

a

b

Fig. 212   a) Mature intact passion fruit from a maypop vine. b) The same fruit in longitudinal section showing the seeds within and the leathery exterior.   *Passiflora incarnata*   Passifloraceae

Fig. 213   The banana is a berry with a leathery exterior. *Musa* x *paradisiaca*   Musaceae

Fig. 214   Mature tomato both intact and in cross section. An example of a berry with soft flesh inside and a soft skin outside. *Solanum esculentum*   Solanaceae

Fig. 215   The grape is a soft-skinned berry. *Vitis* sp.   Vitaceae

a

b

a

b

Fig. 216   Berries whose exterior and interior is more or less of the same consistency. a) Blueberries.   *Vaccinium corymbosum*   Ericaceae b) Infructescence of greenbrier.   *Smilax bona-nox*   Smilacaceae

Fig. 217   a) One of an aggregate of pawpaw berries.   b) Berry in LS showing hard interior and seeds. *Asimina incana*   Annonaceae

Specialized Berries

Hesperidium - A specialized berry with a leathery exterior or rind with oil cavities and a fleshy, often juicy interior that is divided into several to many sections. Characteristic of citrus (Rutaceae) and its relatives.

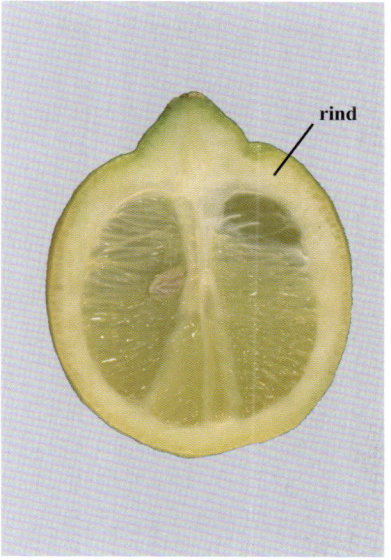

Fig. 218   The navel orange, interior shown in XS and exterior shown intact. Wedge-shaped sections define locules filled with fleshy, juice-containing hairs. *Citrus aurantium*   Rutaceae

Fig. 219   LS of lemon with thick waxy rind clearly visible. *Citrus limon*   Rutaceae

Pepo - A specialized berry with a leathery to hard exterior or rind and a fleshy interior that lacks septa. Characteristic of some members of the Family Cucurbitaceae.

a

b

Fig. 220   a) The cucumber, one of many common fruits classified as a pepo.   b) XS of cucumber showing the dark green leathery rind at the edge. *Cucumis sativus*   Cucurbitaceae

Fig. 221   The yellow squash is a pepo, as are gourds and pumpkins. *Cucurbita pepo*   Cucurbitaceae

Fig. 222   XS of cantaloupe with hard green exterior rind and soft orange flesh and seeds in center. *Cucumis melo*   Cucurbitaceae

Drupe

A fleshy, indehiscent fruit with one or more hard pits (**stones** or **pyrenes**) in the center, each of which contains a seed.

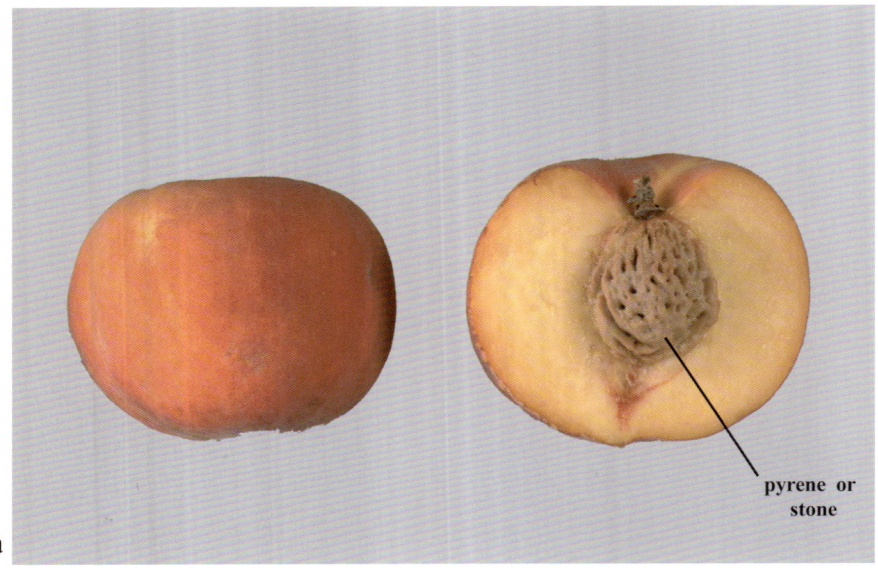

pyrene or stone

seed   pyrene

a

b

Fig. 223 a) The peach, a familiar drupe, is sectioned longitudinally to show the single stone or pyrene within. b) The hard pyrene in LS with one half removed to show the single seed inside. *Prunus persica* Rosaceae

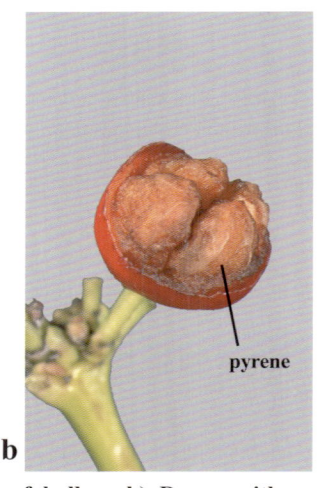

pyrene

a   b   a   b

Fig. 224 a) Intact drupe of wild cherry. b) LS of drupe. *Prunus caroliniana* Rosaceae

Fig. 225 a) Fruiting branch of holly. b) Drupe with four pyrenes exposed. *Ilex* sp. Aquifoliaceae

Fig. 226 Infructescence of immature drupes of dogwood. *Cornus florida* Cornaceae

Fig. 227 Partial infructescence of winged sumac with the pyrenes of many individual drupes exposed. *Rhus copallina* Anacardiaceae

Follicle

A dry dehiscent fruit that originates from a single carpel and splits along a single suture.

Fig. 228   a) Nearly mature follicles of milkweed.   b) Mature follicles beginning to dehisce.   c) Follicle completely split open and seeds being released.   *Asclepias curassavica*   Asclepiadaceae

Fig. 229   a) Immature follicles of periwinkle.   b) Completely dehisced follicle with seeds exposed.   *Catharanthus roseus*   Apocynaceae

Fig. 230   a)   Nearly mature follicle of Carolina larkspur.   b) Mature follicle.   *Delphinium carolinianum*   Ranunculaceae

Fig. 231   Aggregate of follicles.   *Aquilegia* sp.   Ranunculaceae

Capsule

A dry fruit produced from a gynoecium with two or more carpels. Capsules are categorized into sub-groups depending upon how they release their seeds. **Loculicidal capsules** split or dehisce into the locule forming segments or valves that are longitudinally divided by a septum. This septum takes the form of a distinct ridge with a portion (half) of adjacent locules on either side.

Loculicidal Capsules

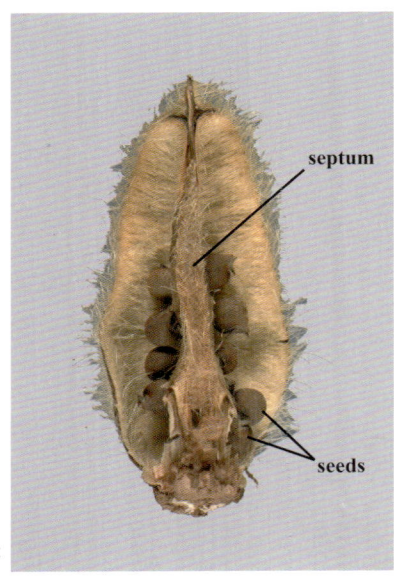

Fig. 232 **a) Loculicidal capsule of swamp hibiscus. b) Same capsule with two of the valves removed. c) A single valve or segment showing the septum with portions of adjacent locules and seeds attached.** *Hibiscus moscheutos* Malvaceae

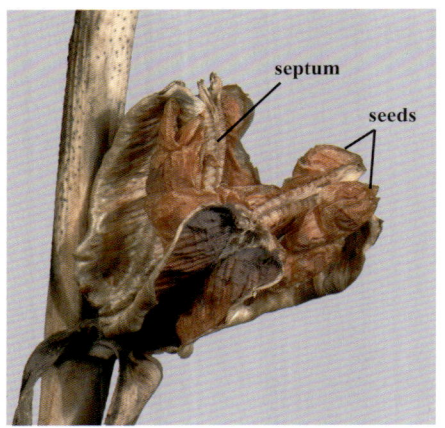

Fig. 233 *Iris* sp. Iridaceae

Fig. 234 *Allium* sp. Alliaceae

Fig. 235 *Ruellia* sp. Acanthaceae

Fig. 236 *Agave* sp.
Agavaceae

Fig. 237 *Biophytum* sp.
Oxalidaceae

Fig. 238 *Amaryllis* sp.
Amaryllidaceae

Capsule

Septicidal Capsule - A capsule that splits along the septa to release the seeds. This results in valves or segments that represent a single entire locule without a septum dividing it.

Poricidal Capsule - A capsule that releases seeds by means of pores such as holes or flaps.

Circumscissile Capsule - A capsule that splits along a horizontal line that circumscribes the entire fruit, resulting in the top portion falling off like a lid.

Septicidal Capsules

Fig. 239   a) Mature capsule of pipevine splitting along septa to form six valves. b) A single valve of pipevine with seeds. *Aristolochia elegans*   Aristolochiaceae

Fig. 240   a) The septicidal capsule of trumpet vine that splits into two valves or segments. b) A single valve with seeds removed. *Campsis radicans*   Bignoniaceae

Poricidal Capsules

Fig. 241   a) Developing capsule of field poppy that releases seeds from holes or pores along the top. *Papaver rhoeas*   Papaveraceae b) Mature capsule of Venus' looking glass which releases seeds through flaps on the side. *Triadanis perfoliata*   Campanulaceae

Circumscissile Capsule

Fig. 242   a) Mature circumscissile capsule of purslane before splitting. b) Capsule after dehiscence with top portion gone to reveal multiple seeds nestled in the persistent bottom portion. *Portulaca* sp.   Portulacaceae

Samara

A dry indehiscent winged fruit containing a single seed.

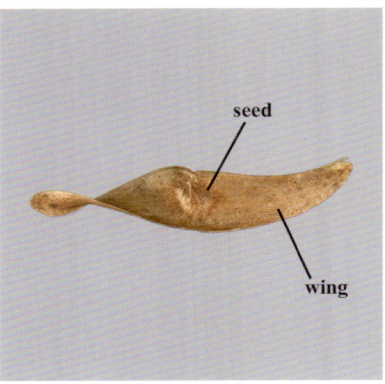

Fig. 243 a) Individual samara of river birch. b) Carpellate catkin from which the samaras are dispersed when mature.
Betula nigra   Betulaceae

Fig. 244 Samara of tree-of-heaven.
Ailanthus altissima   Simaroubaceae

Fig. 245 Cluster of immature samaras of green ash.
Fraxinus pennsylvanica   Oleaceae

Fig. 246 Developing samaras of the winged elm.
Ulmus alata   Ulmaceae

Fig. 247 Aggregate of samaras of the tulip poplar.
Liriodendron tulipifera   Magnoliaceae

Fig. 248 Samaras of green ash.
Fraxinus pennsylvanica   Oleaceae

Fig. 249 Samara of Chinese elm.
Ulmus parvifolia   Ulmaceae

Fig. 250 Samaras of tulip poplar.
Liriodendron tulipifera   Magnoliaceae

Schizocarp

A typically dry but sometimes fleshy fruit that is derived from two or more carpels, and that splits or breaks up into two or more segments called **mericarps** that each contain one seed.

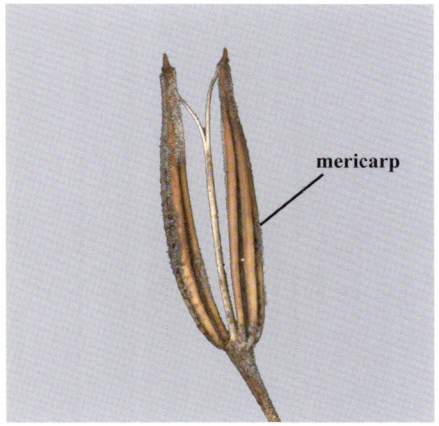

Fig. 251   Intact mature schizocarp of chervil.
Chaerophyllum tainturieri   Apiaceae

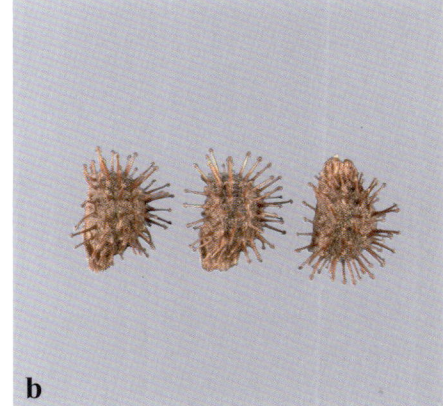

Fig. 252   a) Intact mature schizocarp of caesarweed showing five mericarps with hooked spines that aid in dispersal. b) Three individual mericarps.
Urena lobata   Malvaceae

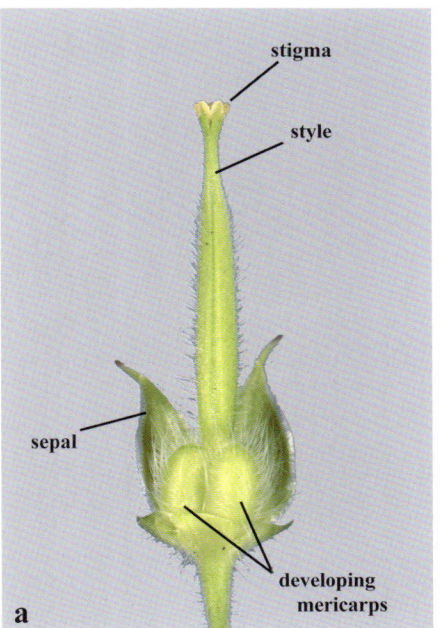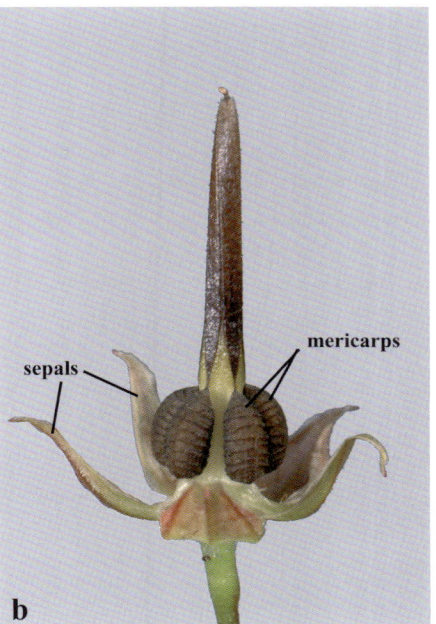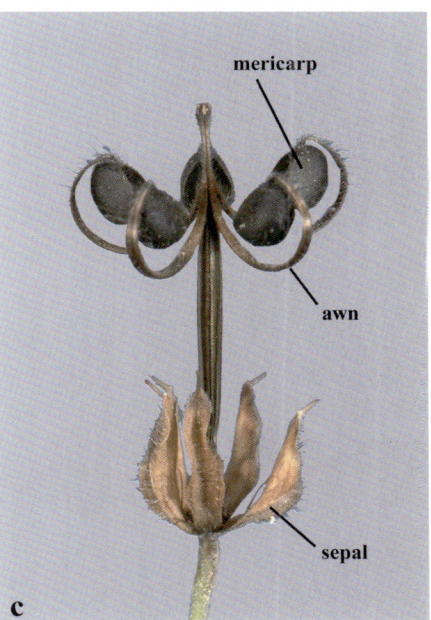

Fig. 253   a) Immature schizocarp of geranium with several sepals removed to show seeds developing at base. b) Mature schizocarp prior to dehiscence. c) Fruit after dehiscence with all five mericarps separated from base and curving up towards the top.
Geranium carolinianum   Geraniaceae

Fig. 254   Schizocarp of sugar maple with two mericarps.
Acer saccharum   Aceraceae

Fig. 255   Intact and dehiscing three-segmented schizocarps of the popcorn tree. *Sapium sebiferum*   Euphorbiaceae

Nut - Nutlet

An indehiscent dry fruit consisting of a single seed surrounded by a hard smooth wall. The distinction between nut and nutlet is subjective with larger fruits termed nuts and smaller fruits termed nutlets.

Fig. 256   Acorn of the shumard oak. An acorn is a nut partially enclosed by a cupule.
Quercus shumardii   Fagaceae

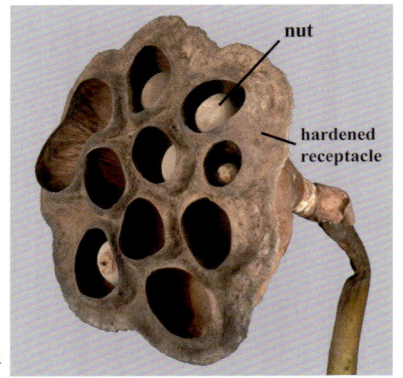

Fig. 257   a) Mature receptacle of the American lotus at time of fruit release. The nuts drop out into the water as the receptacle pulls away. b) CU of American lotus fruit known as a 'pond nut'.
Nelumbo lutea   Nelumbonaceae

Fig. 258   Nutlet and subtending bracts of ironwood.
Carpinus caroliniana   Betulaceae

Fig. 259   Fruit of hornbeam with top half of papery bract removed to expose the nutlet inside.
Ostrya virginiana   Betulaceae

Fig. 260   (L) Fruit of pickerelweed with accrescent perianth present. (R) Nutlet with perianth removed.
Pontederia cordata   Pontederiaceae

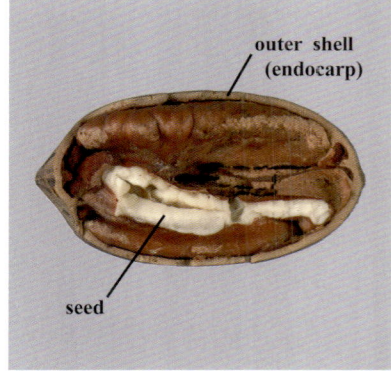

Fig. 261   Pecans are generally thought of as nuts, but in reality are dehiscent drupes with a leathery outer husk. The hard inner portion is actually a pit and the edible portion is the seed within.
a) Cluster of pecans still in the thick green leathery husks. b) Pecan with top half of husk removed to show the hard brown shell. c) Pecan in approximate LS showing thin outer shell and inner edible material.
Carya illinoiensis   Juglandaceae

Indehiscent Pod

A fruit that is typically dry and that does not split open at maturity. Seeds are sometimes liberated by animal dispersers such as in the case of the Brazil nut.

Fig. 262 Indehiscent pod and underground fruit of the peanut.
Arachis hypogaea Fabaceae

Fig. 263 Indehiscent pod of the baobab tree.
Adansonia digitata Bombacaceae

Caryopsis

A grain characteristic of grasses consisting of a small, dry, indehiscent fruit which has a thin coat fused to a single seed.

Fig. 264 Caryopses of wheat.
Triticum aestivum Poaceae

Fig. 265 Caryopses of corn.
Zea mays Poaceae

Fig. 266 Caryopses of rye.
Secale cereale Poaceae

Achene

A usually dry indehiscent fruit with a thin wall surrounding a single seed.

Fig. 267   a) Fruiting head of false dandelion.
b) Achene with pappus of fine hairs attached to apex.
Pyrrhopappus carolinianus   Asteraceae

Fig. 268   a) Fruit of netleaf leather-flower, an aggregate of achenes. b) Single achene with pappus of hairs.
Clematis reticulatus   Ranunculaceae

Fig. 269   a) Fruit of bristly buttercup, an immature aggregate of achenes. b) Mature aggregate of achenes.
c) Single mature achene.   *Ranunculus hispidus* var. *nitidus*   Ranunculaceae

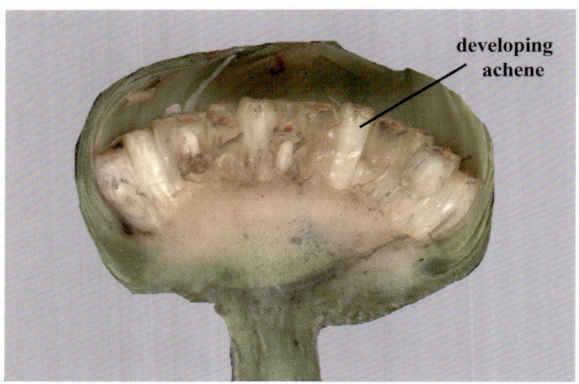
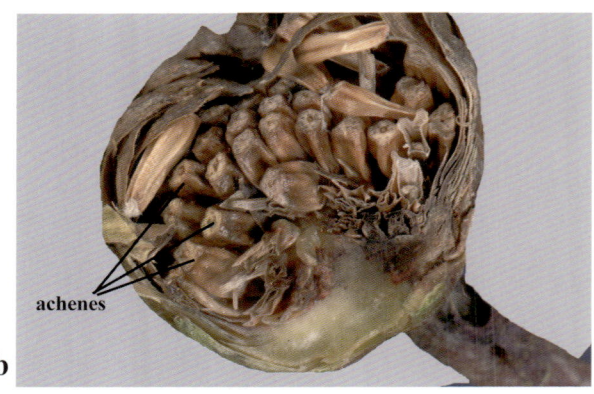

Fig. 270   a) LS of head of Stokes' aster showing developing achenes. b) LS of head showing mature achenes.
Stokesia laevis   Asteraceae

Fig. 271   a) Developing achenes in infructescence of sticktight. b) Achene with barbed spines at apex for dispersal. *Bidens* sp.   Asteraceae

Fig. 272   Submerged coontail achene.
Ceratophyllum demersum Ceratophyllaceae

Silique

Fruit of the Family Brassicaceae. Produced from a gynoecium with two carpels, and seeds attached to a persistent thickened rim(**replum**) with a membrane(**false septum**). The seeds and false septum are not visible until the two valves of the fruit have split away. Many repla often remain attached to the stem.

Fig. 273 Siliques found on plants in the mustard family. From left: bittercress *(Cardamine hirsuta)*, tansy mustard *(Descurainia pinnata)*, shepherd's purse *(Capsella bursa-pastoris)*, pepperweed *(Lepidium virginicum)*. Brassicaceae

Fig. 274 Silique of tansy mustard with valves pulled away. *Descurainia pinnata* Brassicaceae

Fig. 275 Sequence of pepperweed siliques. a) Developing and mature fruits. b) Fruit with one valve detached revealing one seed. c) Fruit with both valves removed to show replum with seeds attached. d) Replum with one seed attached and one seed detached and replum with no seeds. *Lepidium virginicum* Brassicaceae

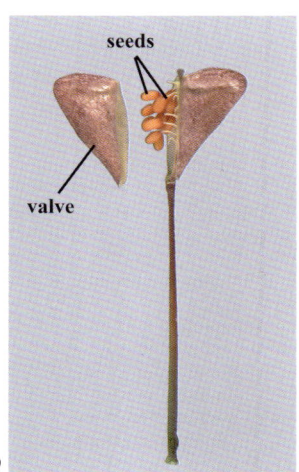

Fig. 276 Sequence of shepherd's purse siliques. a) Developing and mature fruits. b) Fruit with one valve detached and one valve intact. c) Fruit with both valves detached to show seeds attached to replum. d) Replum with seeds attached and replum with seeds removed showing thickened rim with membranous false septum. *Capsella bursa-pastoris* Brassicaceae

Legume

A dry fruit that originates from a single carpel and dehisces by means of two longitudinal sutures. Some split completely and equally along both sutures resulting in a V-shaped fruit after dehiscence. Others split primarily along one suture or minimally to resemble a follicle or an indehiscent pod.
Only found in the Family Fabaceae (= Family Leguminosae).

Fig. 277   Mature legumes of milkpea before and after dehiscence. *Galactia volubilis*   Fabaceae

Fig. 278   Legumes of white leadtree. *Leucaena leucocephala*   Fabaceae

Fig. 279   Immature legumes of mimosa tree. *Albizia julibrissin*   Fabaceae

Fig. 280   *Crotalaria spectabilis*   a) Developing legume, b) Legume in longitudinal section showing seeds.

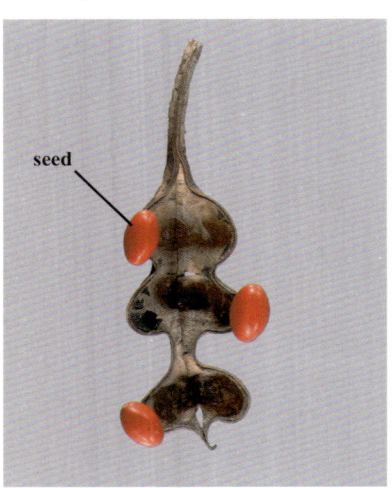

seed

Fig. 281   Legume of coral bean split along a single suture. *Erythrina herbacea*   Fabaceae

seeds

Fig. 282   Legume of bladderpod with very little dehiscence. *Sesbania punicea*   Fabaceae

Fig. 283   Immature legume of wisteria. *Wisteria sinensis*   Fabaceae

Loment

A dry fruit produced from a single carpel that breaks along transverse lines into segments of one seed each. Only found in the Family Fabaceae (= Family Leguminosae).

Fig. 284   a) **Infructescence of sensitive plant consisting of a cluster of loments. b) Developing and mature loments. c) Loment showing a single segment or seed that has broken away and three that remain on the fruit.** *Mimosa pudica*   Fabaceae

Fig. 285   a) **Developing loments on fruiting branch of false moneywort. b) Two mature loments. c) Mature loment that has begun to dehisce along transverse lines.** *Alysicarpus ovalifolius*   Fabaceae

Fig. 286   a) **Two immature loments developing on beggarweed. b) Three seeds broken away from a mature loment and attached to author's pants.** *Desmodium incanum*   Fabaceae

Pome

A fleshy indehiscent fruit only found in the Subfamily Maloideae of the Family Rosaceae. The outer portion is soft and formed from the expansion of the floral structures that surround the ovary and enlarge as the fruit matures. The central portion contains the seeds which are surrounded by a cartilaginous or papery structure that is sometimes indistinct and difficult to see.

 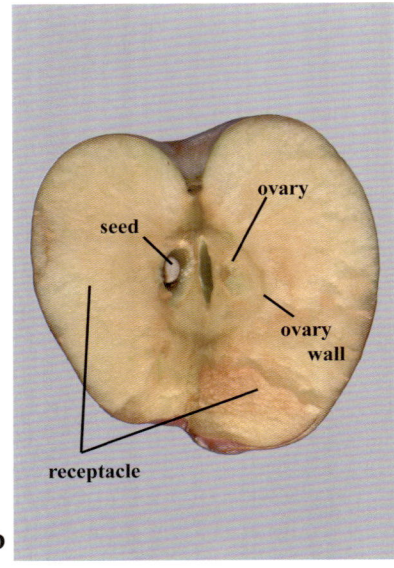 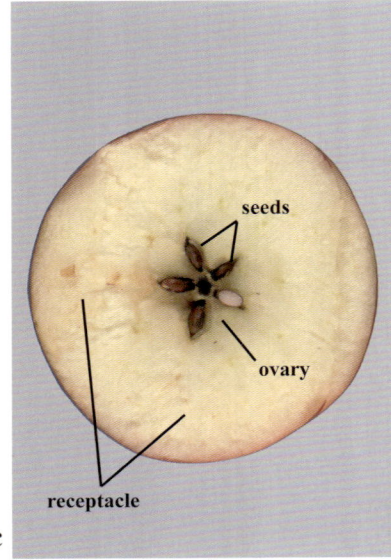

a b c

Fig. 287 a) The apple is a common example of a pome fruit. b) LS of apple showing fleshy receptacle and ovary that is distinguished by a faint line representing the ovary wall. c) XS of apple showing seeds at center of fruit. Ovarian tissue is barely discernible from that of the receptacle. *Malus sylvestris*   Rosaceae

 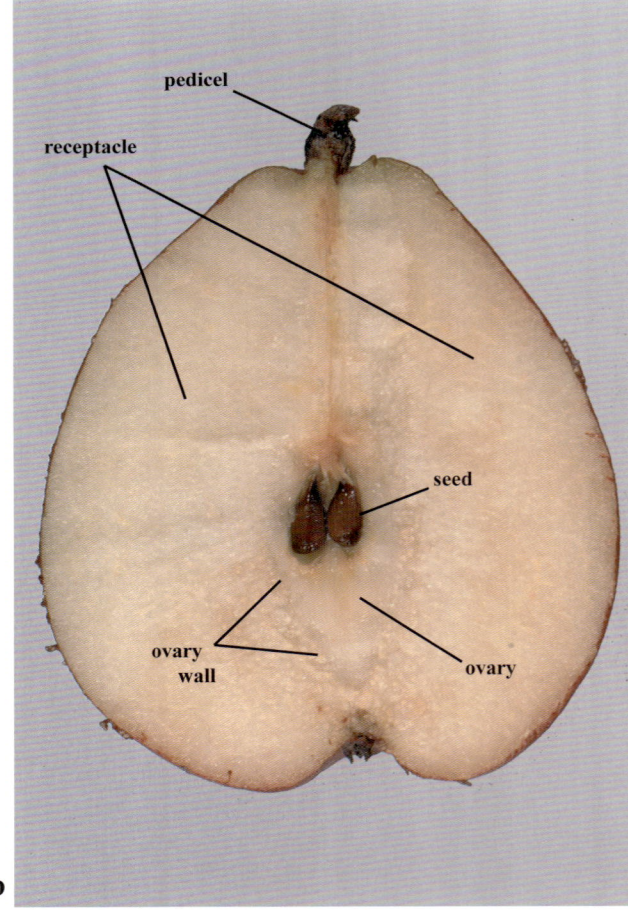

a b

Fig. 288 a) A pear, another example of a pome fruit. b) LS of pear showing ovarian and receptacle tissues and seeds. *Pyrus communis*   Rosaceae

Accessory Fruits

 55

A fruit or cluster of fruits that includes tissue not derived from the ovary.

a **b**

Fig. 289   a) The mature strawberry fruit primarily consists of an edible red flesh. This material is formed from the flower stalk or receptacle. The small, yellow gritty structures on the strawberry surface are the actual fruits or achenes. b) LS of mature strawberry showing receptacle and location of achenes on outside surface.   *Fragaria x ananassa* Rosaceae

a **b**

Fig. 290   The fruit of figs is both an accessory and multiple fruit called a syconium. Both the flowers and the fruits are borne on the inside of this fleshy structure. a) Intact syconium of edible fig. b) LS of syconium showing developing fruits in central hollow area.   *Ficus carica*   Moraceae

Multiple Fruits

A fruit formed from multiple, closely packed flowers where the developing structures coalesce into a single unit.

Fig. 291   a) Inflorescence of a pineapple showing multiple flowers scattered spirally around a central axis.
b) Developing pineapple fruit where the berries derived from multiple flowers have coalesced into a single structure.
Ananas comosus   Bromeliaceae

Fig. 292   a) Carpellate inflorescence of sycamore. b) Mature fruit made up of a ball of tightly packed achenes.
Platanus occidentalis   Platanaceae   c) Carpellate inflorescence or catkin of the red mullberry.   d) Mature multiple fruit made up of many drupes.   *Morus rubra* Moraceae   e) Carpellate inflorescence of sweetgum.   f) Mature fruit made up of a ball of multiple capsules.   *Liquidambar styraciflua* Altingiaceae

Aggregate Fruits

A fruit formed from several or many carpels of only one flower.

a

b

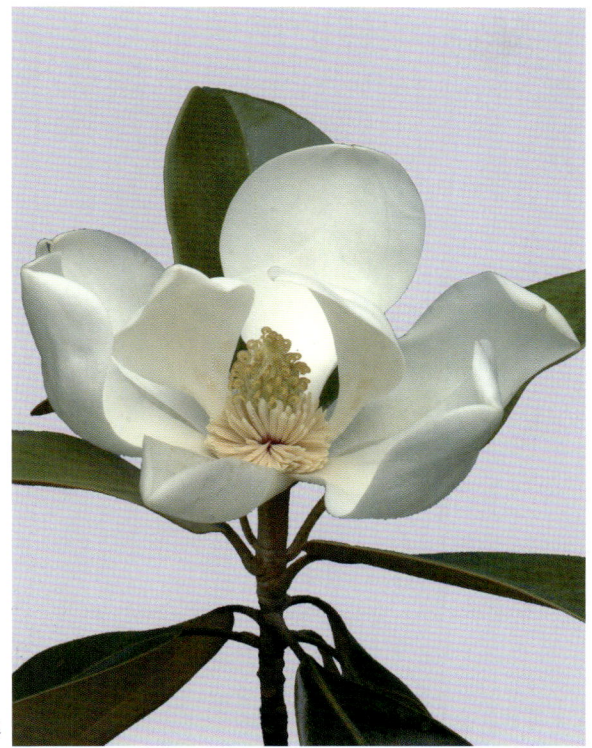

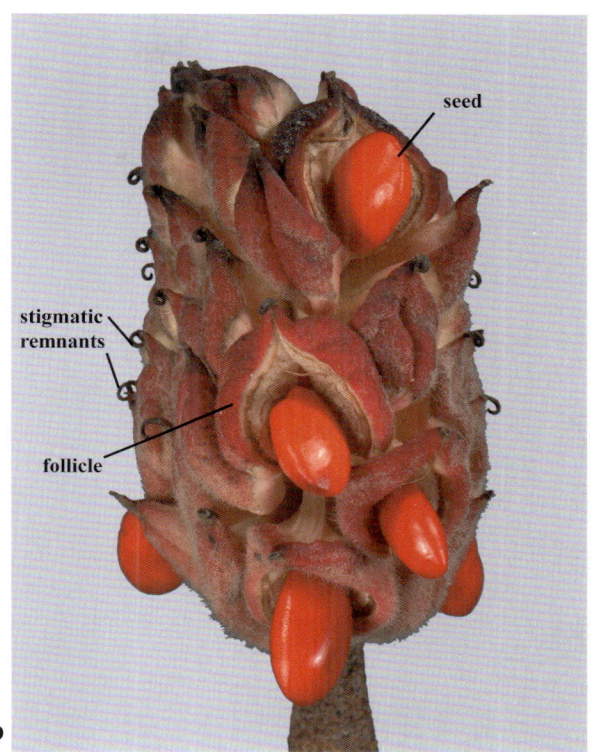

Fig. 293 a) Flower of magnolia. b) Mature fruit consisting of an aggregate of follicles.
Magnolia grandiflora Magnoliaceae

a

c

e

b

d

f

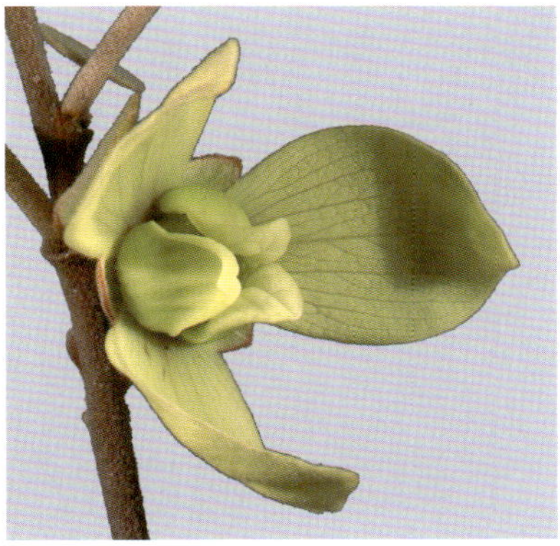

Fig. 294 a) Flower of pawpaw. b) Mature fruit consisting of an aggregate of berries. *Asimina incana* Annonaceae
c) Flower of blackberry. d) Mature fruit consisting of an aggregate of drupes. *Rubus argutus* Rosaceae
e) Flower of illicium. f) Developing fruit consisting of an aggregate of follicles. *Illicium floridanum* Illiciaceae

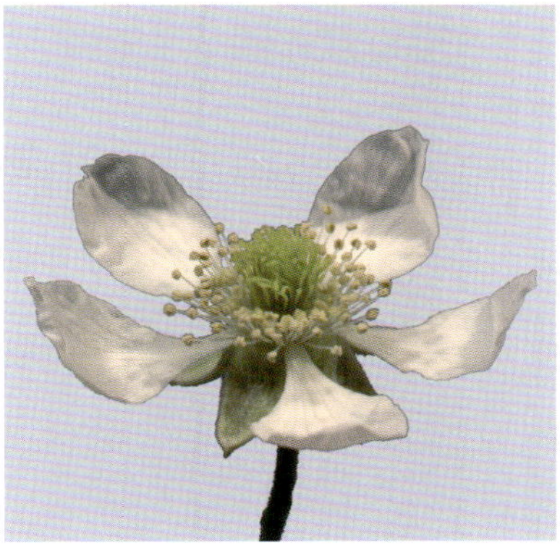

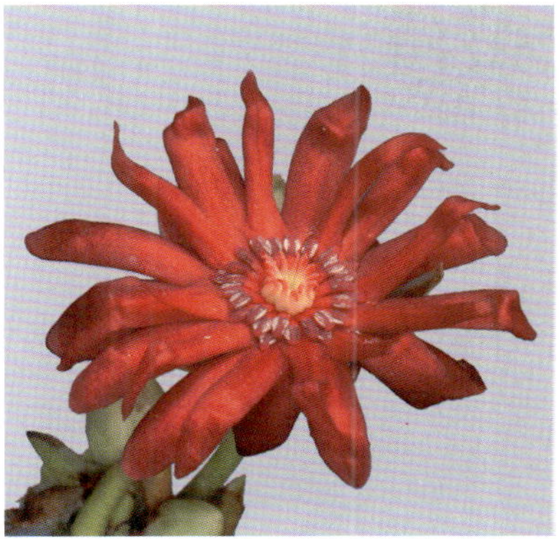

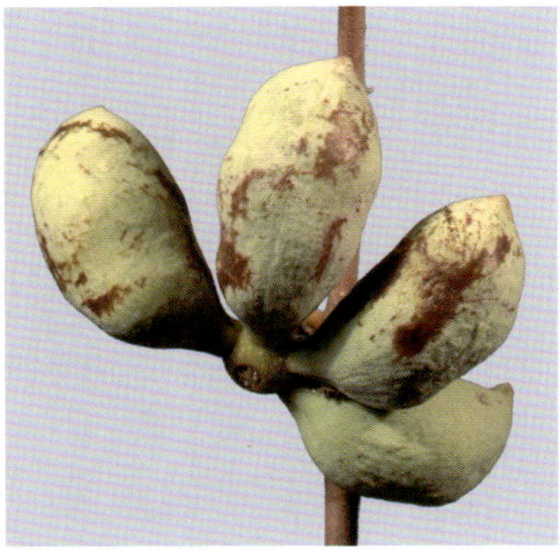

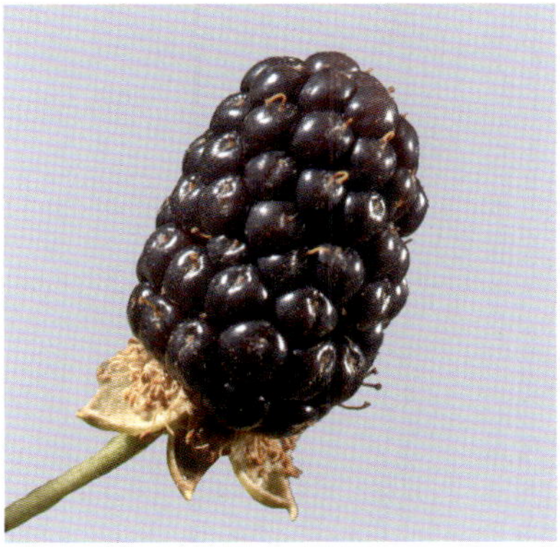

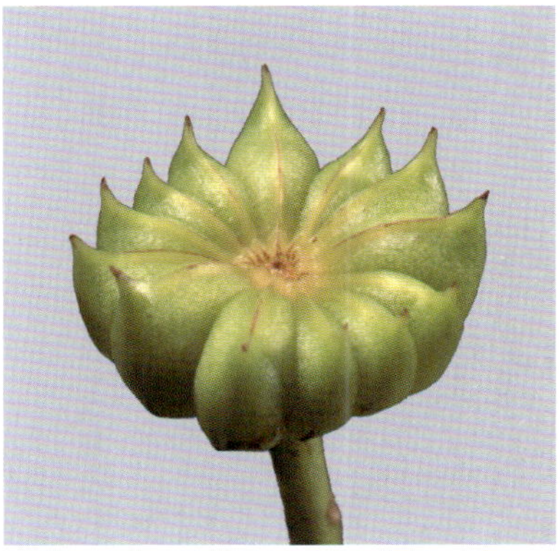

Seed Dispersal

Fig. 295 WIND is used by many plants that have evolved light seeds bearing wings or other associated structures that permit them to be transported on air currents. The achenes of the false dandelion are attached to a fuzzy pappus. *Pyrrhopappus carolinianus*

Fig. 296 WATER can sometimes carry fruits or seeds great distances, such as with the coconut above. If conditions are suitable it will sprout and begin to grow. *Cocos nucifera*

Fig. 297 BIRDS and other animals such as rodents and mammals are often attracted to brightly colored fruits, such as the American beauty berry. Many seeds have evolved to pass through the digestive system of disperser animals before germinating. *Callicarpa americana*

aril

seed

Fig. 298 INVERTEBRATES such as ants sometimes feed on special edible structures called arils, found attached to seeds. They may carry the seed with its aril back to the nest where the seed is later discarded. Larger animals may also be attracted to arils. *Ricinus communis*

Fig. 299 PHORESY refers to the act of carrying. Many seeds have spines, barbs, or tiny hooked hairs that allow them to become attached to the fur of animals and hitch a ride. Three species of phoretic seeds are attached to the author's sock above.
Sandspur *Cenchrus* sp.,
beggarweed *Desmodium incanum*, and
Spanish needles *Bidens alba*.

Fig. 300 EXPLOSIVE DEHISCENCE refers to fruits that explode or pop apart thus propelling the seeds they contain a considerable distance. The two light colored schizocarps of wild poinsettia above are almost at the point where they will explode. *Euphorbia heterophylla*

Plant Taxonomy

The taxonomic coverage of common plant families begins here and continues to page 293. Immediately following this is a list of botany reference and textbooks. This is followed by a glossary of terms which students are encouraged to use to clarify any questions regarding botanical terminology used throughout this book. The very last page lists every plant family treated in this book in alphabetical order. At a glance, the reader should be able to determine if a particular family has been included and on what page to find the coverage.

Seedless Vascular Plants (Ferns & Fern Allies)

Introduction

The term Seedless Vascular Plants has been applied to those plant families commonly referred to in other works as Ferns and Fern Allies. These families do not represent a monphyletic group, but rather a paraphyletic grade of the early land plants. Current interpretations of their evolutionary history indicate that there are two main clades that are monophyletic. These are the Lycopsids and the Monilophytes. There are only three families of extant Lycopsids. These are the Lycopodiaceae or club mosses, the Selaginellaceae or spikemosses, and the Isoetaceae or quillworts.

The Monilophytes include all the other extant families of ferns and fern allies. The non-fern groups are represented by the Equisetaceae or horsetails and the Psilotaceae or whisk ferns. Taxonomic coverage of the true ferns includes the following families: Ophioglossaceae (adder's tongue and grape ferns), Osmundaceae (royal ferns), Cyatheaceae (tree ferns), Marsileaceae (water clovers), Polypodiaceae (polypody ferns), Salviniaceae (water spangles), and the Azollaceae (mosquito ferns).

Lycopodiaceae - *Clubmoss Family*

Most primitive group of vascular plants.
Dichotomously branching stems and roots.
Tiny vegetative leaves (microphylls) with a single vein.
Microphylls typically alternate and arranged in a spiral pattern.
Sporangia either borne by fertile leaves (sporophylls) that are scattered along the stem,
or by leaves modified into a cone-like strobilus.
Homosporous.

Fig. 301   a) Clubmoss with sterile branching stems running along the ground surface and fertile stems that are erect.
b) Section of stem with roots intact showing their dichotomously branching nature.
Lycopodiella appressa

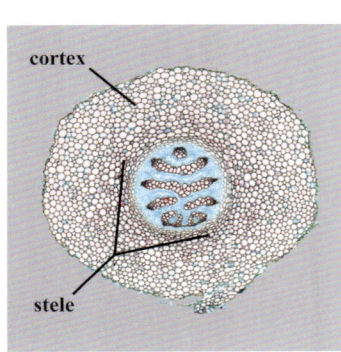
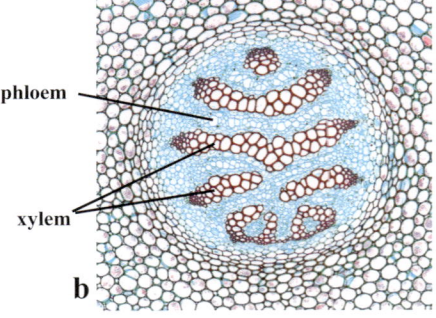

Fig. 302   a) Small spine-like vegetative leaves (microphylls) on a fertile stem.
b) Tip of fertile stem where leaves have formed a strobilus.
Lycopodiella appressa

Fig. 303   a) XS of rhizome.
b) CU of stele.
Lycopodium sp.

Fig. 304   Clubmoss with erect branching sterile stems. *Lycopodium digitatum*

Fig. 305   CU of sterile stems with tiny vegetative leaves (microphylls).

a

b

c

Fig. 306   a) Developing strobili on a single fertile stem. b) Mature dehiscing strobili of two fertile stems. c) Single isolated mature strobilus.   *Lycopodium digitatum*

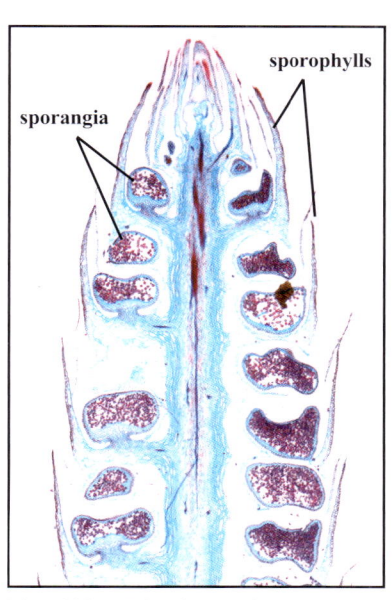

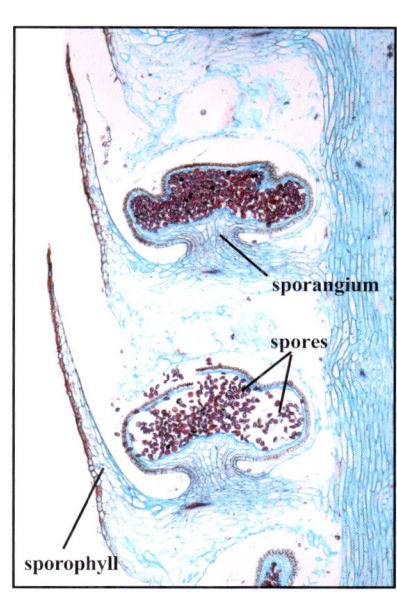

Fig. 307   CU of mature strobilus. *Lycopodium digitatum*

Fig. 308   LS of strobilus. *Lycopodium* sp.

Fig. 309   CU of LS of strobilus. *Lycopodium* sp.

Selaginellaceae - *Spikemoss Family*

Dichotomously branching stems and roots.
Small vegetative leaves (**microphylls**) typically arranged in four rows along stem.
Leaves with a small flap or ligule at their base.
Sporangia typically borne by fertile leaves (**sporophylls**) that are modified into strobili.
Heterosporous.
Only one genus (*Selaginella*).

Fig. 310   Branch and strobilus characters of spikemoss. a) Sterile branch showing patterns of branching and arrangement of leaves. b) Fertile branch with many strobili at tips. c) CU of two strobili. *Selaginella* sp.

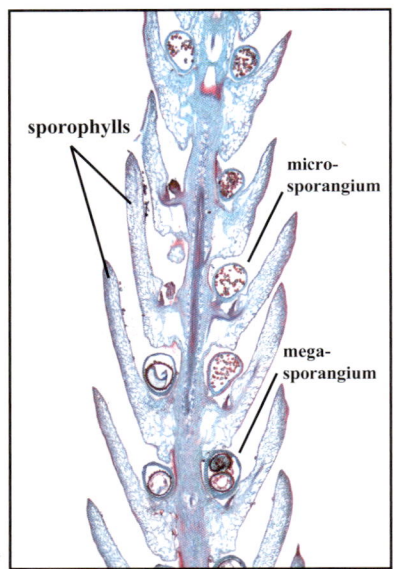 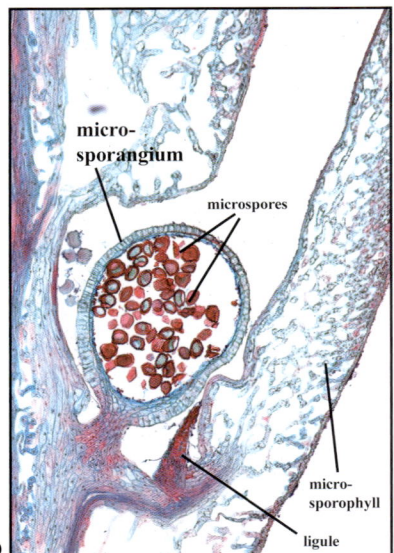 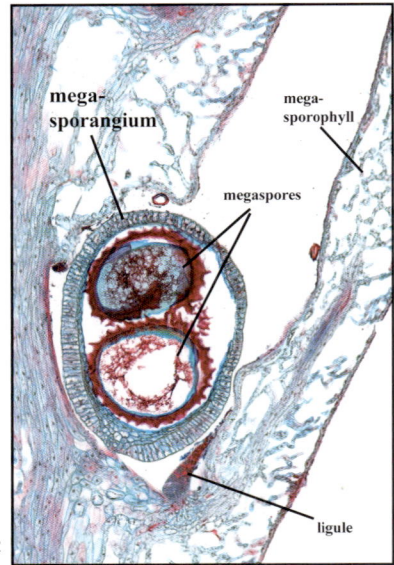

Fig. 311   Strobilus characteristics of spikemoss. a) LS of strobilus. b) LS of microsporangium. c) LS of megasporangium. *Selaginella* sp.

Isoetaceae - *Quillwort Family*

Short, fleshy underground stem (**corm**).
Long, slender quill-like leaves with ligules.
Roots not dichotomously branched.
Sporangia embedded in base of leaves.
Heterosporous.
Grow in wetlands or aquatic situations.
Only one genus (*Isoetes*).

Fig. 312   Plant characters of quillwort. a) Habit, showing long slender leaves. b) Two clusters of quillworts.
Isoetes flaccida

embedded
sporangium

Fig. 313   CU of leaf bases of quillworts showing the embedded sporangia.
Isoetes flaccida

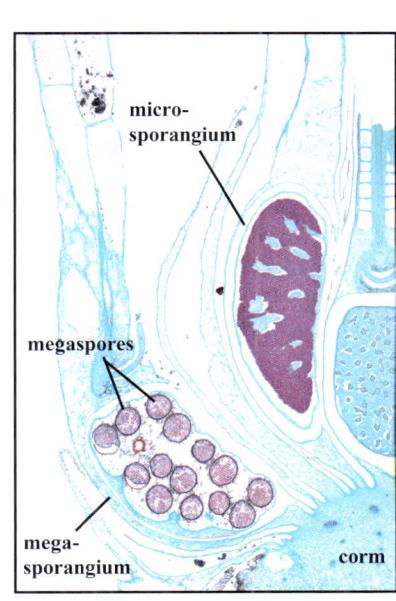

micro-
sporangium

megaspores

mega-
sporangium

corm

Fig. 314   LS of quillwort base.
Isoetes sp.

Psilotaceae - *Whisk Fern Family*

Roots absent, replaced by underground stems.

Dichotomously branching, erect aerial stems.

Tiny scale-like leaves on aerial stems.

Sporangia appear three-chambered, but are actually three fused sporangia forming a synangium.

Sporangia borne on the tip of extremely short spike-like branches.

Homosporous.

Only two genera (*Psilotum* and *Tmesipteris*).

Fig. 315   Plant characters of whisk fern. a) Dense growth habit of whisk fern. b) One aerial stem isolated to show dichotomous branching. c) New growth at ground level.
Psilotum nudum

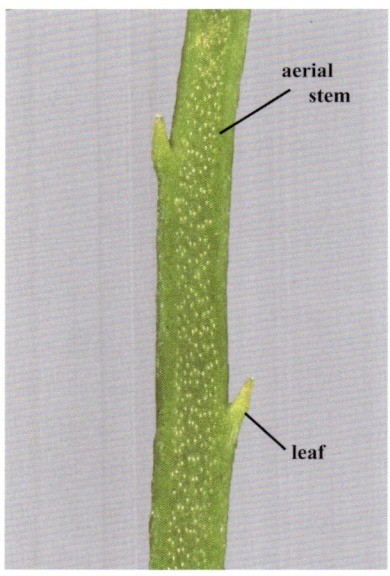  

Fig. 316   Stem and reproductive characters of whisk fern. a) CU of section of aerial stem showing leaves. b) Aerial stems laden with synangia. c) CU of synangia and the spike-like branches that bear them.
Psilotum nudum

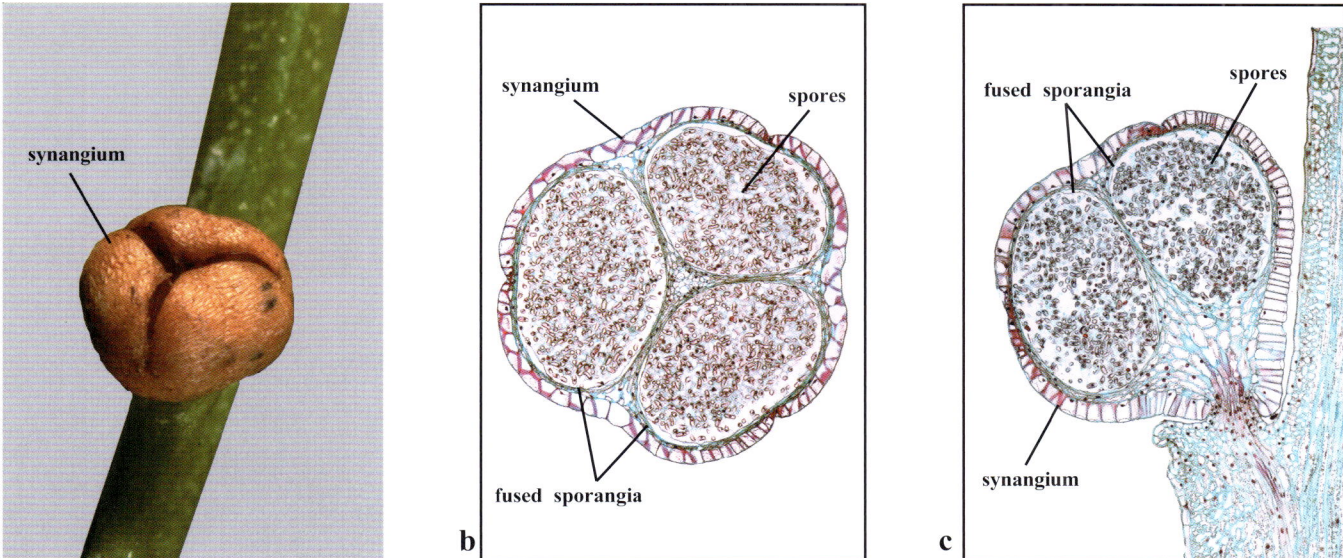

Fig. 317   a) Synangium characters of whisk fern. a) CU of mature synangium. b) XS of synangium. c) LS of synangium. *Psilotum nudum*

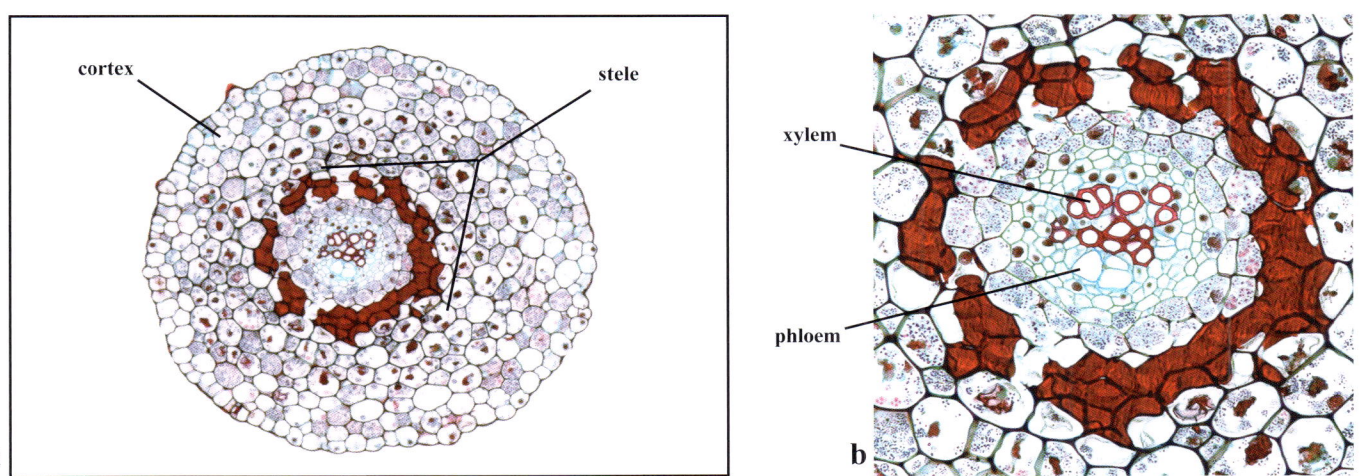

Fig. 318   Rhizome characters of whisk fern. a) XS of rhizome showing vascular core or stele. b) XS of rhizome with CU of stele showing solid core of xylem. *Psilotum nudum*

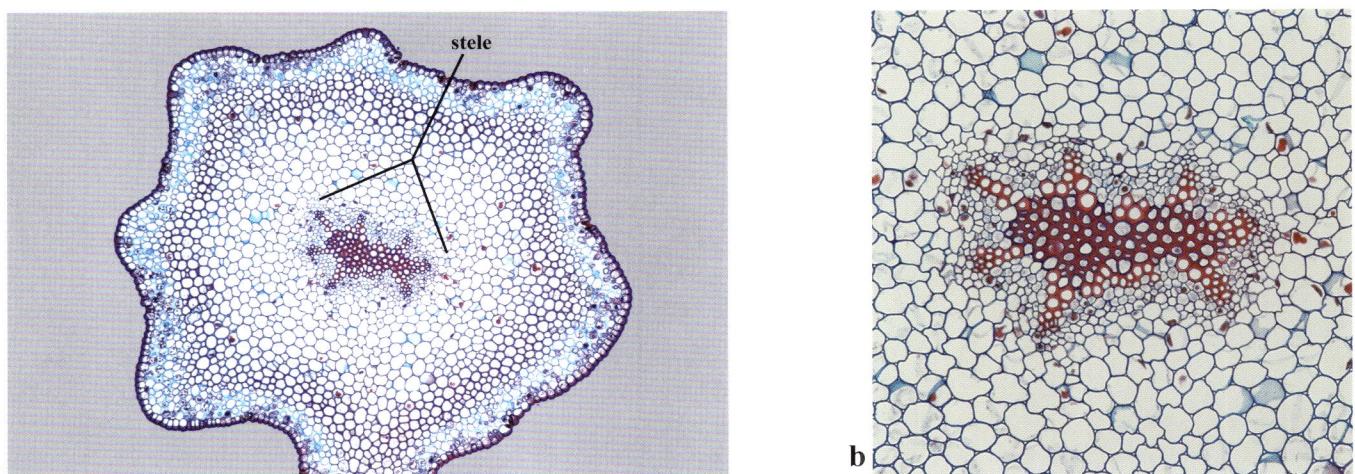

Fig. 319   Aerial stem characters of whisk fern. a) XS of aerial stem. b) XS of aerial stem with CU of stele. *Psilotum nudum*

Equisetaceae - *Horsetail Family*

Small to large herbaceous plants (up to 8 meters).

Roots develop from rhizomes and are not dichotomously branching.

Stems are jointed and have ridges.

Leaves occur in small, dark whorls at the nodes and are connected at the base to form a sheath.

Branches occur in whorls at the nodes.

Sporangia are borne along the margin of umbrella-shaped sporangiophores.

Sporangiophores are clustered into strobili.

Homosporous.

Spores bear thickened bands (elaters) that aid in dispersal.

Only one genus (*Equisetum*).

Fig. 320   Portion of stem of horsetail showing whorls of projecting branches and small whorls of tightly appressed leaves at the nodes.   *Equisetum arvense*

Fig. 321   Whorl of small leaves on the fertile stem of the scouring rush.
Equisetum hyemale

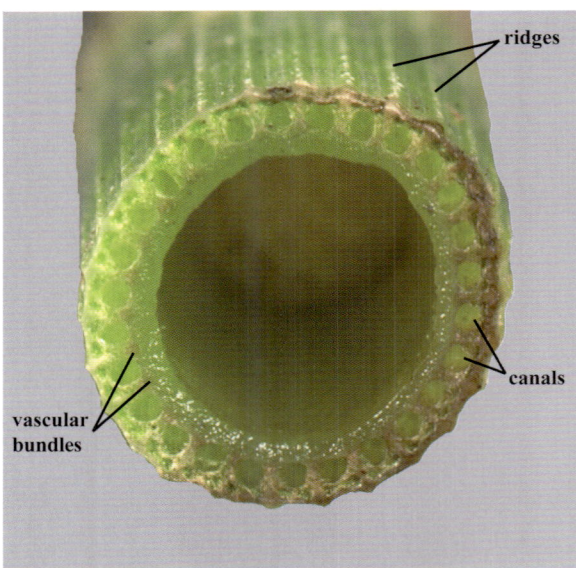

Fig. 322   XS of living fertile stem of scouring rush.
Equisetum hyemale

Fig. 323   Strobili of scouring rush. a) Strobili in various stages of dehiscence. b) CU of mature strobilus showing dark round sporangiophores.   *Equisetum hyemale*

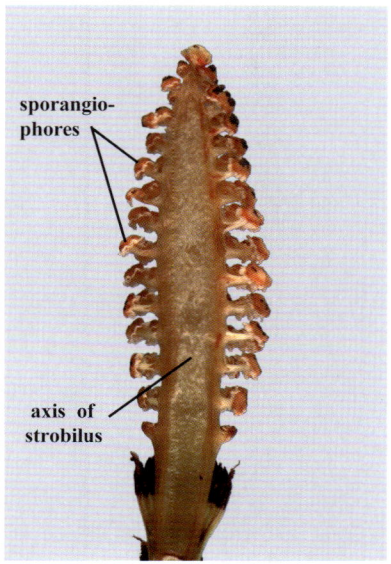

Fig. 324   LS of mature strobilus of the scouring rush. *Equisetum hyemale*

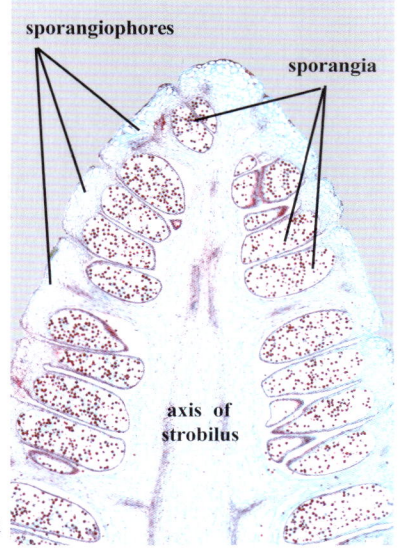

Fig. 325   Characters of horsetail strobilus. a) CU of LS of strobilus showing sporangiophores and the sporangia borne on them. b) XS of strobilus. *Equisetum arvense*

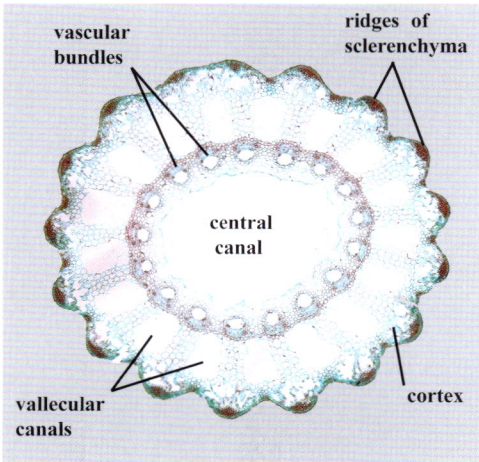

Fig. 326   XS of sterile stem of horsetail. *Equisetum arvense*

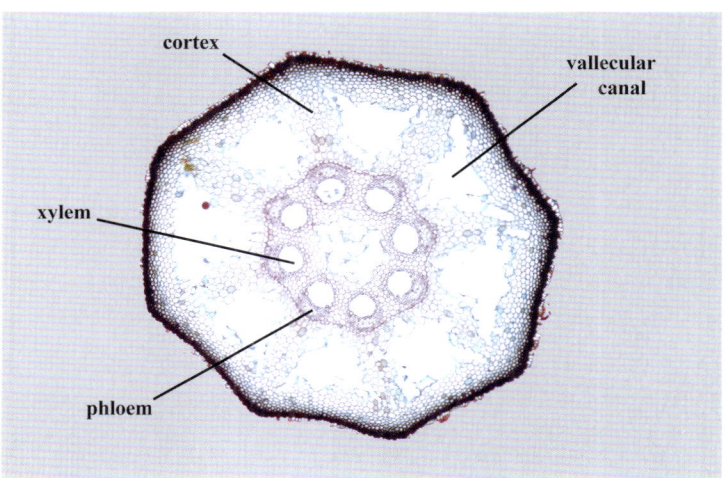

Fig. 327   XS of rhizome of horsetail. *Equisetum arvense*

Ferns

Introduction

Ferns grow from a **rhizome** that may be underground or found on the substrate surface. The leaves or **fronds** vary in size and shape and are typically **circinate,** or coiled into a **fiddlehead** when young. Ferns are divided into two main groups. The more primitive are the eusporangiate ferns (Family Marattiaceae), characterized by having thick-walled **sporangia** without stalks. The more advanced leptosporangiate ferns have have very thin-walled (one cell layer) sporangia that are borne on stalks. Each sporangium has a row or cluster of cells termed an **annulus** that functions in the release of spores. The sporangia of leptosporangiate ferns occur in clusters called **sori** that may or may not be covered by protective tissue called an **indusium.** Leptosporangiate fern families are distinguished by characters of the stalks, annuli, and sori. The Family Ophioglossaceae is a primitive leptosporangiate fern that shows intermediate characteristics.

Haploid spores released by the sporangium divide by mitosis and produce a small heart-shaped haploid **gametophyte** or **prothallus.** Egg-producing **archegonia** and sperm-producing **antheridia** are both produced on the gametophyte. Water provides the medium permitting fertilization that results in the large obvious diploid sporophyte generation that we recognize as ferns. The tiny prothallus or gametophyte generation, although photosynthetic, lacks the vascular tissue that is present in the sporophyte.

Terminology

annulus - Row or cluster of thick-walled cells on a sporangium that function in spore release.

antheridium - Male reproductive structure that is a tiny capsule containing sperm. Antheridia are found on the base and 'wings' of the prothallus.

archegonium - Female reproductive structure that is tiny and flask-shaped and encloses an egg. Many archegonia are found on the central area of the prothallus below the notch.

blade - Leafy, foliaceous portion of a frond.

circinate - Term describing coiled arrangement of leaves and leaflets.

false indusium - Recurved leaf margin that folds over sori developing at the leaf edge. It is not a separate structure from the leaf like a true indusium.

fiddlehead - Coiled immature leaf with apex at center of the roll.

frond - Leaf of a fern consisting of the stipe and blade.

gametophyte - Small, usually photosynthetic, often heart-shaped structure resulting from a germinated spore and bearing the sex organs. The haploid gamete-producing generation of a fern. Also called the prothallus.

indusium - Membranous outgrowth of a leaf that acts as a protective structure over a developing sorus.

pinna - Leaflet or primary division of a leaf.

pinnule - Secondary or greater division of a leaf. Segment of a leaflet.

prothallus - Same as the gametophyte.

rachis - Extension of the petiole forming the central axis of a divided leaf from which the pinnae arise.

rhizome - Creeping stem. May be horizontal or vertical, below the soil surface or above.

sorus - Small cluster of sporangia appearing as brown or yellow spots or lines on the underside of a leaf.

sporangium - Reproductive structure that produces and encases spores.

sporophyte - Large, leaf-bearing phase commonly observed. The diploid spore-producing generation of a fern resulting from fertilization.

stele - Collective term for the vascular bundles.

stem - Horizontally or vertically developing structure from which the leaves arise. Grows on the ground surface or just below. Typically short.

stipe - Petiole or leaf stalk of a frond.

Fig. 328 SIMPLE (with stipe)
Pyrrhosia lingua
Polypodiaceae

Fig. 329 SIMPLE (with very short stipe, not shown)
Asplenium nidus   Polypodiaceae

Fig. 330   PINNATE
Nephrolepis cordifolia
Polypodiaceae

Fig. 331   BIPINNATE
Pteridium aquilinum
Polypodiaceae

Fig. 332   PINNATIFID
Woodwardia areolata
Polypodiaceae

Fig. 333   PINNATE - PINNATIFID
Osmunda cinnamomea
Osmundaceae

Fig. 334   Each tiny leaf-like structure is a young fern sporophyte approximately an inch in length. This is the vascular diploid phase in the alternation of generations that will eventually produce large fronds and rhizomes. Below these obvious young sporophytes is a carpet of small dark green, rounded prothalli. These prothalli are the gametophytes or non-vascular haploid phase in the alternation of generations. The sporophytes sprout from the gametophytes following the fertilization of eggs in the antheridia on the gametophyte surface.

Fern Development

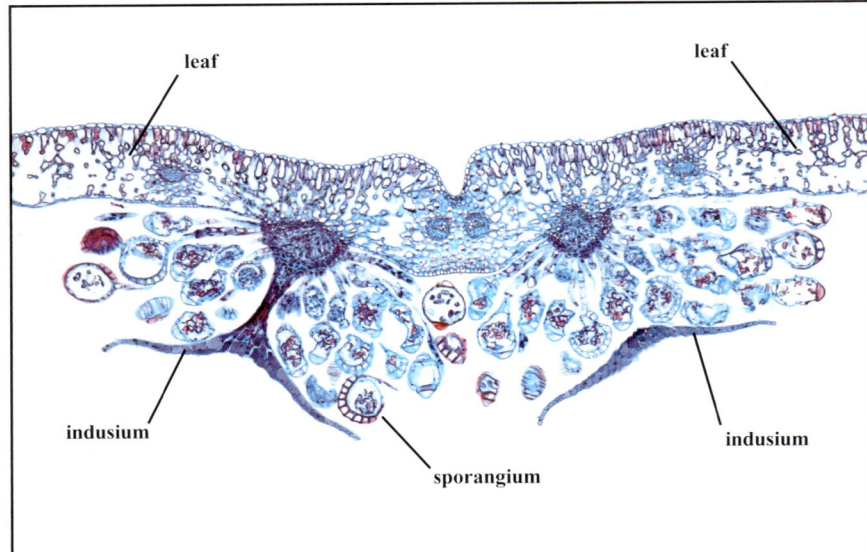

Fig. 335   XS of fern leaf. Two umbrella-shaped indusia, each with clusters of sporangia, project downwards from the lower surface of the leaf.

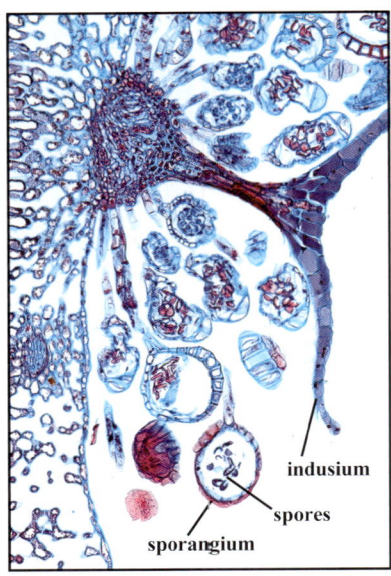

Fig. 336   CU of indusium and its associated sporangia.

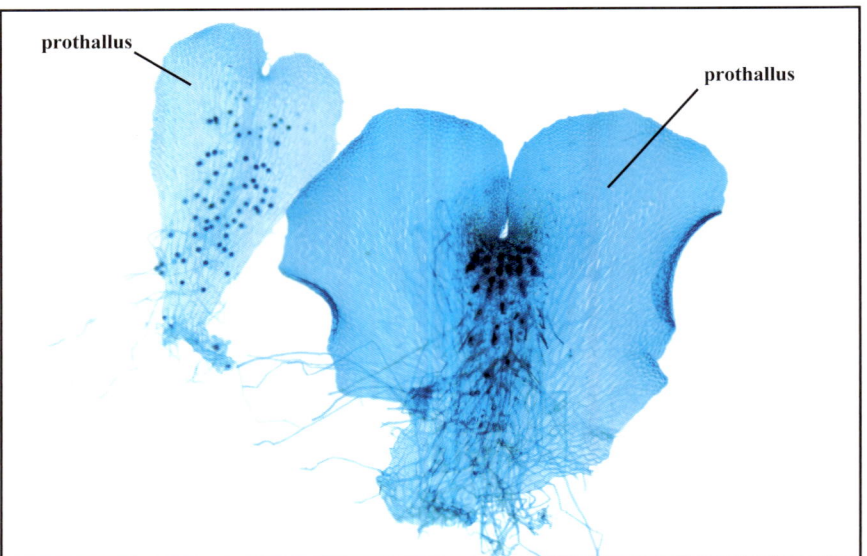

Fig. 337   Two fern prothalli or gametophytes. The dark spots are gametangia, consisting of archegonia and antheridia.

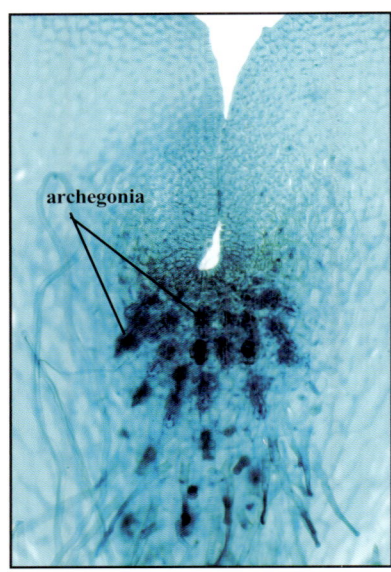

Fig. 338   CU of fern prothallus (gametophyte) showing archegonia.

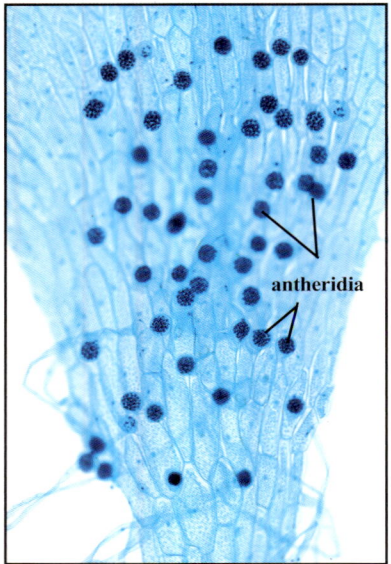

Fig. 339   CU of fern prothallus (gametophyte) showing antheridia.

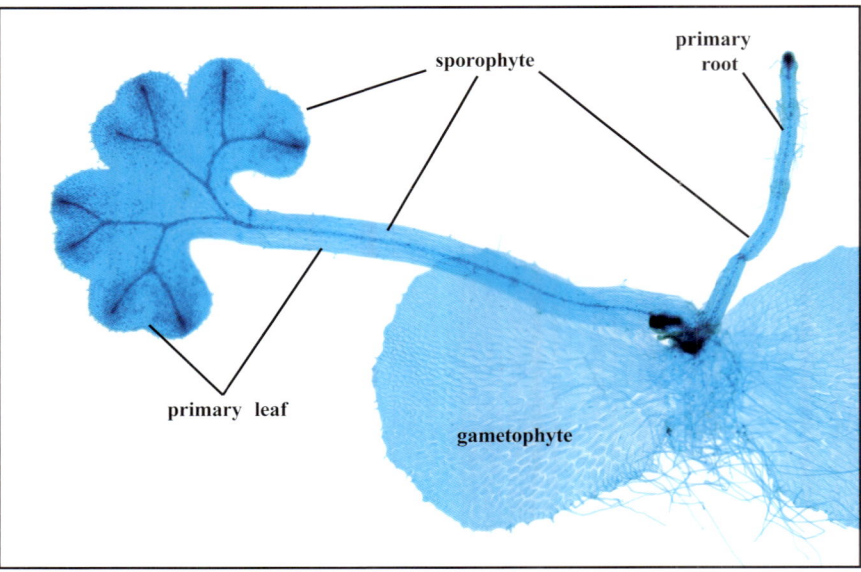

Fig. 340   Fern prothallus (gametophyte) with young sporophyte attached.

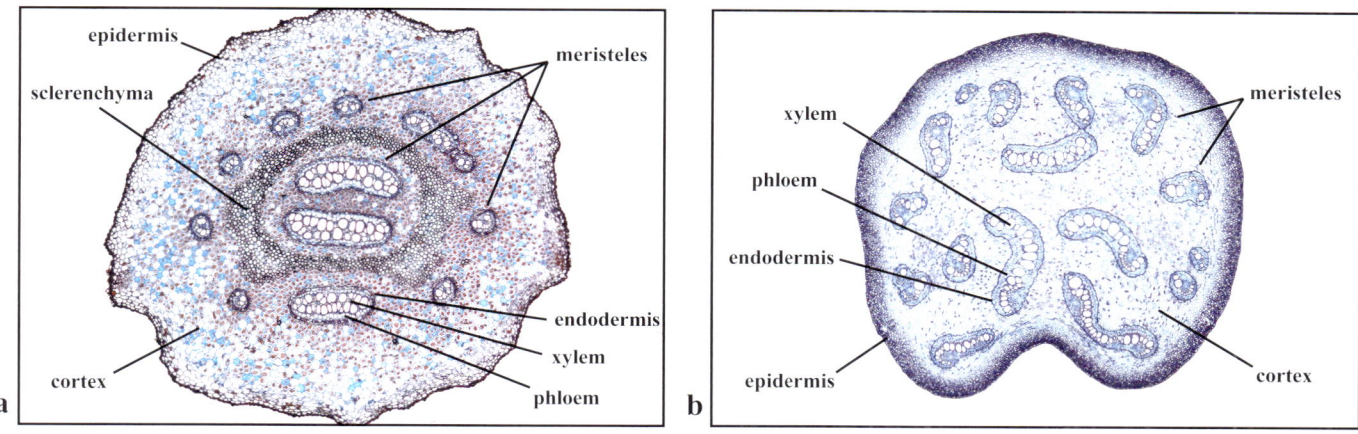

Fig. 341   a) & b) XS of two different rhizomes with internal tissues identified.

Fig. 342   Rhizome of rabbit's-foot fern from which multiple fronds are sprouting.
Davallia fejeensis   Polypodiaceae

Sori

a

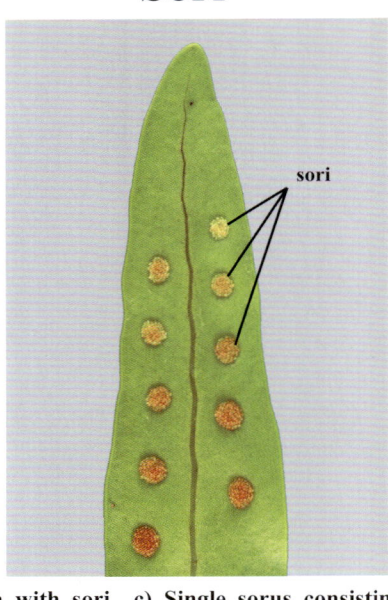

b

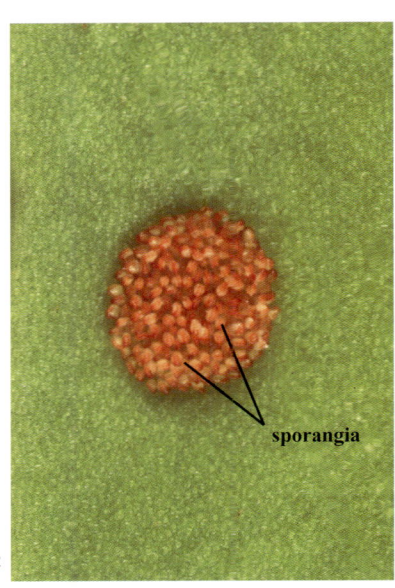

c

Fig. 343   a) Fertile frond. b) Fertile pinna with sori. c) Single sorus consisting of many sporangia and no indusium. Golden Polypody Fern   *Phlebodium aureum*   Polypodiaceae

a

b

c

Fig. 344   a) Fertile pinna. b) Portion of fertile pinna with multiple sori. c) CU of sori with white round indusia. Japanese Holly Fern   *Cyrtomium falcatum*   Polypodiaceae

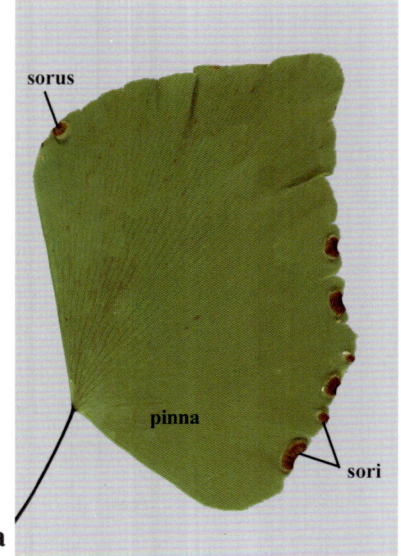

a

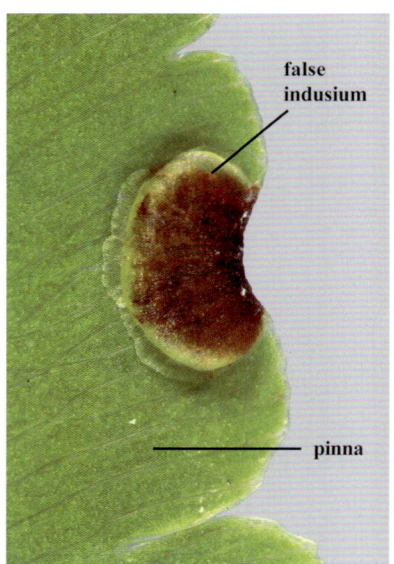

b

c

Fig. 345   a) Fertile pinna with several sori at edge. b) CU of false indusium covering sorus. c) CU of false indusium pulled back revealing clustered sporangia beneath.   Peruvian Maidenhair Fern   *Adiantum peruvianum*   Polypodiaceae

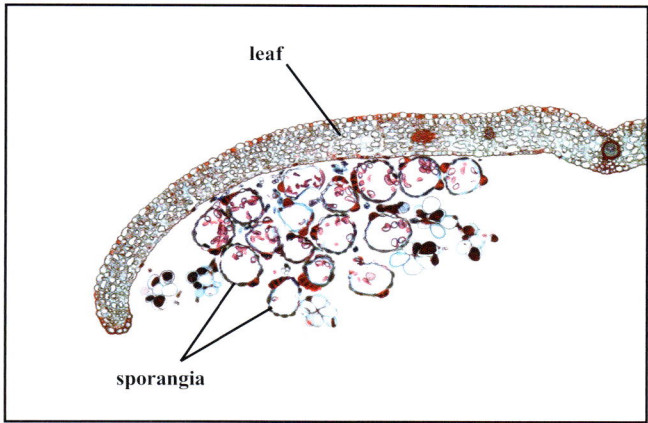

Fig. 346   XS of leaf with sorus that has no indusium.
Polypodium sp.
Polypodiaceae

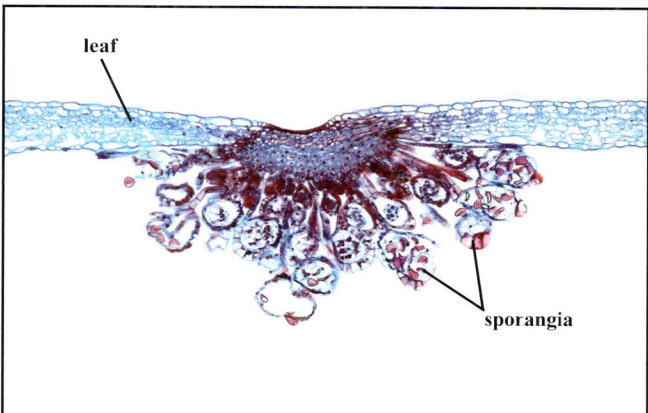

Fig. 347   XS of leaf with sorus that has no indusium.
Polypodium sp.
Polypodiaceae

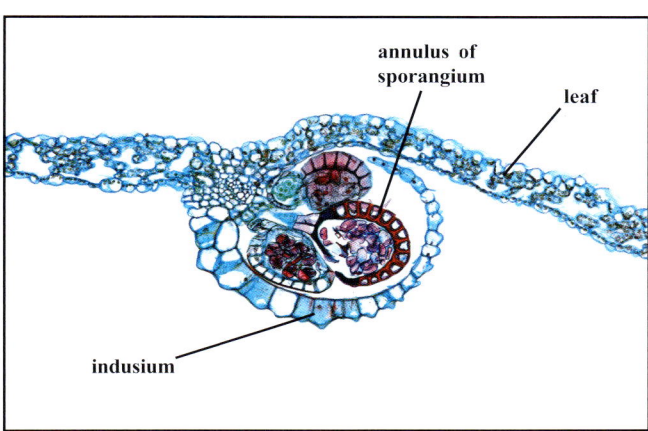

Fig. 348   XS of leaf with a sorus covered by an indusium. Note that this indusium does not have a central stalk.
Asplenium sp.
Polypodiaceae

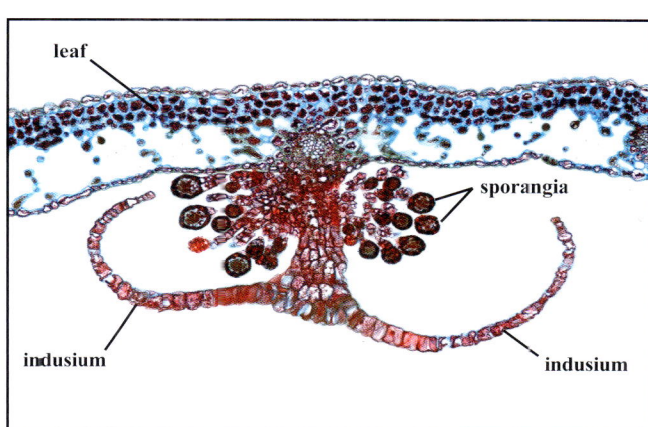

Fig. 349   XS of leaf with a sorus covered by an umbrella-shaped indusium with a central stalk.
Dryopteris sp.
Polypodiaceae

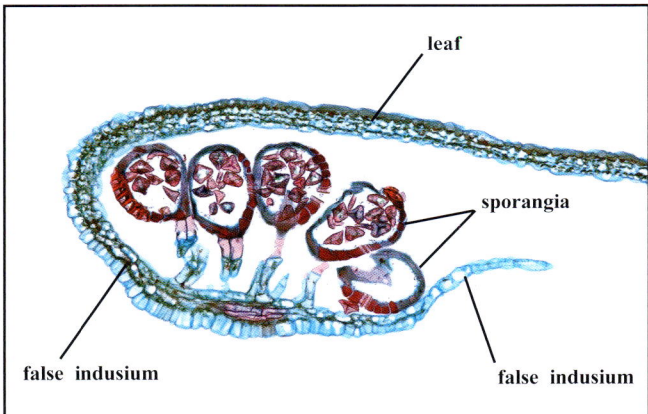

Fig. 350   XS of leaf with a sorus covered by the leaf edge forming a false indusium.
Adiantum sp.
Polypodiaceae

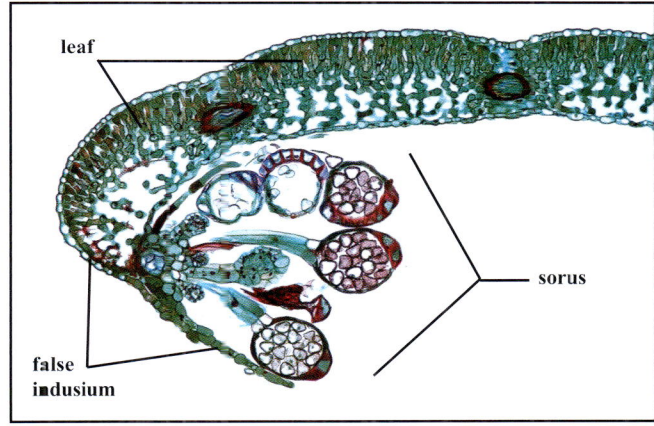

Fig. 351   XS of leaf with a sorus covered by the leaf edge forming a false indusium.
Pteridium sp.
Polypodiaceae

Ophioglossaceae

Genus Ophioglossum - *Adder's Tongue Ferns*

Small, fleshy underground stem.
Leaves dimorphic.
Sterile leaves <40cm.
Sterile leaves simple, ovate, tongue-shaped.
Fertile leaf longer than sterile with two rows of sporangia embedded in tip.
Homosporous.

Fig. 352   a) Sterile simple leaves of an adder's tongue. b) Sterile and fertile leaves. *Ophioglossum petiolatum*

 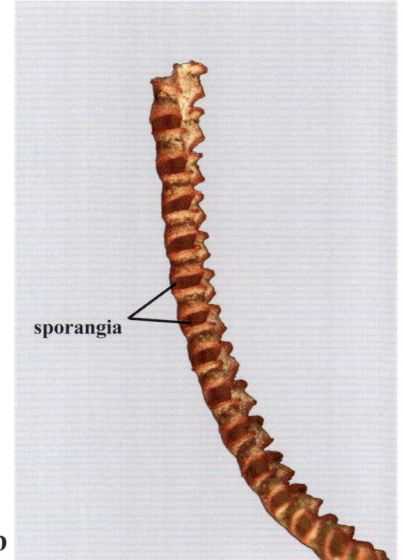 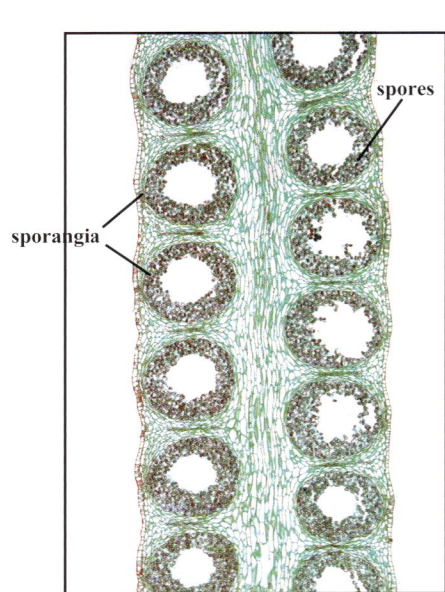

Fig. 353   a) Tip of fertile leaf. b) Edge of tip of fertile leaf showing the embedded sporangia. *Ophioglossum petiolatum*

Fig. 354   LS of sporangia embedded in fertile leaf. *Ophioglossum sp.*

Ophioglossaceae

Genus Botrychium - *Grape Ferns*

Small, fleshy underground stem.
Leaves dimorphic.
Sterile leaves <15cm.
Sterile leaves pinnate to 4-pinnate.
Fertile leaf a spike that branches at the tip with two rows of sporangia on each branch.
Homosporous.

Fig. 355   a) Sterile pinnate leaves of a grape fern. b) Sterile and fertile (upright spike) leaves.
Botrychium biternatum

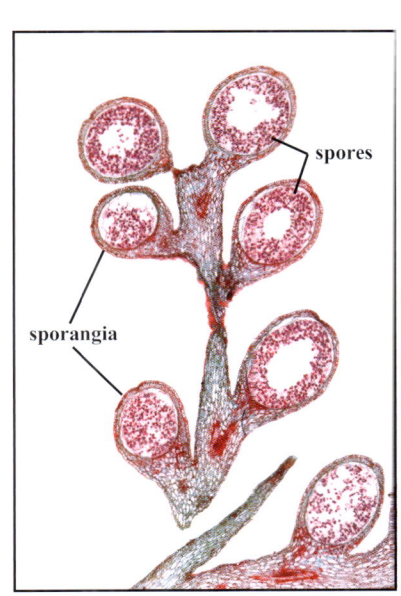

Fig. 356   a) Branching tip of fertile leaf showing the grape-like clusters of sporangia. b) CU of sporangia.
Botrychium biternatum

Fig. 357   LS of fertile leaf with sporangia.
Botrychium sp.

Osmundaceae - *Royal Fern Family*

Rhizomes with persistent stipe bases.

Leaves pinnately compound, 50-150cm long, grow in erect clusters.

Fully dimorphic species with separate sterile and fertile leaves.

Most species partially dimorphic with fertile pinnae near tip or middle of an otherwise sterile leaf.

Sporangia large, globose, and borne in clusters on short stalks.

Sporangium wall is thicker than other leptosporangiate ferns, but thinner than eusporangiate ferns.

Annulus limited to a lateral cluster of cells.

Fig. 358   Sterile and fertile leaves of the royal fern.
Osmunda regalis

Fig. 359   Green sterile leaves and brown fertile stalk of cinnamon fern in-situ.
Osmunda cinnamomea

Fig. 360   Immature fertile leaf of cinnamon fern.
Osmunda cinnamomea

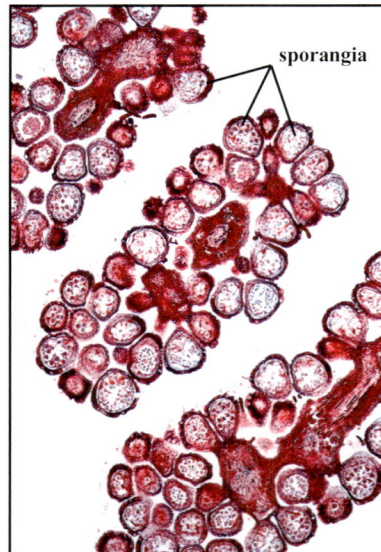

Fig. 361   Sporangia characters of cinnamon fern. a) Nearly mature fertile leaf. b) Single pinna with clusters of cinnamon-colored sporangia. c) LS of clusters of sporangia from section of fertile pinna.
Osmunda cinnamomea

Cyatheaceae - *Tree Fern Family*

Tree-like. Stems unbranched, reaching 20m in height.
Leaves pinnately or bipinnately compound, typically 2-3m long.
Leaves scaly.
Sporangia borne in sori.
Annulus equator-like, complete and uninterrupted by stalk.
Homosporous.
Genera: All genera tropical with no native North American species.
Sometimes planted as an exotic ornamental (ex. *Cyathea*).

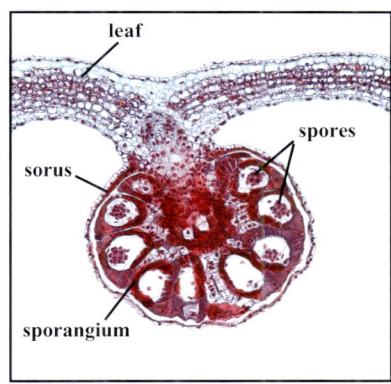

Fig. 362   LS of sporangium.
Cyathea furfuracea

Fig. 363   a) & b) Tree ferns growing in Monteverde Cloud Forest in Costa Rica.
Cyathea sp.

Fig. 364   Fiddlehead of tree fern.
Cyathea sp.

Fig. 365   Sori on underside of frond of tree fern.
Cyathea sp.

Polypodiaceae - *Polypody Fern Family*

Extremely diverse family.

Terrestrial (predominantly) with some epiphytic, aquatic, and climbing species.

Leaves primarily from underground stems or rhizomes.

Leaves typically pinnately to multiply pinnately compound, <2m in length.

Sporangia typically borne in sori.

Annulus vertical and interrupted by stalk.

Homosporous.

Genera: *Acrostichum, Adiantum, Asplenium, Blechnum, Cyrtomium, Dryopteris, Nephrolepis, Phlebodium, Pleopeltis, Polypodium, Pteris, Thelypteris, Woodwardia.*

Fig. 366   Resurrection fern demonstrating its ability to become lush when exposed to moisture following periods of desiccation.
Polypodium polypoides

  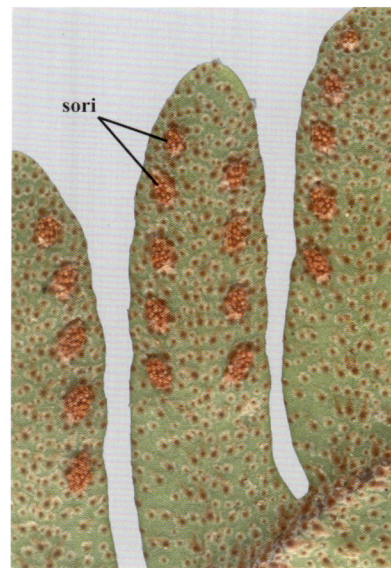

sori

Fig. 367   Characters of resurrection fern. a) Top surface of frond. b) Under surface of frond showing sori. c) CU of pinna showing sporangia contained within each sorus.
Polypodium polypoides

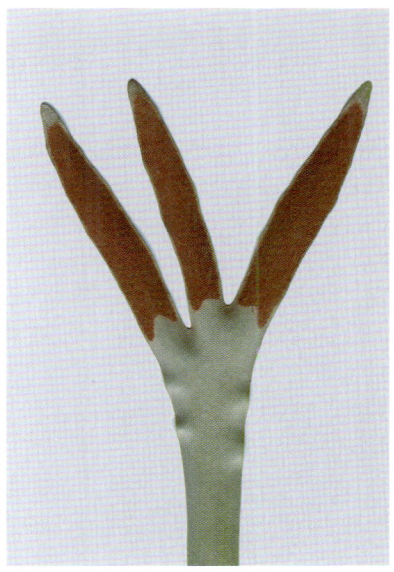

Fig. 368   Characters of a staghorn fern. a) Habit showing dimorphic leaves. b) CU of flat basal leaf modified for attachment. c) Underside of frond or erect leaf showing sori at tips.   *Platycerium bifurcata*

Fig. 369   Leaf characters of American bird's-nest fern. a) Upper surface of frond. b) Lower surface of frond showing sori arranged in parallel rows. c) CU of sori.   *Asplenium nidus*

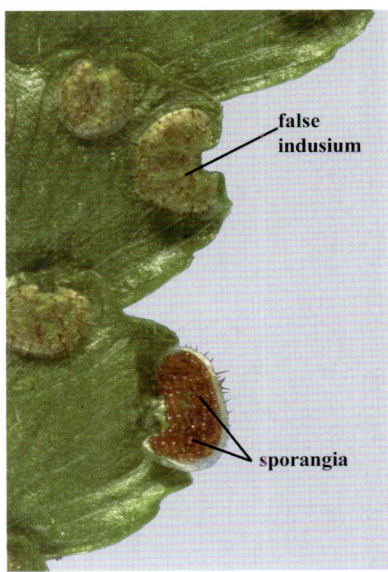

Fig. 370   Leaf characters of the rough maidenhair fern. a) Lower surface of frond. b) CU of frond showing reflexed leaf margins forming false indusia. c) CU of false indusium bent away from leaf to show sporangia.   *Adiantum hispidulum*

Marsileaceae - *Water Clover Family*

Aquatic or in muddy areas bordering water.
Stipe is long (<10") and slender.
Leaf resembles a four-leaf clover or is absent.
Heterosporous.
Sporangia lack an annulus.
Sori borne in **sporocarps** (hardened pea-shaped
structures with short stalks at base of stipes).
Three genera: *Marsilea*, *Pilularia*, and *Regnellidium*.

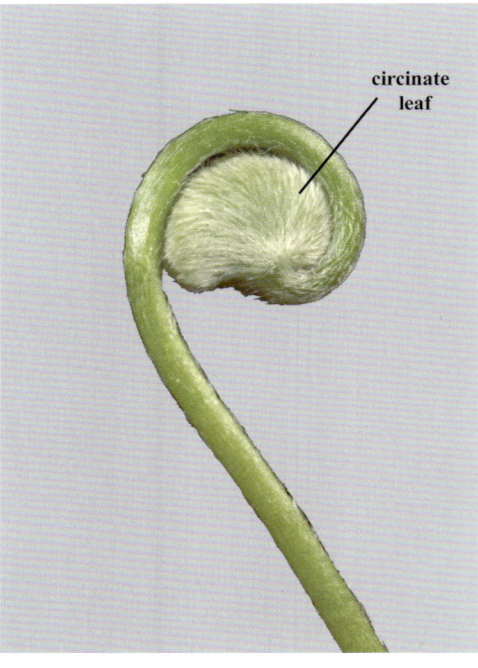

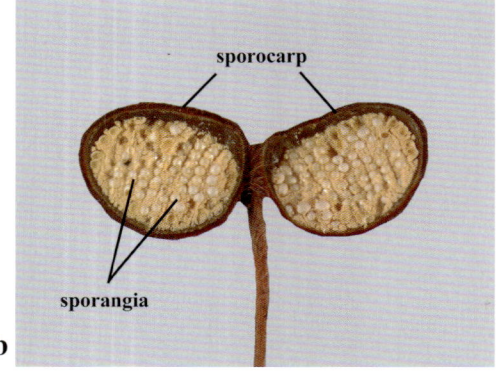

a

b

Fig. 371   Characters of sporocarp of water clover. a) Mature sporocarp.
b) LS of sporocarp showing sporangia within.
Marsilea sp.

Fig. 372   Young water clover leaf still in
the circinate stage.
Marsilea sp.

Fig. 373   Mat of water clover leaves floating on the water's surface.
Marsilea mutica

Salviniaceae - *Water Spangle Family*

Aquatic floating ferns.
Dimorphic leaves from long thin rhizomes.
Floating oval leaves 5-15mm long with short hairs on upper surface.
Dissected irregular leaves hang in water and act as roots.
Heterosporous.
Sporocarps produced on irregular dangling leaves.
One Genus: *Salvinia*.

Fig. 374　Plant characters of water spangles.　a) Floating mat.　b) Leaf showing the white hairs that keep it afloat. c) Profile of floating leaves at water's surface and rhizomes with dissected leaves beneath the water. *Salvinia minima*

Azollaceae - *Mosquito Fern Family*

Aquatic floating ferns.
Moss-like in appearance.
Leaves grow from branched, thin rhizomes.
Individual leaves less than 1mm long, but
overlap forming small mats that appear larger.
Roots grow from rhizomes.
Heterosporous.
Sporocarps minute, produced in leaf axils.
Leaves may have reddish tint.
One Genus: *Azolla*.

Fig. 375　Plant characters of mosquito fern.　a) Habit showing many overlapping floating leaves.　b) Dense mat of plants. *Azolla caroliniana*

Gymnosperms
(Non-Flowering Seed Plants)

Introduction

The term *gymnosperm* means 'naked seed' and refers to the fact that the seed of such plants is not enclosed within ovarian tissue. Most evidence supports that gymnosperms refers to a paraphyletic grade of early seed plants. Extant seed plants can be classified into five major monophyletic groups. These are the Cycads, Ginkgo, Gnetopsids, and Conifers that make up the gymnosperms or non-flowering seed plants; and the Angiosperms or flowering seed plants. Cycads are palm-like plants that flourished with greatest diversity during the Mesozoic Era. The Ginkgo family (Ginkgoaceae) contains the sole species *Ginkgo biloba*. It is a true living fossil and was first described from the fossil record. The Gnetopsids are represented by three families, of which the Ephedraceae or Mormon tea is the most common. The conifers are all cone-bearing, with five families covered here: Pinaceae (pines), Cupressaceae (junipers and cedars), Taxaceae (yews), Araucariaceae (Araucarias), and the Podocarpaceae (podocarps).

Fig. 376 Older cycads, such as this sago palm, may show multiple trunks. *Cycas revoluta*

Cycads

Characteristics of Cycads

Evergreen and palm-like in appearance.

Short trunks, usually unbranched.

Leaves pinnately compound, spirally arranged, and forming a crown at the apex of the trunk.

Leaves stiff and leathery.

Wood soft and pithy, with slime canals present.

Pollen (microspores) produced in pollen strobili.

Ovules (megaspores) produced from megasporophylls that may be leaf-like or packed into strobili.

Roots associated with blue-green algae.

Vascular bundles form an inverted 'omega' pattern.

Venation of Cycads

Fig. 377 Family Cycadaceae has stiff, slender leaflets each with only a single central vein.
Cycas revoluta

Fig. 378 Family Zamiaceae has leaflets that vary in shape, but all with parallel linear veins.
Zamia floridana

Fig. 379 Family Stangeriaceae has dichotomously branched pinnate veins on the leaflets.
Stangeria eriopus

Stangeriaceae - *Stangeria Family*

a

b

microsporophylls

micro-sporangia

c

Fig. 380 Plant characters of stangeria. a) Compound leaf. b) Male strobilus. c) CU of male strobilus.
Stangeria eriopus

Cycadaceae - *Sago 'Palm' Family*

Leaflets with a single vein.

More than two ovules per megasporophyll.

Megasporophylls pinnate and leaf-like, not compacted into strobili.

Represented by the single genus *Cycas*.

Fig. 381   Characters of male sago palm. a) Habit. b) Strobilus.
Cycas revoluta

Fig. 382   Characters of female sago palm. a) Habit. b) Megasporophylls are light in color and clustered in center of the plant.
Cycas revoluta

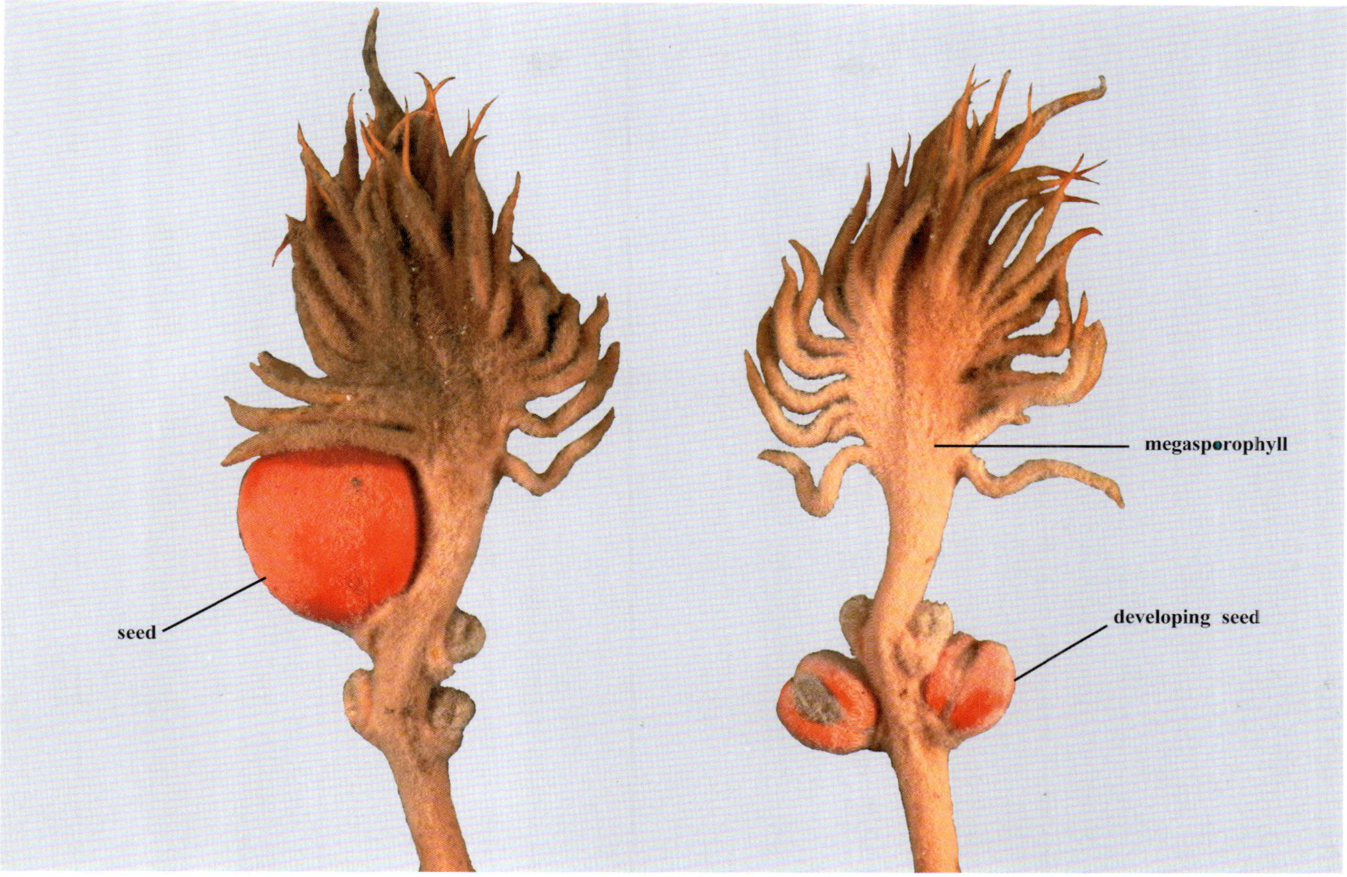

megasporophyll

seed

developing seed

Fig. 383   Two megasporophylls of a sago palm that have been isolated to show the developing seeds attached. *Cycas revoluta*

Fig. 384   Cataphylls are spine-like modified leaves. Here they protect the central area of the plant from which the new leaves will sprout. *Cycas revoluta*

Fig. 385   Flush of growth of new leaves from center of sago palm. *Cycas revoluta*

Zamiaceae - *Coontie Family*

Leaflets with many parallel veins.
Two ovules produced per megasporophyll.
Megasporophylls reduced and clustered into strobili.
Very short unbranched trunks (leaves may appear to come from ground).
Eight genera, including *Ceratozamia*, *Dioon*, and *Zamia*.

Fig. 386   Male coontie with dehisced pollen strobili on ground about base.
Zamia floridana

Fig. 387   Male strobilus consisting of tightly-packed microsporophylls.
Zamia floridana

Fig. 388   Female coontie with one large erect seed strobilus in the center.
Zamia floridana

Fig. 389   Female strobilus made of tightly-packed megasporophylls.
Zamia floridana

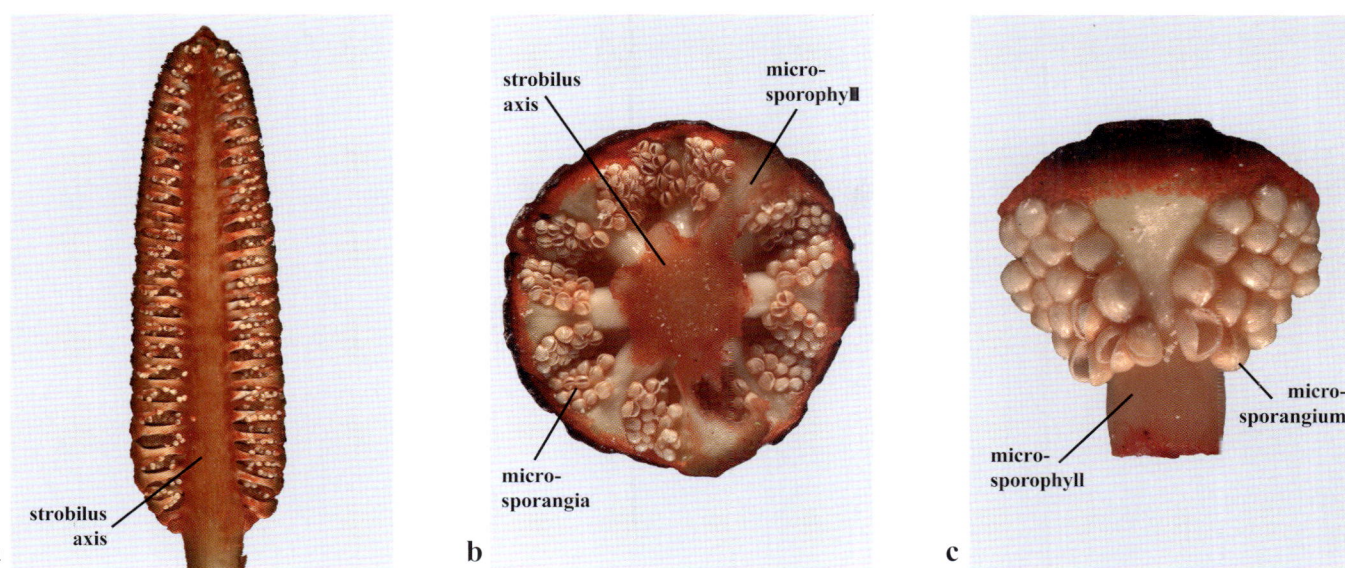

Fig. 390   Characters of male strobilus of coontie. a) LS of strobilus. b) XS of strobilus. c) Abaxial surface of microsporophyll with mircrosporangia.   *Zamia floridana*

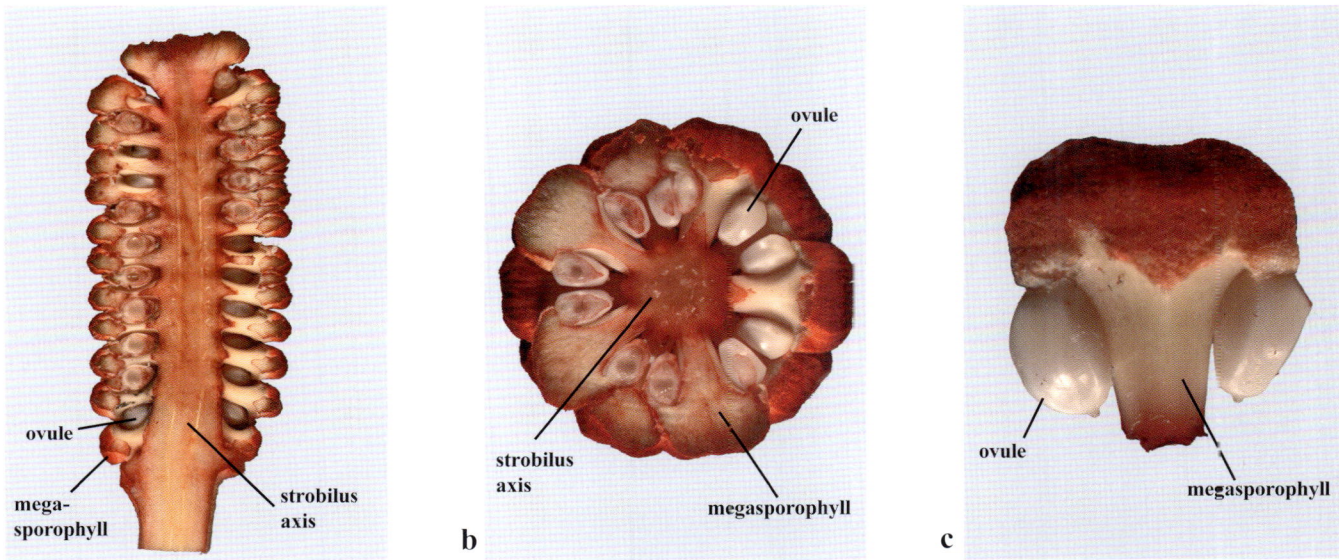

Fig. 391   Characters of female strobilus of coontie. a) LS of strobilus. b) XS of strobilus. c) Adaxial surface of megasporophyll with two ovules.   *Zamia floridana*

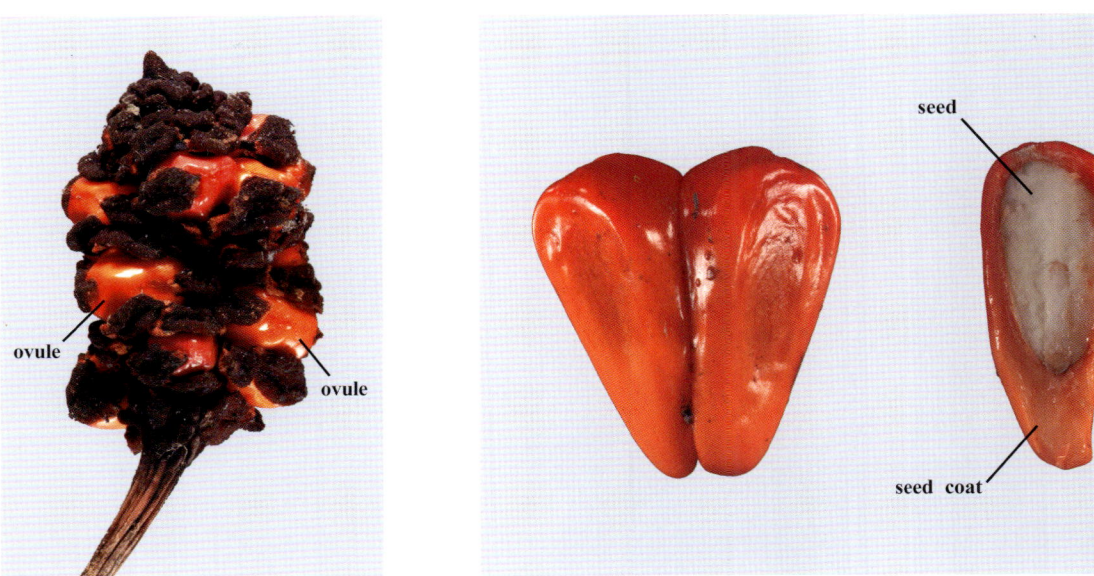

Fig. 392   Dehiscing female strobilus.
Zamia floridana

Fig. 393   Mature intact ovules (left) and LS of ovules (right). Note fleshy brightly-colored seed coat.   *Zamia floridana*

Ginkgoaceae - *Ginkgo Family*

Deciduous tree reaching 30 meters. Many branches.
Simple, fan-shaped leaves with palmate venation.
Two vascular bundles supply leaf, leaving a distinctive leaf scar.
Leaves alternate, but may appear whorled on short shoots.
Dioecious.
Pollen produced from strobili.
Ovules usually produced in pairs on short shoots.
Seeds have a thick, fleshy outer coat.
Living fossil that was first described from the fossil record.
Represented by one species, *Ginkgo biloba*.

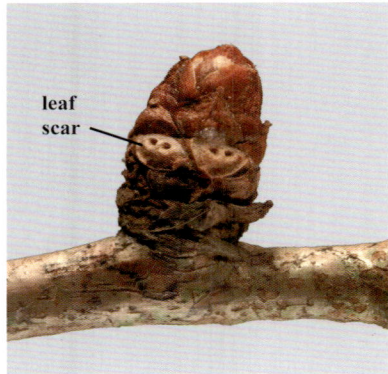

Fig. 394   Bud on short shoot that bears two obvious leaf scars.

Fig. 395   Typical fan-shaped ginkgo leaf.

Fig. 396   Fossil imprint of prehistoric ginkgo leaf.

Fig. 397   Ginkgo branch coming into foliage and showing long shoots and short shoots.

Gnetopsids

Seeds exposed and not enclosed in an ovary.
Vessels present in the wood.
Flower-like reproductive structures borne from strobili.
Opposite leaves.
Represented by three families, each of which has one genus.

Family Welwitschiaceae - *Genus Welwitschia*

Found in desert regions of southwestern Africa.
Large plant with strap-like leaves several meters in length.

a

Photo by Richard Abbott

b

Photo by Richard Abbott

Fig. 398   a) Two young plants in cultivation. Left = female, right = male. b) CU of staminate strobili. The Welwitschiaceae is a monotypic family with the single species *Welwitschia mirabilis*.

Family Gnetaceae - *Genus Gnetum*

Tropical lianas, trees, and shrubs.
Leaves simple and broad.
Seeds surrounded by fleshy, colorful material.

a

Photo by Richard Abbott

b

Photo by Richard Abbott

c

Photo by Richard Abbott

Fig. 399   Plant characters of *Gnetum gnemon*. a) Immature ovulate strobilus. b) Mature seeds. c) Shrubby habit.

Family Ephedraceae - *Genus Ephedra*

Found in arid regions of southwestern United States.
Shrubs or vines.
Leaves scale-like, often shed soon after developing leaving branches bare.
Pollen and ovulate strobili consist of opposite or whorled bracts.
Known as the Mormon Tea or Joint Fir Family.

a

Photo by Richard Abbott

b

Photo by Richard Abbott

c

ovules

bracts

Photo by Richard Abbott

Fig. 400   Plant characters of Mormon tea. a) Bushy habit.   *Ephedra* sp. b) Pollen strobili.   *Ephedra trifurca* c) Ovulate strobili.   *Ephedra viridis*

Conifers

Characteristics of Conifers

Medium to extremely large trees, mostly evergreen.
Leaves simple and variable. Often needle-like, awl-like, or scale-like.
Pollen produced in strobili.
Seeds produced in woody or fleshy cones.
Monoecious or dioecious.

Conifer Terminology

bract - Modified leaf associated with reproductive structures.
cone - Female reproductive structure, typically consisting of woody ovulate scales arranged spirally around an axis.
fascicle - Unit of two or more pine leaves that are joined at the base.
long shoot - Long, woody branches that bear the short shoots.
needle - Common term for the stiff, slender, pointed leaves of many conifers.
ovulate scale - Woody scale that produces the seeds or ovules and makes up the basic unit of a pine cone.
short shoot - Short woody branches that bear the leaves and arise from the long shoots.
wing - Flat, papery extension of tissue associated with a seed that permits dispersal by wind.

Fig. 401   **Various structures of the slash pine including a cone, a seed (resting on the cone), two seedlings, and a dehisced pollen strobilus (on ground beneath wing of seed), and fallen pine needles.**

Pinaceae - *Pine Family*

Mostly evergreen trees, conical in shape.
Leaves are needle-like and produced in bundles (fascicles) or singly.
Branches with long shoots and short shoots.
Resin canals in leaves and wood resulting in strong, characteristic odor.
Ovulate scales spirally arranged into cones that are hard at maturity.
Bracts are free from ovulate scales in female cones.
Two seeds produced from each ovulate scale.
Seeds each have a large wing (extension of the scale).
Pollen grains with two bladder-like wings.
Genera: *Abies, Larix, Picea, Pinus, Pseudotsuga, Tsuga.*

Fig. 402   Scaly, plate-like bark typical of many pine trees. *Pinus elliotii*

a

b

Fig. 403   Shoot characters of deodar cedar. a) Leafing branch showing long and short shoots. b) CU of short shoot. *Cedrus deodara*

a

b

Fig. 404   Pine needles and fascicles. a) Two fascicles of spruce pine each consisting of two needles or leaves.   *Pinus glabra* b) Base of fascicle of longleaf pine, which contains three needles or leaves per fascicle.   *Pinus palustris*

ovulate
scale

umbo

ovulate
scale

umbo

a   b   c

Fig. 405   Cone development of spruce pine.   a) Very young cone.   b) Immature cone.   c) Mature cone.
Pinus glabra

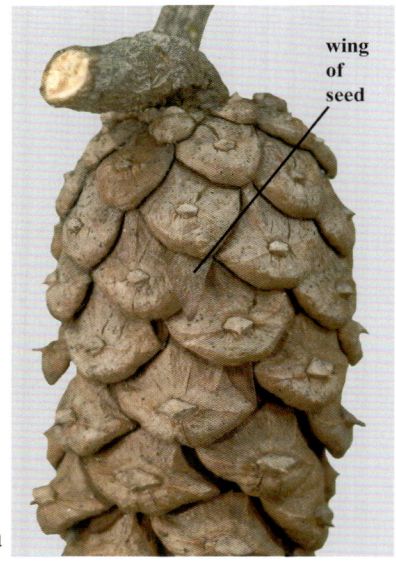

wing
of
seed

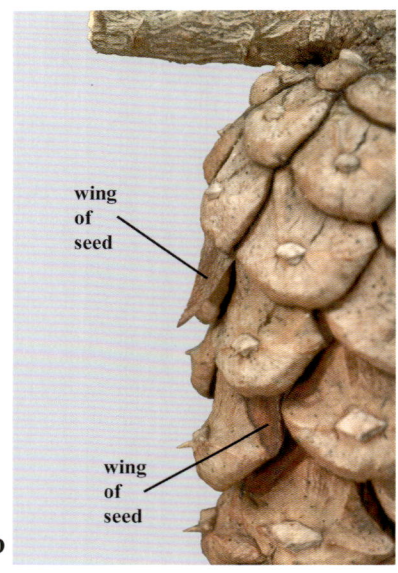

wing
of
seed

wing
of
seed

papery
wing

seed

a   b   c

Fig. 406   Pine seeds.   a) & b) Seed protruding from cone of spruce pine.   *Pinus glabra*
c) Two seeds showing attached wing that functions in aerial dispersal.   *Pinus* sp.

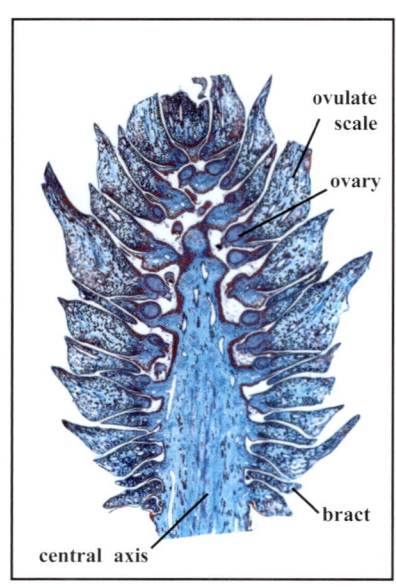

ovulate
scale

ovary

bract

central axis

Fig. 407   Cone of Norway spruce.
Picea abies

Fig. 408   Cone of eastern hemlock.
Tsuga canadensis

Fig. 409   LS of cone.
Pinus sp.

Fig. 410   Pine pollen strobili. a) Pollen strobili the majority of which have not dehisced. b) Pollen strobili the majority of which are brown and have already dehisced.   *Pinus* sp.

Fig. 411   Pine pollen strobili. a) Pollen strobili before dehiscence. b) Pollen strobili during dehiscence. *Pinus* sp.

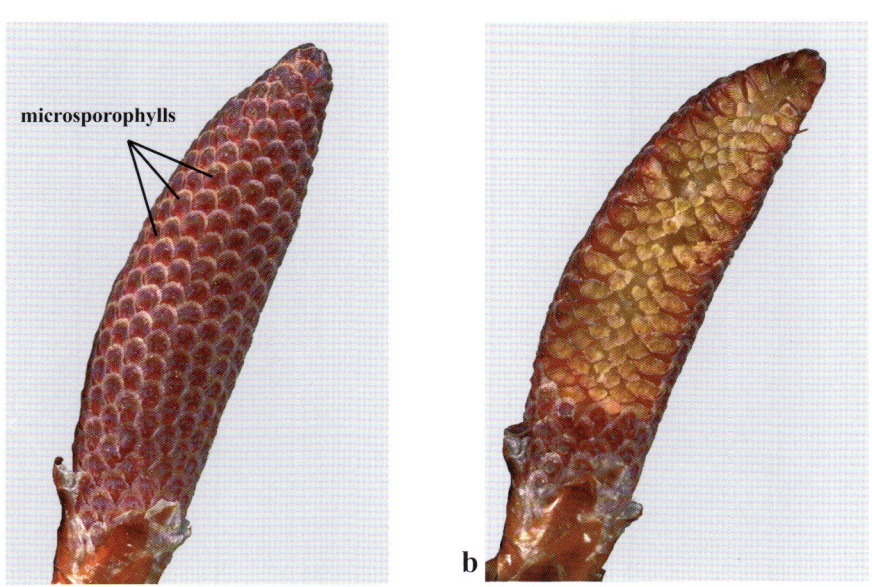

Fig. 412   CU of pollen strobilus. a) Strobilus prior to dehiscence. b) LS of strobilus prior to dehiscence.   *Pinus* sp.

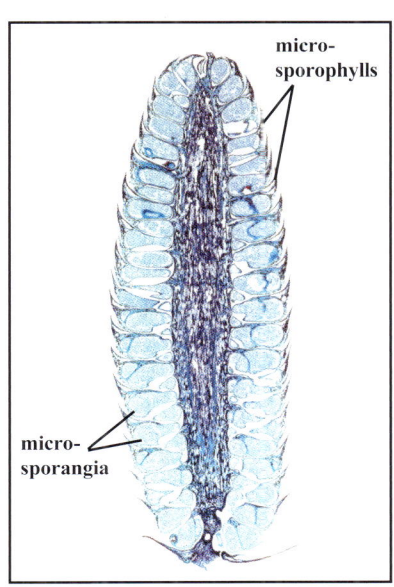

Fig. 413   LS of pollen strobilus. *Pinus* sp.

Cupressaceae - *Juniper & Cedar Family*

Shrubs and trees that may be evergreen or deciduous.
Bark typically has vertical furrows.
Leaves simple, spirally arranged or opposite and decussate.
Leaves scale-like, awl-like, or needle-like with dimorphic examples on the same tree.
Monoecious or dioecious.
Bracts are fused to ovulate scales in the female cones.
Ovulate scales are round to oval, producing 12-15 ovules.
Seeds are usually wingless.
Pollen is wingless.
Genera: *Callitris, Chamaecyparis, Cryptomeria, Cunninghamia, Cupressus, Juniperus, Platycladus, Sequoia, Taxodium, Thuja.*

Fig. 414   Plant characters of China fir. a) Ovulate cone. b) Pollen strobili (post-dehiscence). c) Underside of leaf showing white stomata.   *Cunninghamia lanceolata*

Fig. 415   Reproductive characters of Arizona juniper. a) Ovulate cone and pollen strobili. b) Ovulate cone dehiscing. c) Pollen strobili.   *Cupressus arizonica*

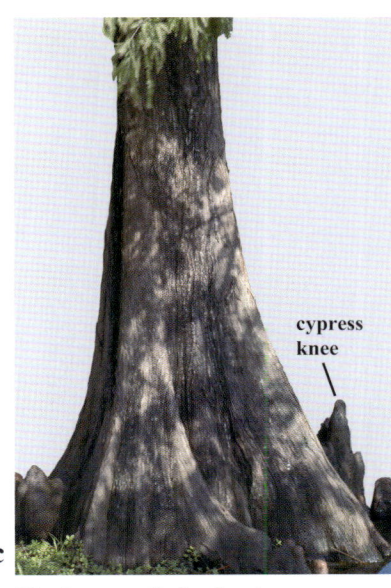

Fig. 416   Plant characters of bald cypress. a) Ovulate cone. b) Deciduous branchlet with dozens of slender leaves. c) Cypress trunk showing characteristic widening at base and several small knees.   *Taxodium distichum*

Fig. 417   Reproductive characters of southern red cedar. a) Ovulate cone. b) Pollen strobili.   *Juniperus silicicola*

Fig. 418   Needle and scale leaves. *Juniperus chinensis*

Fig. 419   Plant characters of arbor-vitae. a) Flat foliage of scale-like leaves. b) Microsporangiate or pollen strobili. c) CU of showy microsporophylls subtended by scale leaves.   *Platycladis orientalis*

Taxaceae - *Yew Family*

Evergreen trees or shrubs.
Leaves simple, alternate, spirally arranged but appearing two-ranked.
Leaves with two white or light-colored longitudinal bands (of stomata) on abaxial surface.
Dioecious.
Seeds hard and bony, with fleshy colorful outgrowths (arils) surrounding them.
Pollen wingless and produced in small strobili.
Genera: *Taxus*, *Torreya*.

Fig. 420   Plant characters of English yew. a) Fruiting branch. b) CU of fruit showing a red aril surrounding a hard dark seed. c) Leaves at point of attachment to branch.
Taxus baccata

Fig. 421   Plant characters of Florida yew. a) Branch with pollen strobili. b) CU of pollen strobili. c) Underside of leaves showing white stomatal bands.
Taxus floridana

Araucariaceae - *Araucaria Family*

Large evergreen trees with very symmetrical appearance.
Leaves simple, spirally arranged or two-ranked.
Leaves awl-shaped or broad and leathery.
Monoecious or dioecious.
Pollen produced in small cones.
Pollen wingless.
Female cones large and woody.
Ovulate scales are small and fused to a large bract.
Seed is very large and embedded in the scale.
Two Genera: *Agathis* and *Araucaria*.

Photo by F.W. Howard

Fig. 422　Norfolk Island pine tree.
Araucaria heterophylla

a

b

Fig. 423　Plant characters of araucaria. a) Fruiting branch. b) CU of ovulate cones and branches of needle-like leaves.
Araucaria subulata

a

b

Fig. 424　Plant characters of agathis. a) Fruiting branch showing broad leathery leaves. b) CU of ovulate cone.
Agathis vitiensis

Fig. 425　CU of leaves of *Araucaria bidwillii*.

Fig. 426　CU of leaves of Norfolk Island pine.
Araucaria heterophylla

Podocarpaceae - *Podocarp Family*

Evergreen trees or shrubs.

Leaves simple and highly variable (scale-like to ovate and leathery).

Ovulate scales are reduced and not fused to the bracts.

Ovulate scales bear one seed which is embedded in the scale.

Female cones modified into a fleshy, drupe-like structure.

Pollen produced in small, cylindrical strobili.

Pollen grains have 2-4 wings.

Genus: *Podocarpus* (including many segregate genera).

Fig. 427   Female reproductive characters of podocarpus. a) Fruiting branch. b) CU of ovulate cone that is here reduced to a single seed. At its base is an aril-like structure.
Podocarpus macrophyllus

Fig. 428   Male reproductive characters of podocarpus. a) Branch with pollen strobili. b) CU of pollen strobili.
Podocarpus macrophyllus

Fig. 429 Plant characters of podocarpus. a) Fruiting branch. b) CU of ovulate cone that is reduced to a single ovule and thus produces one seed. *Podocarpus nagi*

fleshy
seed
coat

hard
protective
layer

nutritive layer
with embryo

Fig. 430 XS of ovulate cone.
Podocarpus nagi

strobilus

Fig. 431 Podocarpus branch with pollen strobili.
Podocarpus nagi

strobilus

strobilus

strobilus

Fig. 432 Characters of pollen strobili. a) Developing strobili prior to dehiscence. b) Strobili post-dehiscence.
Podocarpus nagi

Angiosperms
(Flowering Plants)

Introduction

Angiosperms are defined as flowering plants that have seeds enclosed within a fruit which is a matured ovary. Although the angiosperms are strongly supported as a monophyletic group, hypotheses of relationships among all families have been hotly debated in recent years. For ease of discussion, angiosperms can be divided into three major groups. These are the primitive angiosperms, the monocots, and the eudicots (true dicots). In this book, the following families have been treated as primitive angiosperms: Illiciaceae, Nymphaeaceae, Magnoliaceae, Annonaceae, Lauraceae, Piperaceae and Aristolochiaceae. Recognizing a paraphyletic group such as the primitive angiosperms allows for convenient discussion of flowering plant families that are neither monocots or eudicots. However, it should be noted that some of the members of the primitive angiosperms are more closely related to the monocots or the eudicots than they are to each other. Members of the primitive angiosperms represent unique evolutionary lineages and are not a single, natural, diagnosable group.

The monocot angiosperms are considered to be a monophyletic group. Their defining characteristics are an embryo with a single cotyledon, pollen grains bearing a single groove or pore (monosulcate), leaves with parallel venation, vascular bundles scattered throughout the stem, and a fibrous root system. Most monocots also have 3-merous flowers resulting in perianth parts occurring in threes or multiples of threes. As hinted above, many of these features can also be found in some members of the primitive angiosperms.

The eudicot angiosperms are also considered to be a monophyletic group. This is primarily based on having pollen grains with three grooves or pores (tricolpate). For this reason the eudicots are also referred to as the 'tricolpates'. Other traditional characteristics used to define the eudicots are an embryo with two cotyledons, leaves with reticulate venation, vascular bundles arranged in rings within the stem, and a taproot type root system. Most eudicots have 4-merous or 5-merous flowers.

Illiciaceae - *Star Anise Family*

Habit: Shrubs, small trees.
Leaves: Alternate, simple, entire. Pinnate venation.
Pellucid dots. Aromatic like licorice.
Inflorescences: Solitary to 2-3 flowers in axillary clusters.
Flowers: Radial, sometimes showy, bisexual.
Perianth: Numerous, distinct tepals.
Stamens: Numerous, distinct, spirally arranged on short receptacle.
Carpels: 5-many in a single whorl.
Ovary: Superior.
Placentation: Basal.
Fruit: Radial aggregate of follicles (each with one seed). Star shaped.
Seeds: Flat, smooth.
Genus: *Illicium* (monogeneric family).

Fig. 433   Flower of star anise.
Illicium parviflorum

Fig. 434   Flowering branch of the star anise.
Illicium parviflorum

 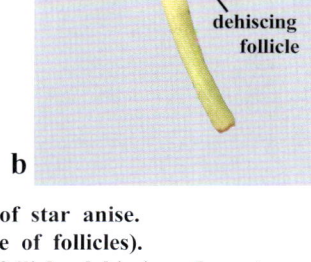

Fig. 435   Fruit characters of star anise.
a) Immature fruit (aggregate of follicles).
b) Mature fruit with some follicles dehiscing along top.
Illicium parviflorum

 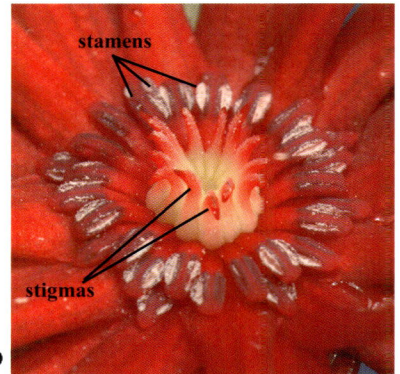 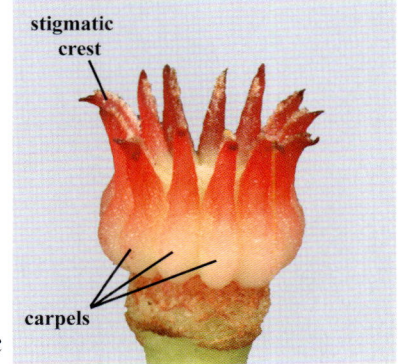

Fig. 436   Floral characters of Florida anise. a) Flowering branch. b) CU of stamens and stigmas. c) Ring of carpels after tepals and stamens have been removed.   *Illicium floridanum*

Nymphaeaceae - *Water Lily Family*

Habit: Aquatic herbs. Grow from rhizomes.
Leaves: Alternate, simple, usually entire. Typically with long petiole and floating blade. Palmate to pinnate venation. Heterophylly.
Inflorescences: Solitary flowers.
Flowers: Large, showy, radial, perfect. Long pedicel. Floats on or emerges above the water surface.
Calyx: 4-12 sepals, distinct, petaloid.
Corolla: 8 to numerous petals in several whorls.
Stamens: Numerous, arranged spirally.
Carpels: 3-numerous.
Ovary: Superior or inferior.
Placentation: Parietal.
Fruit: Various. Fleshy capsule, aggregate of nuts, berry, and indehiscent pod.
Seeds: Small, usually arillate and operculate.
Other: Stem with milky sap, mucilage, and prominent air canals.
Genera: *Brasenia, Cabomba, Nuphar, Nymphaea.*
Note: Genus *Cabomba* is sometimes treated as a distinct family.

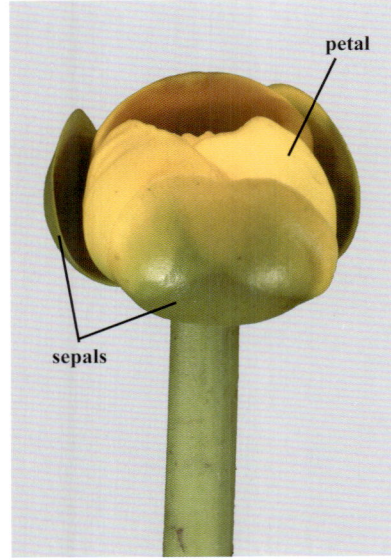

Fig. 437   Flower of spatterdock.
Nuphar luteum

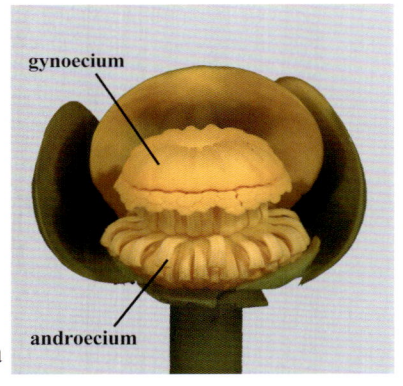

a

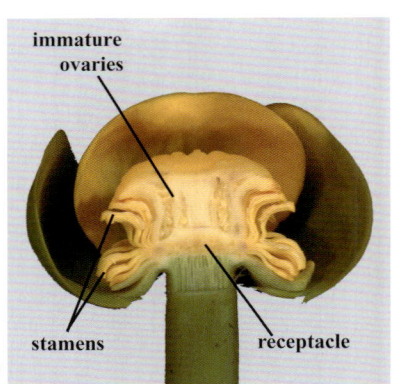

b

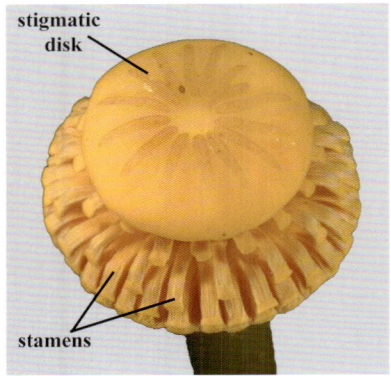

c

Fig. 438   Floral characters of spatterdock. a) Portion of perianth removed to show androecium and gynoecium. b) LS of flower. c) All sepals and petals removed to show stamens and stigmatic disk.
Nuphar luteum

a

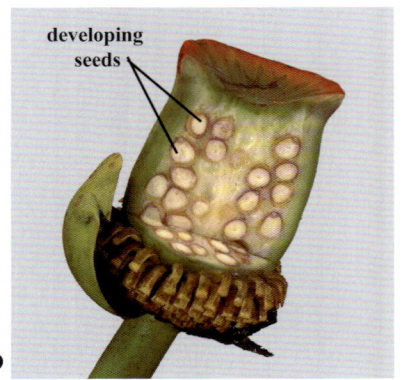

b

c

Fig. 439   Fruit characters of spatterdock. a) Immature fruit with some persistent sepals and stamens. b) LS of fruit. c) XS of fruit.
Nuphar luteum

Fig. 440   Plant characters of water lily. a) Floating leaves. b) Flower showing multiple petals.
Nymphaea odorata

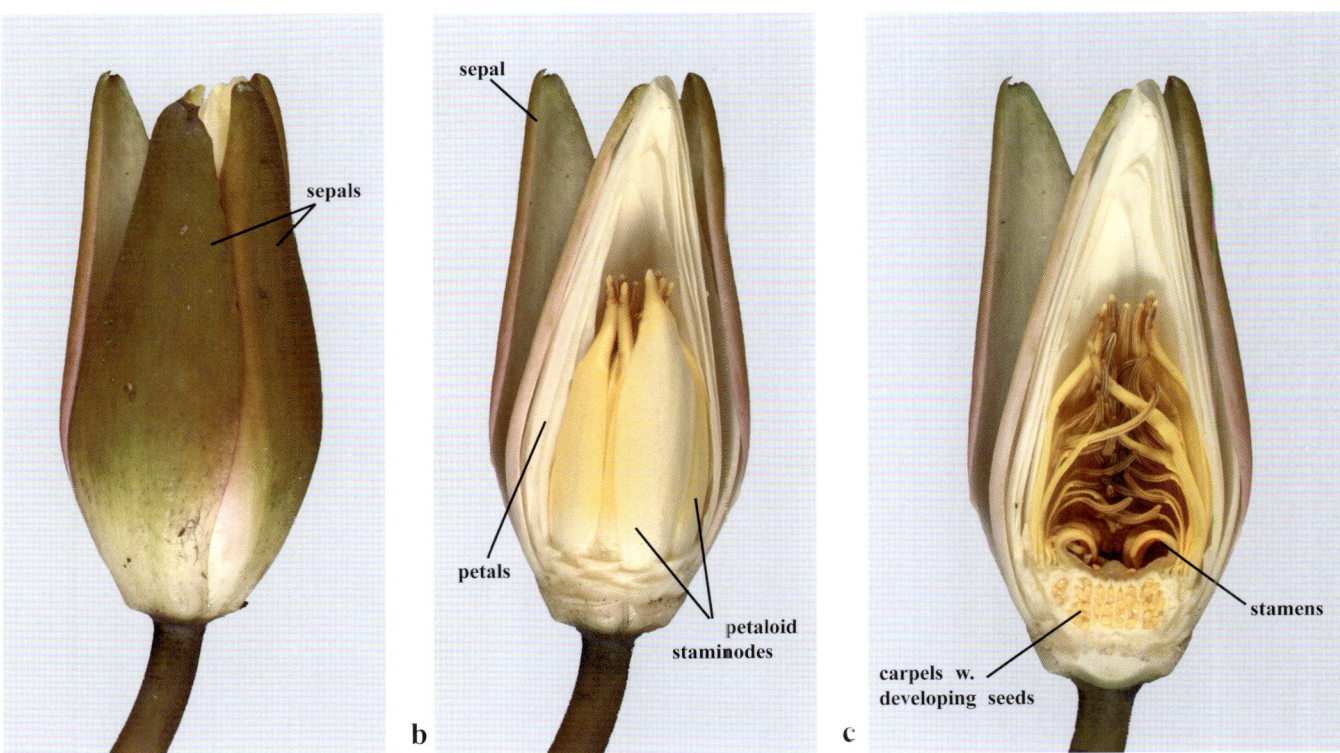

Fig. 441   Flower bud characters of water lily. a) Intact bud. b) Front sepals and petals removed to show petaloid staminodes. c) LS of bud.   *Nymphaea odorata*

Fig. 442   Plant characters of water lily. a) LS of gynoecium. b) Petaloid staminodes. c) XS of petiole showing air canals.
Nymphaea odorata

Magnoliaceae - *Magnolia Family*

Habit: Trees and shrubs.
Leaves: Alternate, simple, entire or lobed (*Liriodendron*).
Sometimes leathery (*Magnolia*). Pinnate venation, aromatic.
Inflorescences: Terminal, solitary flowers.
Flowers: Large, showy, radial, bisexual. On elongate receptacles.
Perianth: 6-many petaloid tepals.
Stamens: Numerous, spirally arranged around receptacle base.
Carpels: Numerous, spirally arranged around receptacle axis.
Each with elongate stigmatic crest.
Ovary: Superior.
Placentation: Parietal.
Fruit: Cone-like aggregate of follicles (*Magnolia*), berries,
or aggregate of samaras (*Liriodendron*).
Seeds: With a fleshy coat (**sarcotesta**)(*Magnolia*), red-orange,
and attached to fruit by a slender thread (**funicle**).
Other: Protective stipules cover leaf/flower buds, leaving ring-
like stipule scars that encircle the twig or stem.
Genera: *Liriodendron, Magnolia.*

Fig. 443   Leaf of tulip tree.
Liriodendron tulipifera

Fig. 444   Stem and floral characters of tulip tree. a) Branch tip showing new leaf and valvate bud scales.
b) Flower bud with stipule. c) Flower.   *Liriodendron tulipifera*

Fig. 445   Fruiting structures of tulip tree. a) Immature fruit. b) Mature fruit (aggregate of samaras). c) Samara.
d) Ring of persistent bracts and central receptacle.   *Liriodendron tulipifera*

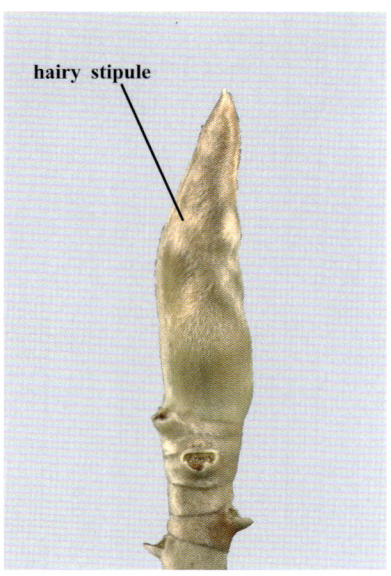
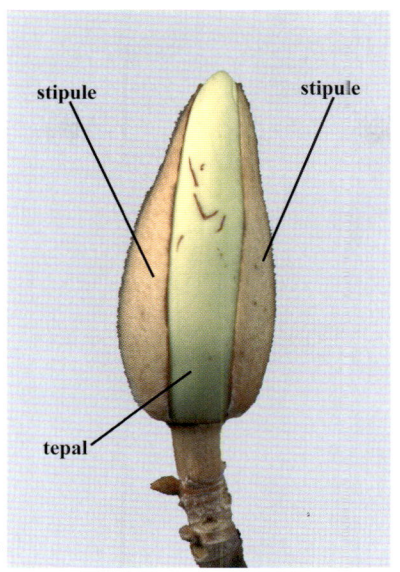

Fig. 446 Stem and floral characters of magnolia. a) Leaf bud covered by densely hairy protective stipule. b) Flower bud with protective stipule splitting to reveal tepals. c) Branch showing series of stipule scars.   *Magnolia grandiflora*

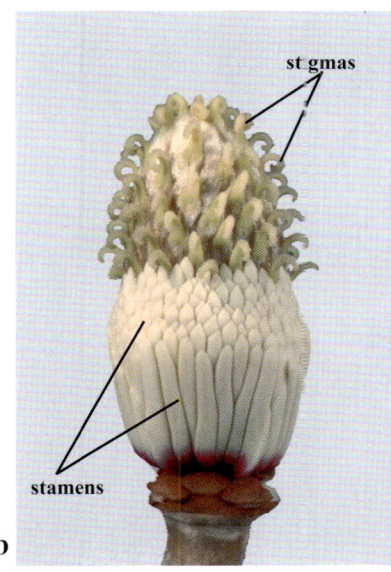
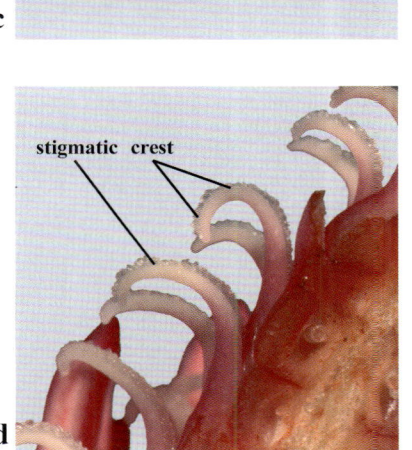

Fig. 447 Floral characters of magnolia. a) Flowering branch. b) Receptacle bearing intact stamens and stigmas. c) Stamen. d) CU of stigmas with stigmatic crests on their upper surfaces.   *Magnolia grandiflora*

Fig. 448 Fruit and leaf characters of magnolia. a) Mature fruit (aggregate of follicles). b) CU of one dehisced follicle with seeds suspended by thread-like funicles. c) Leaf.   *Magnolia grandiflora*

Annonaceae - *Pawpaw Family*

Habit: Shrubs, trees, woody vines.

Leaves: Alternate, simple, entire. Typically with short petiole.
2-ranked. Pinnate venation. Aromatic.

Inflorescences: Often solitary or small clusters of cymes.

Flowers: Showy, radial, bisexual. Produced on a short, globose
receptacle. Increase in size after opening. Drooping in appearance.

Calyx: 3 sepals, distinct.

Corolla: 6 petals, distinct. Outer petals tend to be larger than inner.

Stamens: Numerous, arranged spirally around receptacle and
clustered into a tight ball.

Carpels: 3-numerous, spirally arranged on top of receptacle.

Ovary: Superior.

Placentation: Parietal.

Fruit: Aggregate of free berries (*Asimina*) or fused berries (*Annona*).

Seeds: Large, often arillate. Contain ruminate endosperm.

Genera: *Annona, Asimina, Cananga, Deeringothamnus.*

Fig. 449   Annona fruit (aggregate of fused berries).
Annona muricata

Fig. 450   Flower of ylang-ylang with small sepals and long pendent petals.
Cananga odorata

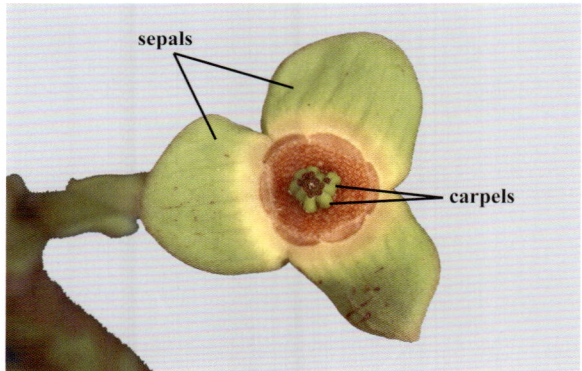

Fig. 451   Ylang-ylang flower after petals have fallen showing persistent sepals and cluster of carpels that will develop into fruit.
Cananga odorata

Fig. 452   Fruit (aggregate of unfused berries) developing from a single flower.
Cananga odorata

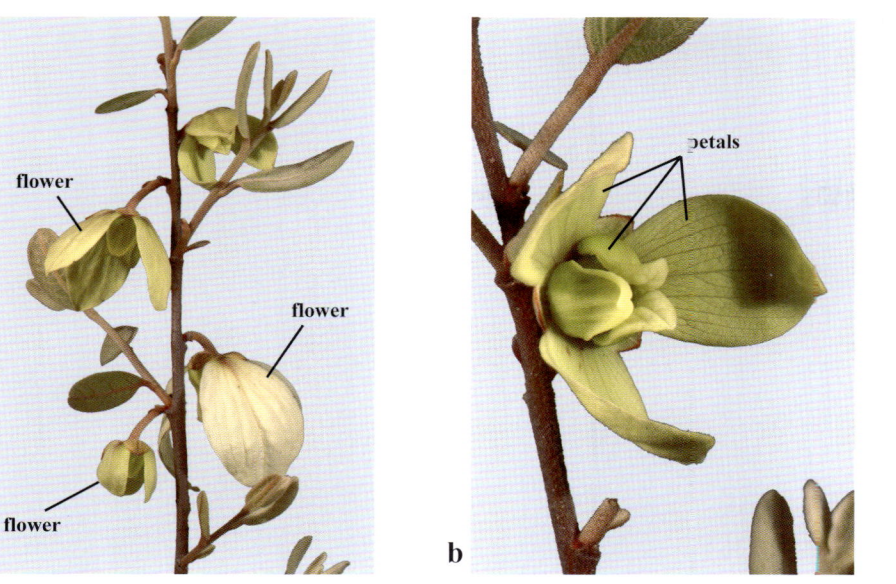

Fig. 453 Floral characters of pawpaw. a) Flowering branch. b) Flower with all six petals intact. c) Flower with one of the smaller inner whorl petals removed to show ball of stamens. *Asimina incana*

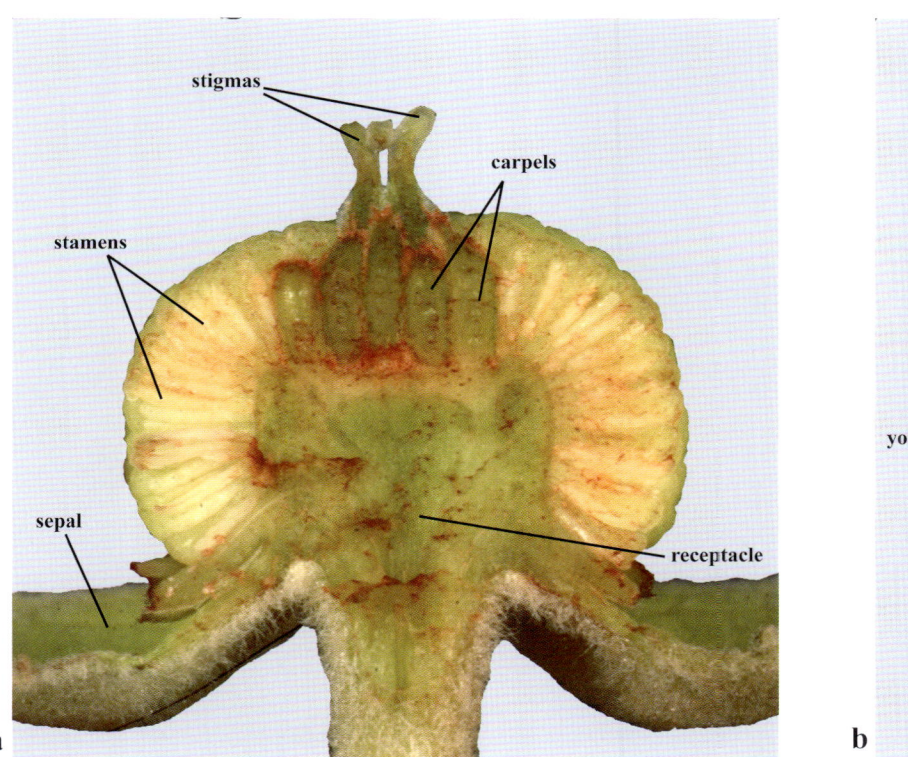

Fig. 454 Floral and fruit characters of pawpaw. a) LS of flower with all petals removed to show attachment of stamens and carpels to short globose receptacle. b) Fruiting branch with fruits of different age. *Asimina incana*

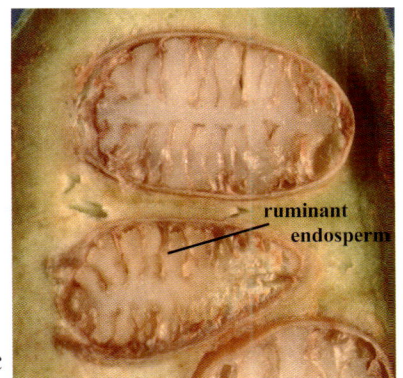

Fig. 455 Fruit of pawpaw. a) One intact fruit (aggregate of unfused berries). b) LS of one berry containing four seeds. c) LS of two seeds showing ruminant endosperm. *Asimina incana*

Lauraceae - *Bay Family*

Habit: Shrubs, trees.
Leaves: Alternate (occasionally opposite), simple, entire (rarely lobed).
Arranged spirally. Pinnate to palmate venation, aromatic.
Inflorescences: Cymes.
Flowers: Small, pale, radial. Bisexual or unisexual.
Perianth: 6 tepals, distinct to slightly connate.
Stamens: 3-12, anthers open by flaps. Some filaments w. nectar glands.
Carpels: 1.
Ovary: Superior.
Placentation: Parietal.
Fruit: Drupe or berry with persistent woody receptacle (**cupule**).
Large embryo with fleshy cotyledons.
Genera: *Cinnamomum, Lindera, Litsea, Persea, Sassafras.*

Fig. 456   Partial inflorescence and leaf of the avocado tree. *Persea americana*

a

b

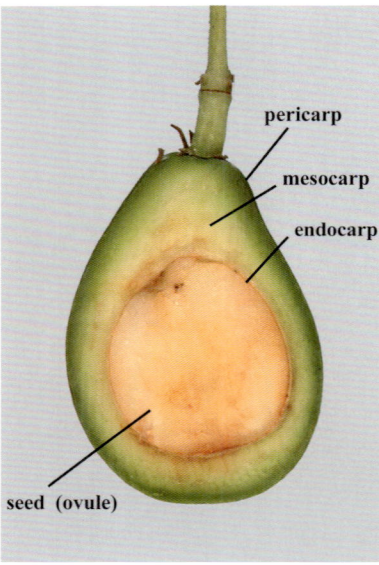

c

Fig. 457   Flower and fruit characters of avocado. a) Staminate flower showing flaps on dehisced anthers. Yellow nectar glands at base of stamens. b) Developing fruit. c) LS of developing fruit.   *Persea americana*

Fig. 458   Immature fruit of the camphor tree. *Cinnamomum camphora*

Fig. 459   Mature fruits of red bay. *Persea borbonia*

Fig. 460   Spirally arranged leaves of avocado. *Persea americana*

Piperaceae - *Pepper Family*

Habit: Herbs, vines, small trees. Stem with swollen nodes.
Leaves: Alternate, simple, entire. Palmate or pinnate venation.
Petioles sometimes sheathe stems. Aromatic.
Inflorescences: Minute flowers clustered on a thick erect spike.
Flowers: Inconspicuous, minute (resemble tiny dots).
Bisexual or unisexual. Subtended by a peltate bract.
Perianth: Absent, flowers apetalous.
Stamens: 1-19, usually 6.
Carpels: 1-4.
Ovary: Superior.
Placentation: Basal.
Fruit: Drupes.
Genera: *Peperomia, Piper.*

Fig. 461   Peperomia w. flowering branch.   *Peperomia obtusifolia*

Fig. 462   Leaf of rootbeer plant showing a common piper leaf shape.   *Piper auritum*

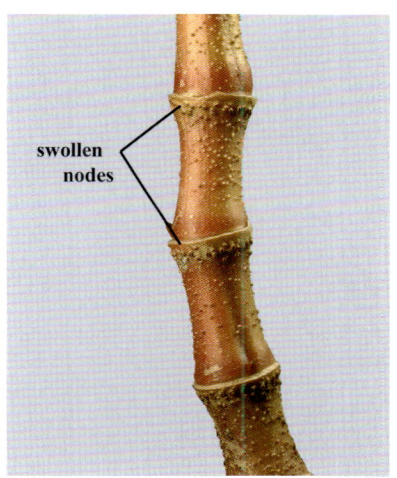

Fig. 463   Swollen nodes of stem. *Piper* sp.

Fig. 464   Stem showing sheathing petiole and inflorescence opposite leaf.   *Piper auritum*

Fig. 465   Peperomia inflorescence. a) Flower spike. b) CU of flowers on spike.   *Peperomia obtusifolia*

Aristolochiaceae - *Dutchman's Pipe Family*

(Characters given for the Genus *Aristolochia*)
Habit: Vines, shrubs, herbs.
Leaves: Alternate, simple, entire. Sometimes lobed and cordate at base. Palmate venation, pellucid dots, aromatic. Bases broadly sheathing.
Inflorescences: Various.
Flowers: Large, showy, often with curvy shapes and the color of carrion to trap fly pollinators. Bisexual.
Perianth: Large and petaloid, with 3 fused differentiated sepals.
Stamens: 6-12, free or fused to style.
Carpels: 4-6.
Ovary: Usually inferior.
Placentation: Axile.
Fruit: Septicidal capsule opening basally.
Seeds: Flat or 3-sided.
Genera: *Aristolochia, Asarum, Saruma*.

Fig. 466   *Saruma henryi*
This uncommon genus has cordate leaves and 3-merous flowers.

a   b

Fig. 467   Plant characters of Virginia snakeroot. a) Leafing branch. b) Pipe-like profile of flower. *Aristolochia serpentaria*

seed   aril

a   b   c

Fig. 468   Fruit characters of Virginia snakeroot. a) Intact mature fruit capsule. b) Dehiscing fruit capsule showing seeds. c) Triangular-shaped seeds with white arils.   *Aristolochia serpentaria*

Fig. 469 Floral characters of Dutchman's pipe. a) Carrion-colored flower and cordate leaves. b) Flower in profile with half of petaloid tepal removed to show the trap-like base or calyx. c) LS of flower base that traps fly pollinators. *Aristolochia elegans*

Fig. 470 Fruit characters of Dutchman's pipe. a) Immature capsules of progressive ages. b) LS of capsule showing stacked seeds. c) XS of capsule with six locules. d) Mature septicidal capsule after dehiscence. *Aristolochia elegans*

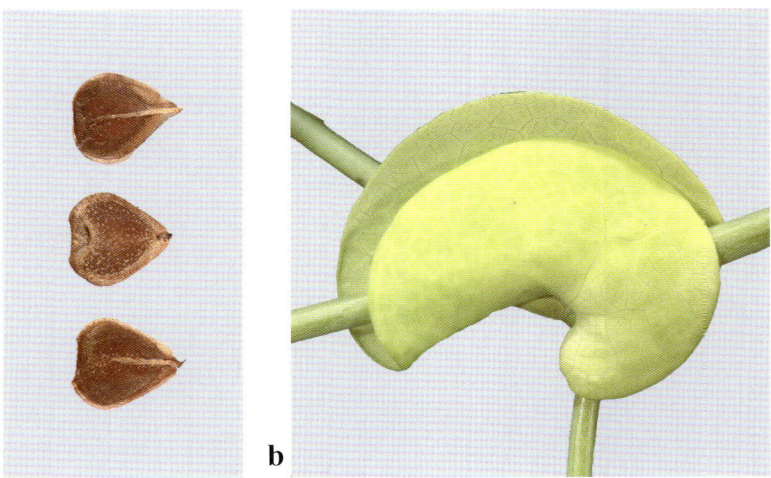

Fig. 471 a) Flattened triangular seeds. b) Base of leaf that is broadly sheathing. *Aristolochia elegans*

Fig. 472 Dutchman's pipe flower with fringed limb. *Aristolochia fimbriata*

Ranunculaceae - *Buttercup Family*

Habit: Herbs, shrubs, and vines.
Leaves: Alternate, simple or compound, lobed or dissected.
Margins dentate or serrate. Pinnate to palmate venation.
Inflorescences: Cyme, raceme, panicle, or a single flower.
Flowers: Often showy, radial to bilateral, bisexual.
Receptacle varies from short to elongate.
Perianth (if non-differentiated): 4-numerous tepals, distinct.
Calyx: 5 sepals, distinct, often petaloid.
Corolla: 5 petals, distinct, usually with nectar glands.
Stamens: Numerous, arranged spirally around receptacle.
Carpels: 5-numerous, arranged spirally around receptacle.
Ovary: Superior.
Placentation: Parietal.
Fruit: Aggregate of follicles, aggregate of achenes, berry.
Genera: *Actaea, Anemone, Clematis, Delphinium, Hepatica, Ranunculus, Thalictrum.*

Fig. 473   Columbine leaf.
Aquilegia canadensis

a   b   c   d

**Fig. 474   Floral and fruit characters of columbine. a) Flowering branch. b) Sepal and tubular petal with nectar spur.
c) Developing fruit. d) Mature fruit (aggregate of follicles).**
Aquilegia canadensis

a   b   c

**Fig. 475   Floral and fruit characters of delphinium. a) Flower showing large single nectar spur. b) Developing fruit.
c) Mature fruit (follicle) that has dehisced.**
Delphinium sp.

Fig. 476   a) Anemone flower. b) Anemone fruit (aggregate of achenes).
Anemone berlandieri

Fig. 477   Netleaf leather flower.
Clematis reticulata

Fig. 478   Floral and fruit characters of clematis. a) Staminate flower. b) Mature fruit (aggregate of achenes).
c) Achene with persistent plumose style.
Clematis catesbyana

Fig. 479   Floral and fruit characters of buttercup. a) Flower. b) Developing fruit. c) Mature fruit (aggregate of achenes).
Ranunculus hispidus

Berberidaceae - *Barberry Family*

Habit: Herbs or shrubs.
Leaves: Alternate (typically), simple or compound, entire to lobed to dissected. Pinnate to palmate venation.
Inflorescences: Cyme, raceme, panicle, or single flower.
Flowers: Radial, bisexual.
Perianth: 6-9 tepals, distinct.
Stamens: 4-numerous, typically opposite petals. Release pollen by flaps or slits.
Carpels: One.
Ovary: Superior.
Placentation: Parietal to basal.
Fruit: Berry.
Seeds: Arillate (typically).
Other: Inner wood yellow.
Genera: *Berberis, Caulophyllum, Jeffersonia, Mahonia, Nandina, Podophyllum.*

Fig. 480   Flowering barberry branch.
Berberis julianae

a

b

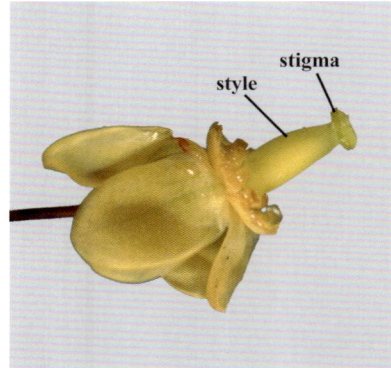

c

Fig. 481   Floral characters of barberry. a) Intact flower. b) Flower with half the petals removed to show stamens. c) Flower with all petals and all stamens removed to show pistil.
Berberis julianae

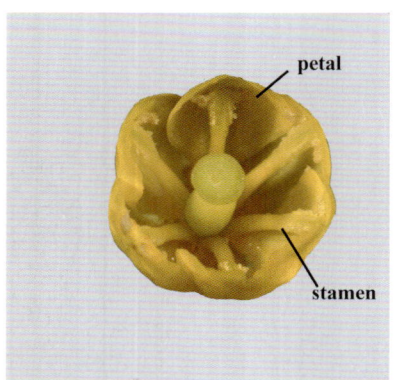

a

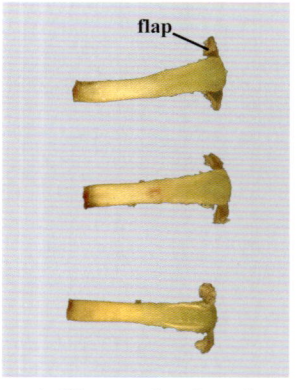

b

Fig. 482   Floral characters of barberry. a) Flower showing the stamens positioned opposite petals. b) Stamens with flaps extended indicating that anthers have dehisced.
Berberis julianae

Fig. 483   Woody grape-holly stem with bark shaved away to show yellow inner wood.
Mahonia bealei

Fig. 484   Plant characters of grape-holly. a) Compound leaf. b) Terminal inflorescence. c) Infructescence of berries.
Mahonia bealei

Fig. 485   Floral and fruit characters of heavenly bamboo. a) Inflorescence. b) Infructescence with developing fruits.
c) Infructescence with mature fruits.
Nandina domestica

Fig. 486   CU of single flower and fruit of heavenly bamboo. a) Flower. b) Fruit (berry). c) Fruit with half of outer
material removed to show seed and tissues.
Nandina domestica

Papaveraceae - *Poppy Family*

Habit: Herbs, shrubs.
Leaves: Alternate (typically), simple or compound, entire to lobed to dissected. Sometimes spiny. Pinnate venation.
Inflorescences: Cyme, panicle, umbel, or single flower.
Flowers: Radial to bilateral, bisexual.
Calyx: 2-3 sepals, distinct. Small to large, ephemeral.
Corolla: 4-6 to numerous petals, distinct, often wrinkled.
Stamens: Numerous and distinct or 6 and connate.
Carpels: 2-numerous.
Ovary: Superior.
Placentation: Parietal.
Fruit: Capsule (typically) or nut.
Seeds: Arillate (typically).
Other: Colored milky sap or clear mucilaginous sap.
Genera: *Argemone, Chelidonium, Eschscholzia, Papaver.*

Fig. 487   Mexican poppy flower.
Argemone mexicana

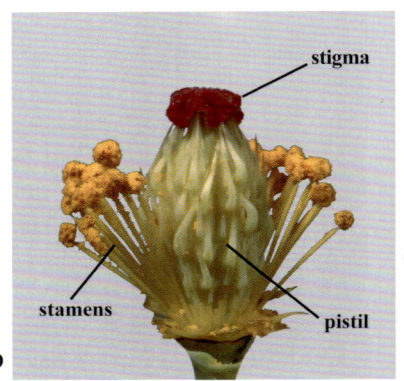

a   b   c

Fig. 488   Floral and fruit characters of Mexican poppy. a) Flower with one petal removed. b) Gynoecium and androecium (with all petals and some stamens removed). c) Intact capsule covered with spines.
Argemone mexicana

a   b   c

Fig. 489   Fruit characters of Mexican poppy. a) LS of developing capsule showing seeds. b) XS of developing capsule illustrating parietal placentation. c) Mature dehisced capsule.
Argemone mexicana

Fig. 490   Flower and fruit characters of California poppy. a) Flower showing multiple stamens. b) Immature capsule on fruiting branch. c) Intact capsule (L) and capsule in LS (R) revealing small black seeds within. *Eschscholzia californica*

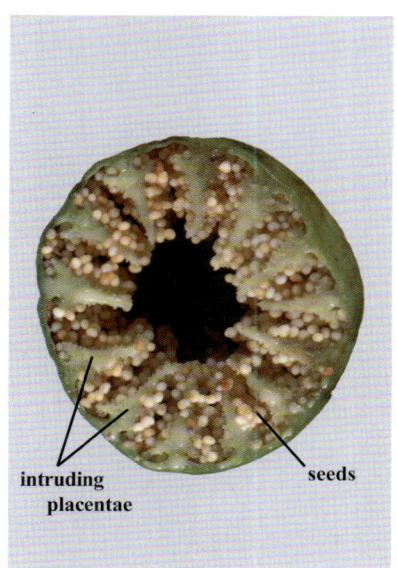

Fig. 491   Fruit characters of opium poppy. a) Developing capsule with large radiate persistent stigma at apex. LS of immature capsule showing developing seeds. c) XS of immature capsule showing developing seeds arranged along sides of deeply intruded placentae. *Papaver somniferum*

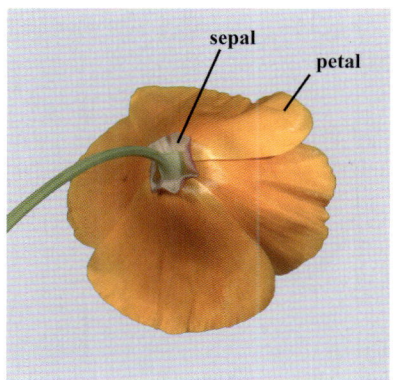

Fig. 492   Mature poricidal capsule. *Papaver rhoeas*

Fig. 493   Colored sap from broken leaf.   *Argemone mexicana*

Fig. 494   Rear view of flower showing small sepals. *Eschscholzia californica*

Platanaceae - *Sycamore Family*

Habit: Trees.

Leaves: Alternate, simple, palmately lobed, serrate, deciduous. Palmate venation. Petiole enlarged at base, enclosing axillary bud.

Inflorescences: Axillary, indeterminate, pendulous. Raceme of globose heads or a single head.

Flowers: Very reduced to minute, radial, unisexual.

Calyx: 3-7 sepals, distinct to slightly connate.

Corolla: Only present in staminate flowers. 3-7 petals, distinct, fleshy.

Stamens: 3-7, distinct, opposite sepals.

Carpels: 5-9.

Ovaries: Superior.

Placentation: Apical.

Fruit: Aggregate of achenes clustered into a globose head. Achenes subtended by long bristles.

Other: Large stipules encircle stem. Patchy, mottled bark on trunk.

Single Genus: *Platanus*.

Fig. 495   Patchy bark of sycamore trunk.
Platanus occidentalis

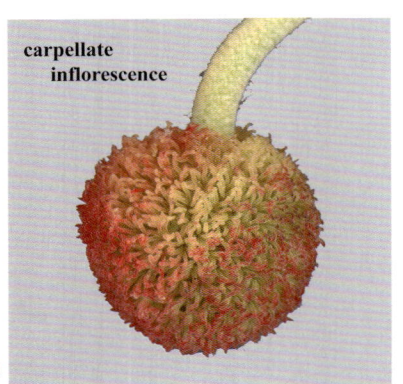

Fig. 496   Female floral characters of sycamore. a) Carpellate inflorescence. b) LS of carpellate inflorescence. c) Portion of LS through carpellate inflorescence showing individual carpellate flowers.
Platanus occidentalis

Fig. 497   Male floral characters of sycamore. a) Staminate inflorescence. b) LS of staminate inflorescence. c) Portion of LS of staminate inflorescence showing individual staminate flowers.
Platanus occidentalis

Fig. 498   Stem and branch characters of sycamore. a) Leafing branch with carpellate and staminate inflorescences present as pendent globose heads. b) Large leafy stipule encircling stem.   *Platanus occidentalis*

Fig. 499   Fruit characters of sycamore. a) Mature infructescence (aggregate of achenes).
b) Portion of LS of infructescence showing cluster of tightly packed achenes.
c) Achene subtended by long bristles for dispersal.   *Platanus occidentalis*

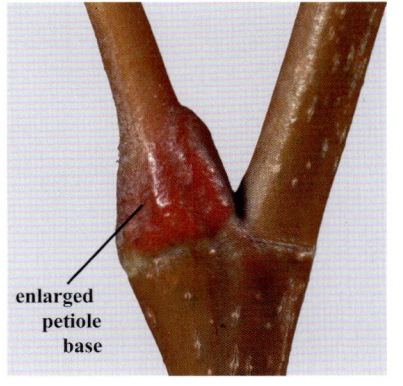

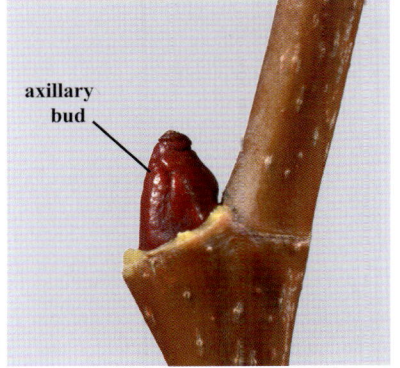

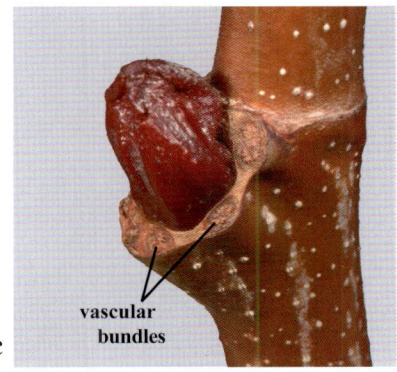

Fig. 500   Bud and leaf characters of sycamore. a) Enlarged petiole base of leaf. b) Axillary bud that was previously enclosed by petiole base. c) CU of axillary bud showing vascular bundles within leaf scar.   *Platanus occidentalis*

Nelumbonaceae - *Lotus Family*

Habit: Aquatic herb. Grows from rhizome.

Leaves: Alternate, simple, entire, peltate. Large, round, and emergent with long petiole. Palmate venation.

Inflorescences: Solitary flowers on scape.

Flowers: Large, showy, radial, bisexual. Emergent with long pedicel.

Perianth: 2 sepals and numerous petals.

Stamens: Numerous.

Carpels: Numerous, embedded in receptacle.

Ovary: Superior.

Placentation: Apical.

Fruit: Aggregate of nuts, each of which grows in its own cavity in the enlarged receptacle.

Seeds: Long lived.

Other: Stem with milky sap.

Single Genus: *Nelumbo.*

Fig. 501   Young leaf of American lotus prior to unfurling. *Nelumbo lutea*

Fig. 502   **Stand of American lotus growing in a water-filled ditch. Shown are the large round peltate leaves, yellow flowers, green buds, and green developing receptacles.**   *Nelumbo lutea*

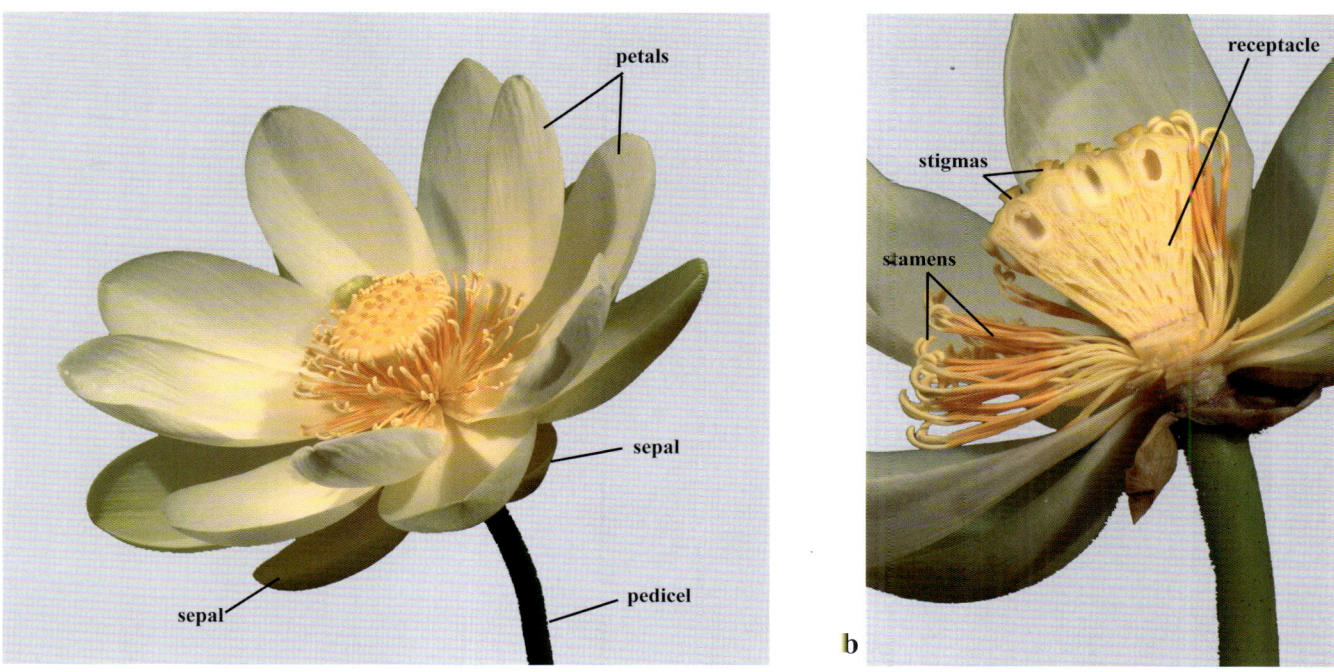

Fig. 503   Floral characters of the American lotus. a) Flower showing numerous stamens arranged around a fleshy central receptacle. b) LS of flower showing spongy nature of receptacle with embedded carpels at the top.   *Nelumbo lutea*

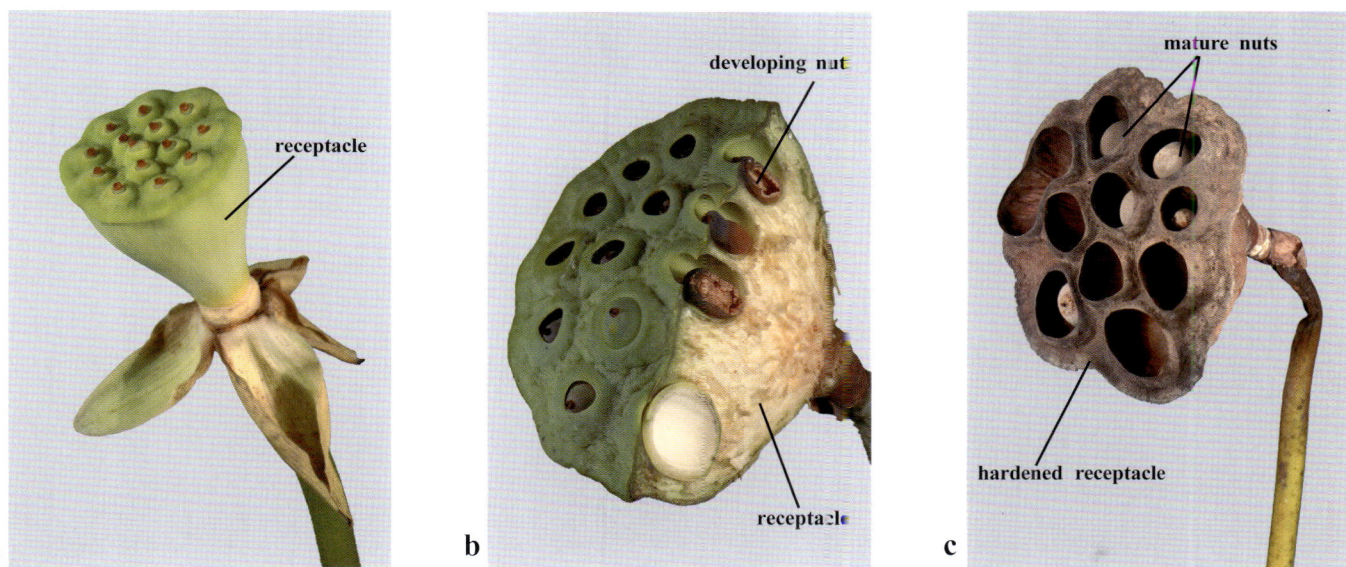

Fig. 504   Fruit characters of the American lotus. a) Young fruit as receptacle begins to enlarge. b) LS of nearly mature fruit showing cavities in which the nuts are developing. c) Mature fruit consisting of hardened receptacle with enlarged cavities that allow nuts to drop in water and disperse.   *Nelumbo lutea*

Fig. 505   Various characters of the American lotus. a) Mature nut. b) Flower bud. c) XS of stem showing air canals and milky sap.   *Nelumbo lutea*

Vitaceae - *Grape Family*

Habit: Vines, sometimes shrubs.

Leaves: Alternate, simple or compound, frequently lobed. 2-ranked. Pinnate to palmate venation. Tendrils opposite leaves, attach by twining or sucker disks. Nodes swollen.

Inflorescences: Cyme, raceme, panicle, appearing opposite leaf. Composed of numerous flowers.

Flowers: Inconspicuous, small to minute, radial, bisexual or unisexual.

Calyx: 4-5 small sepals, sometimes reduced to a rim around ovary.

Corolla: 4-5 petals.

Stamens: 4-5, opposite petals.

Carpels: 2.

Ovary: Superior.

Placentation: Axile.

Fruit: Berry.

Seeds: 4, distinctive shape and features: **raphe** (groove along the sides) and **chalazal knot** (bump or upraised area).

Other: Nectar disk in form of ring between stamens and ovary.

Genera: *Ampelopsis, Cissus, Parthenocissus, Vitis.*

Fig. 506   Virginia creeper vine with palmately compound leaf and cluster of buds.   *Parthenocissus quinquefolia*

a   b   c

Fig. 507   Floral and fruit characters of Virginia creeper. a) Inflorescence. b) CU of flower with recurved petals and erect stamens. c) Infructescence with young green fruits. *Parthenocissus quinquefolia*

a   b

Fig. 508   Floral characters of pepper vine. a) Flowering branch. b) CU of flowers. *Ampelopsis arborea*

Fig. 509   Cluster of maturing muscadine grapes.
Vitis rotundifolia

Fig 510   Cultivated grapes intact and in
LS to show seeds within the berry.
Vitis vinifera

seed

Fig. 511   Flowering branch of grape with three inflorescences.
Vitis cinerea

tendril

Fig. 512   Branch of grape vine showing
tendrils opposite leaves.   *Vitis* sp.

Fig. 513   Inflorescence of grape with many
open flowers.   *Vitis* sp.

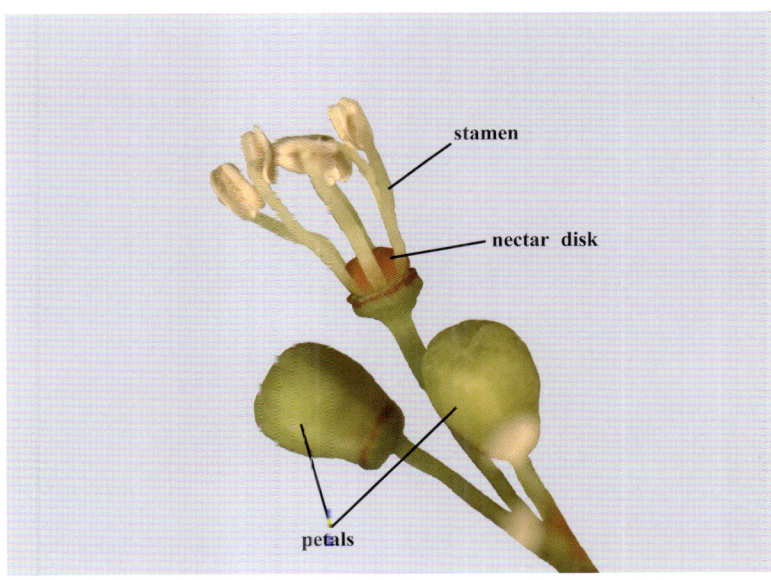

stamen

nectar disk

petals

Fig. 514   CU of grape flower. (Note caducous petals of flower have
fallen off upon opening and are gone.)   *Vitis* sp.

Caryophyllaceae - *Pink Family*

Habit: Herbs. Often with swollen nodes.
Leaves: Opposite, simple, entire. Frequently basally connate.
Often long and thin. Parallel venation.
Inflorescences: Cyme or a single flower.
Flowers: Showy, radial, bisexual.
Calyx: 4-5 sepals.
Corolla: 4-5 petals, often notched or 2-lobed.
Petal may be modified into a slender basal part (claw)
and an extended apical part (limb). Often pink or white.
Stamens: 4-10, sometime adnate to petals.
Carpels: 2-5.
Ovary: Superior.
Placentation: Free-central or basal.
Fruit: Typically loculicidal capsule, sometimes utricle.
Seeds: Sculptured surface.
Genera: *Cerastium, Dianthus, Silene, Stellaria.*

Fig. 515   Flower of a pink.
Dianthus sp.

Fig. 516   Flowering branch of mouse-ear chickweed.
Cerastium glomeratum

a   b

Fig. 517   a) Flowering branch of common chickweed.
b) CU of flower.
Stellaria media

style

ovary

a   b   c   d

Fig. 518   Flower-Fruit sequence of a pink. a) Flower bud. b) Flower in profile. c) LS of carpellate flower. d) Capsule.
Dianthus sp.

Phytolaccaceae - *Pokeweed Family*

Habit: Herbs, shrubs. Sometimes succulent.
Leaves: Alternate, simple, entire. Pinnate venation.
Inflorescences: Raceme or spike, appearing opposite leaves.
Flowers: Small and inconspicuous, radial, bisexual.
Nectar disk present.
Perianth: 4-5 tepals, distinct.
Stamens: 10-numerous.
Carpels: 3.
Ovary: Superior.
Placentation: Axile.
Fruit: Berry.
Seeds: Reniform, typically arillate.
Other: Stems often red or purple (due to betalain compounds), and
with concentric rings of vascular tissue. Anomalous 2ndary growth.
Genera: *Phytolacca*.

Fig. 519   Inflorescence of pokeweed.
Phytolacca dioica

Fig. 520   Fruit characters of pokeweed. a) Infructescence with older fruits at bottom. b) Infructescence with developing
fruits. c) Mature infructescence and stem showing red betalain compounds.
Phytolacca americana

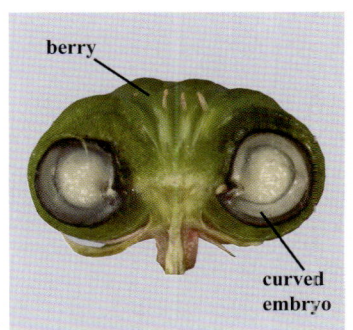

Fig. 521   Fruit characters of pokeweed. a) Berries: Left showing point of attachment and right showing apex.
b) XS of berry showing ten seeds. c) LS of berry with seeds showing curved embryo.   *Phytolacca americana*

Amaranthaceae - *Pigweed Family*

Habit: Herbs or herbaceous-woody shrubs.
Sometimes succulent.
Leaves: Alternate or opposite, simple, entire.
Inflorescences: Various. Small densely clustered flowers.
Flowers: Inconspicuous, minute, radial, bisexual.
Subtended by persistent papery bracts.
Perianth: 4-5 tepals, dry and papery, white to pink.
Stamens: 5, connate, opposite tepals.
Carpels: 2-3.
Ovary: Superior.
Placentation: Basal.
Fruit: Achene, utricle, or circumscissile capsule.
Seeds: Spherical or lenticular. Often shiny.
Other: Stems may be red to purple (betalain compounds).
Often with concentric rings of vascular tissue.
Genera: *Alternanthera, Amaranthus, Celosia, Gomphrena.*

Fig. 522   Flowering amaranthus.
Amaranthus sp.

Fig. 523   Floral characters of alligator weed. a) Flowering branch.
b) CU of cluster of flowers.   *Alternanthera philoxeroides*

Fig. 524   Inflorescence of froelichia.
Froelichia floridana

Fig. 525   Floral characters of celosia.   a) Inflorescence. b) CU of flower.
Celosia sp.

Fig. 526   Flowers of iresine.
Iresine rhizomatosa

Chenopodiaceae - *Goosefoot Family*

Habit: Herbs to shrubs, frequently succulent.
Typically in saline or disturbed habitats.
Leaves: Alternate, simple, entire or lobed or serrate.
May be succulent or reduced. Frequently covered
with hairs causing a powdery or flaky appearance.
Inflorescences: Various.
Flowers: Inconspicuous, minute, greenish, radial, bisexual or unisexual.
Perianth: 5 tepals, frequently enclosing fruit.
Typically green and fleshy.
Stamens: 5, opposite tepals.
Carpels: 2-3.
Ovary: Superior.
Placentation: Basal.
Fruit: Achene, utricle, or circumscissile capsule.
Frequently subtended or enclosed by persistent perianth.
Genera: *Atriplex, Chenopodium, Salicornia, Salsola.*
Other: Stems may be reddish due to betalains.

Fig. 527   Mexican tea in flower.
Chenopodium ambrosioides

a

b

Fig. 528   Floral characters of Mexican tea. a) Flowering branch. b) CU of portion of inflorescence.
Chenopodium ambrosioides

a

b

Fig. 529   Common beet. a) Large, round edible portion is a modified taproot. b) XS of modified taproot showing
concentric circles of vascular tissue.   *Beta vulgaris*

Nyctaginaceae - *Four O' Clock Family*

Habit: Herbs, shrubs, or trees.
Leaves: Typically opposite, simple, entire. Pinnate venation.
Inflorescences: Cyme or head.
Flowers: Showy or inconspicuous, radial, typically bisexual. Subtended by conspicuous leaf-like or petaloid bracts that may form an involucre. Nectar disk present.
Perianth: 5 tepals, connate, funnelform. Often divided into a persistent green base and colored petaloid distal part.
Stamens: Typically 5.
Carpels: 1.
Ovary: Superior (may appear inferior, see 531-a).
Placentation: Basal.
Fruit: Achene or nut, typically enclosed by perianth base.
Other: Various vegetative structures may be red or purple (betalain compounds). Concentric rings of vascular tissue.
Genera: *Abronia, Boerhavia, Bougainvillea, Mirabilis.*

Fig. 530   Flowering and fruiting branch of boerhavia. *Boerhavia diffusa*

a   b

Fig. 531   a) Portion of inflorescence. b) Anthocarp fruits consisting of an achene enclosed by a persistent perianth.   *Boerhavia diffusa*

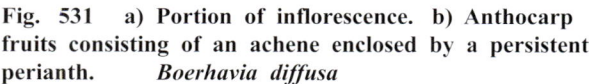

a   b

Fig. 532   Floral structures of bougainvillea. a) Flowers and bracts. b) One flower and its subtending petaloid bract.   *Bougainvillea spectabilis*

a   b   c

Fig. 533   Floral and fruit characters of four o'clocks. a) Flowering branch. b) Flower bud. c) Fruit with two sepals of persistent calyx removed.   *Mirabilis jalapa*

Portulacaceae - *Purslane Family*

Habit: Succulent herbs to shrubs. Recumbent.
Grows from fleshy taproot.
Leaves: Opposite or alternate or basal rosette. Simple, entire,
flat to cylindrical. Succulent. Pinnate venation, often obscure.
Inflorescences: Various, sometimes solitary.
Flowers: Showy, radial, bisexual. Subtended by 2 unequal bracts.
Perianth: 4-6 tepals, petaloid.
Stamens: 4-6 (or numerous), opposite tepals.
Carpels: 2-3.
Ovary: Superior to inferior.
Placentation: Free central to basal.
Fruit: Circumscissile or loculicidal capsule.
Seeds: Lenticular, sometimes arillate.
Other: Red to purple stems (betalain compounds).
Mucilage frequently present.
Genera: *Calandrina, Claytonia, Portulaca, Talinum.*

Fig. 534   Flowering branch of
purslane.
Portulaca oleracea

a   b   c

Fig. 535   Floral characters of purslane. a) Flowering branch. b) Flower with one petal removed showing multiple stamens clustered in the center. c) LS of flower.   *Portulaca* sp.

a   b   c

Fig. 536   Fruit characters of purslane. a) Mature capsule before dehiscence. b) Capsule with top half beginning to dehisce. c) Circumscissile capsule after dehiscence showing bottom half filled with shiny black seeds.   *Portulaca* sp.

Cactaceae - *Cactus Family*

Habit: Herbs, vines, or shrubs with succulent spiny stems. Often ribbed, tuberculate, conical, cylindrical, flattened, or globose. Long and short shoots produced. Typically grow in dry habitats.

Leaves: Photosynthetic leaves produced from long shoots. Alternate, simple, entire. Pinnate venation. Sometimes reduced or lacking. Short shoot leaves modified into spines and spine clusters often with nettling bristles or hairs (glochids) in the Opuntia group.

Inflorescences: Typically a single flower.

Flowers: Large, showy, radial, bisexual. Sunk into tip of modified branch. Hypanthium present with nectar ring.

Perianth: Numerous tepals, petaloid, spirally arranged.

Stamens: Numerous.

Carpels: 3-numerous.

Ovary: Inferior. Embedded in tip of stem.

Placentation: Parietal.

Fruit: Berry, typically spiny. Sometimes purple due to betalains.

Seeds: Numerous, embedded in pulp.

Other: Watery to mucilaginous sap.

Genera: *Cereus, Lophophora, Opuntia, Pereskia.*

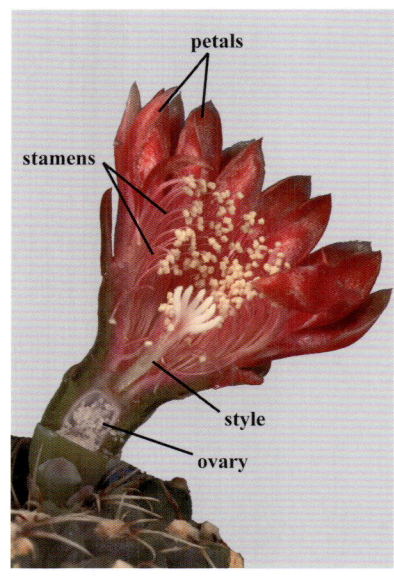

Fig. 537   LS of cactus flower. *Notocactus* sp.

Fig. 538   Cactus habit diversity. a) Columnar growth habit of *Mammillaria* sp. b) Columnar ribbed growth habit. c) Flattened stem of *Schlumbergera* sp.

Fig. 539   CU of spines at nodes. *Mammillaria* sp.

Fig. 540   Flower of Christmas cactus. *Schlumbergera bridgesii*

Fig. 541   LS of flower base showing ovary with developing seeds. *Notocactus* sp.

Fig. 542   Growth characters of paddle cactus. a) Habit. b) Flattened photosynthetic stem or pad bearing highly modified scale-like leaves.
Opuntia sp.

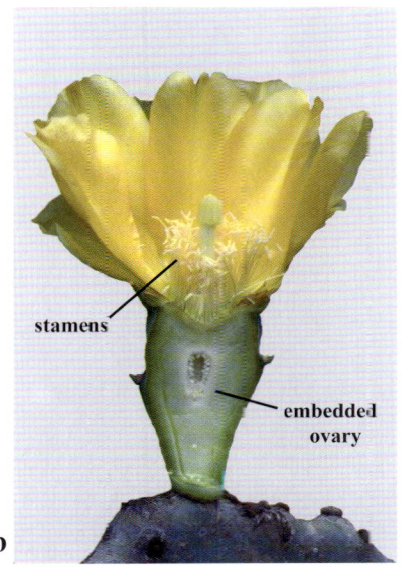

Fig. 543   Floral characters of paddle cactus. a) Intact flower. b) LS of flower. c) CU of embedded ovary.
Opuntia sp.

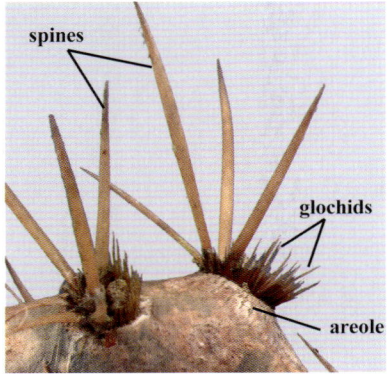

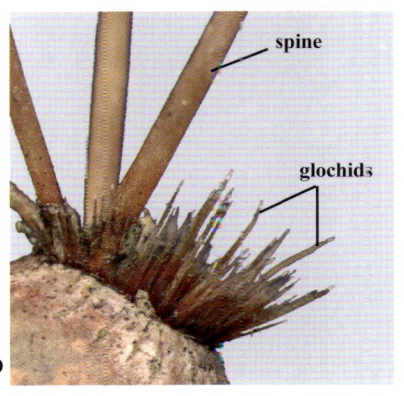

Fig. 544   Characters of paddle cactus. a) & b) Spines and glochids. c) Fruit.
Opuntia sp.

Polygonaceae - *Knotweed Family*

Habit: Herbs, shrubs, vines, trees.

Leaves: Alternate, simple, entire. Pinnate venation.
Membranous sheathing stipule (ocrea) at base of petiole.
Nodes swollen.

Inflorescences: Various.

Flowers: Small, radial, typically bisexual.
Subtended by persistent bracts.

Perianth: 5-6 tepals, petaloid. Persistent.

Stamens: 3-9.

Carpels: 2-3.

Ovary: Superior. Nectar disk at base.

Placentation: Basal.

Fruit: Achene or nutlet. Often enclosed by persistent tepals.

Genera: *Antigonon, Coccoloba, Eriogonum, Polygonella,
Polygonum, Rumex.*

Fig. 545   CU of stem of sea
grape showing sheathing stipules.
Coccoloba uvifera

Fig. 546   Fruiting branch of sea grape. Fruits are achenes
covered with a persistent perianth.
Coccoloba uvifera

Fig. 547   Floral and fruit characters of large-flower
jointweed. a) Budding and flowering branch.
b) Fruiting branch.   *Polygonella robusta*

Fig. 548   Floral and fruit characters of swamp dock. a) Flowering branch. b) CU of flowers. c) Fruiting branch.
d) CU of achenes with winged perianths.   *Rumex verticillatus*

Fig. 549   Floral characters of knotweed. a) Flowering branch. b) Inflorescence. c) CU of flowers.
Polygonum densiflorum

Fig. 550   CU of sheathing stipule or ocrea of knotweed.
Polygonum sp.

Fig. 551   Flowering branch of coral vine.
Antigonon leptopus

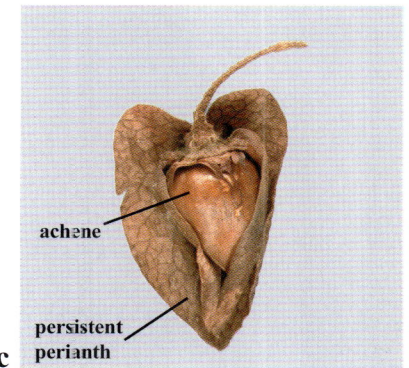

Fig. 552   Fruit characters of coral vine. a) Developing fruits (achenes with persistent perianths). b) Immature fruit with
portion of perianth removed. c) Mature fruit with portion of perianth removed.   *Antigonon leptopus*

Droseraceae - *Sundew Family*

Habit: Insectivorous herbs. Moist-wet to dry-arid habitats.
Leaves: Alternate or basal rosette. In *Drosera*, covered with sticky gland-headed hairs, spatulate or linear, circinate. In *Dionaea*, blade modified into a hinged trap.
Inflorescences: Cyme or solitary flower.
Flowers: Radial, bisexual.
Calyx: 5 sepals.
Corolla: 5 petals.
Stamens: 5, sometimes numerous.
Carpels: 3.
Ovary: Superior.
Placentation: Basal or parietal.
Fruit: Loculicidal capsule.
Other: Insects captured by sticky hairs (sundews) or by snap-trap (Venus' flytrap), then digested.
Genera: *Dionaea* and *Drosera*.

Fig. 553   Habit of linear leaf sundew showing circinate new growth.   *Drosera filliformis*

Fig. 554   Leaf and floral characters of the round leaf sundew. a) Habit (diameter of plant = 3cm). b) Leaves with sticky glandular hairs that trap insects. c) Inflorescense.   *Drosera capillaris*

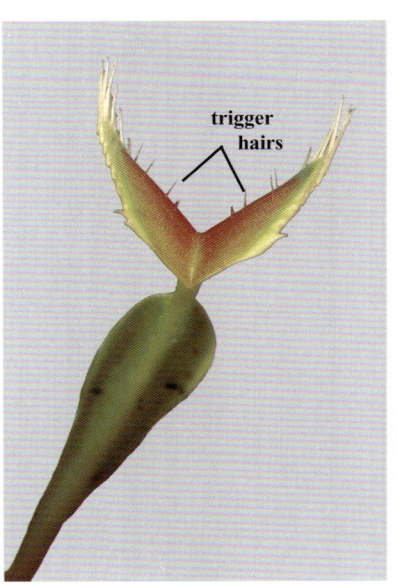

Fig. 555   Leaf and floral characters of Venus' flytrap. a) Modified leaves showing hinged traps. b) Leaf with open trap and touch-sensitive trigger hairs. c) Inflorescence.   *Dionaea muscipula*

Saxifragaceae - *Saxifrage Family*

Habit: Herbs.
Leaves: Alternate or basal rosette. Simple or compound. Entire to dentate. Palmate to pinnate venation.
Inflorescences: Various, often panicles.
Flowers: Radial to bilateral, bisexual or unisexual. Hypanthium present.
Calyx: 4-5 sepals.
Corolla: 4-5 petals, distinct. Frequently clawed.
Stamens: 3-10.
Carpels: 2-4.
Ovary: Superior to inferior.
Placentation: Axile or parietal.
Fruit: Septicidal capsule or follicle.
Other: Nectar disk frequently around base of ovary.
Genera: *Heuchera, Mitella, Saxifraga, Tiarella.*
Note: The Saxifragaceae are traditionally more broadly defined to include many other genera including both shrubs and trees.

Fig. 556   Inflorescence of saxifrage.   *Mukdenia rossii*

Fig. 557   Floral characters of a saxifrage. a) Flower showing clawed petals. b) LS of flower with corolla removed showing two styles and stamens. *Mukdenia rossii*

Fig. 558   Leaves and habit of an alumroot.   *Heuchera* sp.

Fig. 559   Partial inflorescence of a saxifrage.   *Mitella* sp.

Fig. 560   Floral characters of an alumroot. a) Partial inflorescence. b) Flower. c) LS of flower.   *Heuchera* sp.

Altingiaceae - *Sweet Gum Family*

Habit: Shrubs or trees. Tissues with aromatic resins.
Leaves: Alternate, simple, spirally arranged. Entire to serrate, often palmately lobed. Palmate to pinnate venation. Stipules at petiole base.
Inflorescences: Staminate a spherical cluster of stamens. Carpellate a spherical head.
Flowers: Inconspicuous, small, radial, unisexual.
Perianth: Numerous tiny lobes or scales (only on carpellate flowers).
Stamens: Numerous.
Carpels: 2.
Ovary: Half inferior.
Placentation: Axile.
Fruit: Multiple of septicidal capsules.
Genera: *Altingia, Liquidambar*.
Note: Many botanists include these genera in the Hamamelidaceae.
Note: All photos on this page represent a sweetgum tree, *Liquidambar styraciflua*.

Fig. 561   Seed.

Fig. 562   Star-shaped leaf.

a

b

c

Fig. 563   Floral characters of sweetgum. a) Flowering branch before male flowers have dehisced. b) Flowering branch after male flowers have dehisced. c) CU of staminate inflorescences before dehiscence.

a

b

c

Fig. 564   Floral and fruit characters of sweetgum. a) Carpellate inflorescence. b) LS of carpellate inflorescence. c) Infructescence (ball of capsules).

Hamamelidaceae - *Witch Hazel Family*

Habit: Shrubs, trees.
Leaves: Alternate, simple, entire to serrate. 2-ranked. Palmate or pinnate venation. Stipule near base of petiole.
Inflorescences: Solitary or clustered axillary flowers.
Flowers: Showy to inconspicuous, radial, bisexual or unisexual.
Calyx: 4-5 sepals.
Corolla: 4-5 petals, distinct. Frequently wrinkled.
Stamens: 4 stamens and 4 staminodes or numerous. Anthers open by 2 flaps.
Carpels: 2.
Ovary: Half inferior.
Placentation: Axile.
Fruit: Loculicidal or septicidal capsule, leathery-woody.
Other: Aromatic compounds present. Hairs stellate.
Genera: *Fothergilla, Hamamelis, Loropetalum.*

Fig. 565   Loropetalum flowering branch.   *Loropetalum chinense*

a

b

Fig. 566   Floral characters of witch hazel. a) Partial inflorescence with two flowers. b) CU of flower showing short stamens and long linear wrinkled petals.   *Hamamelis virginiana*

a

b

Fig. 567   Witch hazel characters. a) Two unfurling flowers and a cluster of persistent calices. b) Seeds. *Hamamelis virginiana*

Fig. 568   Developing capsule. *Hamamelis molis*

Crassulaceae - *Stonecrop Family*

Habit: Herbs to shrubs, frequently succulent.
Leaves: Fleshy. Alternate or opposite or basal rosette. Entire to serrate-dentate. Pinnate venation.
Inflorescences: Cyme, corymb, panicle, or solitary.
Flowers: Small, showy, radial, bisexual.
Calyx: 4-5 sepals.
Corolla: 4-5 petals.
Stamens: 4-10.
Carpels: 4-5, each with a nectar gland.
Ovary: Superior.
Placentation: Parietal.
Fruit: Aggregate of follicles.
Genera: *Crassula, Echeveria, Kalanchoe, Sedum.*

Fig. 569   Habit of clay rose.
Graptopetalum paraguayense

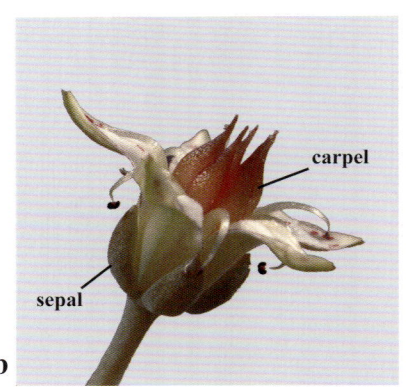

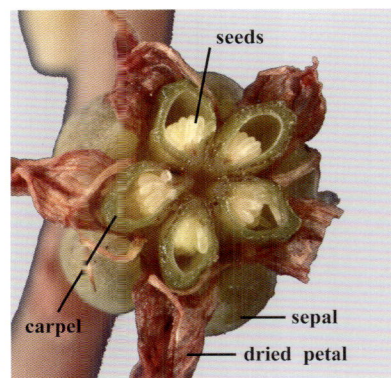

a   b   c

Fig. 570   Floral characters of clay rose. a) Flower. b) Flower in profile showing carpels. c) XS of ovary showing developing seeds.
Graptopetalum paraguayense

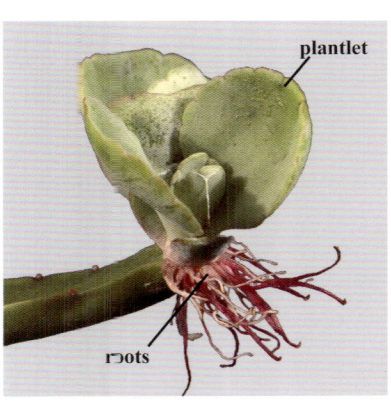

a   b

Fig. 571   Fruit characters of gray clay rose. a) Mature fruit (aggregate of follicles) after dehiscence. b) Fruit with two of five follicles removed.
Graptopetalum paraguayense

Fig. 572   Vivipary in kalanchoe with vegetative reproduction occurring at tip of leaf.
Kalanchoe sp.

Viscaceae - *Christmas Mistletoe Family*

Habit: Herbaceous to shrubby epiphytic parasites. Grow from branches of woody host trees. Stems brittle, break apart easily Dichotomously branched, jointed nodes. Typically green.

Leaves: Opposite, simple, entire, decussate. Leathery to slightly succulent. Pinnate venation. Sometimes greatly reduced or absent.

Inflorescences: Small spikes or cymes.

Flowers: Inconspicuous, small, radial, unisexual. Often dioecious.

Perianth: 3-4 tepals.

Stamens: 3-4, opposite and adnate to tepals.

Carpels: 3-4.

Ovary: Inferior.

Placentation: Basal.

Fruit: Berry containing very sticky material.

Seed: 1 per fruit, no seed coat.

Other: Roots modified into haustoria that permit the invasion of host vascular system.

Genera: *Arceuthobium, Phoradendron.*

Fig. 573 Parasitic growth habit of mistletoe on host branch. *Phoradendron leucarpum*

Fig. 574 Floral and fruit characters of mistletoe. a) Flowering branch. b) Carpellate inflorescence. c) Mature berries. *Phoradendron leucarpum*

Fig. 575 Floral characters of mistletoe. a) Flowering branch with staminate inflorescences. b) Staminate inflorescence. c) CU of portion of staminate inflorescence. *Phoradendron leucarpum*

Geraniaceae - *Geranium Family*

Habit: Herbs or shrubs. Jointed nodes.
Leaves: Usually alternate. Simple and lobed/dissected or pinnately/palmately compound. Usually palmate venation.
Inflorescences: Cyme, umbel, sometimes 1 flower.
Flowers: Showy, radial or bilateral, bisexual.
Calyx: 5 sepals, sometimes 1 modified as a nectar spur.
Corolla: 5, distinct.
Stamens: 10.
Carpels: 5.
Ovary: Superior.
Placentation: Axile.
Fruit: 5-segmented schizocarp. Each segment with 1 seed and attached to a persistent central column.
Other: Glandular hairs with aromatic oils present. Nectar glands alternate with petals on receptacle.
Genera: *Erodium, Geranium, Pelargonium.*

Fig. 576   Geranium characters. a) Seeds.
b) Flowering branch.
Geranium carolinianum

Fig. 577   Floral characters of geranium. a) Flower. b) Flower with some sepals and petals removed to show androecium and gynoecium. c) Flower shortly after petals and stamens have fallen.
Geranium carolinianum

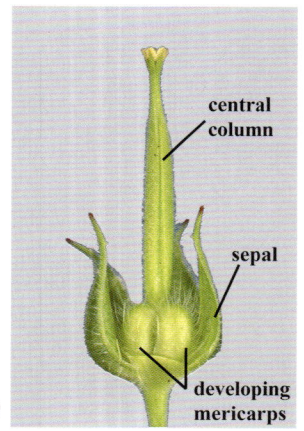

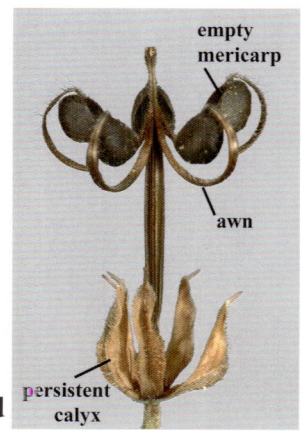

Fig. 578   Fruit characters of geranium. a) Immature schizocarp almost totally covered by persistent calyx. b) Immature schizocarp with portion of calyx removed to show developing mericarps. c) Mature schizocarp just prior to dehiscence. d) Mature schizocarp after seeds have been ejected.   *Geranium carolinianum*

Oxalidaceae - *Wood Sorrel Family*

Habit: Mostly herbs, from rhizomes or tubers. Sometimes shrubs or trees.

Leaves: Alternate, entire. Pinnately or palmately compound, often trifoliate. Sometimes in basal rosettes. Pulvinus at base of petioles and petiolules. Leaflets droop at night

Inflorescences: Umbel or a solitary flower.

Flowers: Showy, small, radial, bisexual. Sometime cleistogamous.

Calyx: 5 sepals.

Corolla: 5 petals.

Stamens: 10, basally connate. Outer whorl shorter than inner.

Carpels: 5, often heterostylous.

Ovary: Superior.

Placentation: Axile.

Fruit: Loculicidal capsule (frequently ridged) or berry.

Seeds: Typically arillate, explosively dispersed.

Other: Oxalate crystals present.

Genera: *Averrhoa, Biophytum, Oxalis.*

Fig. 579   Flowering branch of sorrel.   *Oxalis debilis*

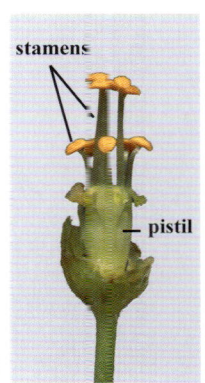
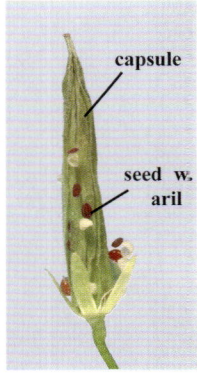

a   b   c   d

Fig. 580   Floral and fruit characters of sorrel. a) Habit, leaves, and flowers. b) CU of flower and bud. c) Androecium and gynoecium after calyx and corolla removed. d) Capsule during dehiscence. Brown seeds with white arils on side. *Oxalis corniculata*

a   b   c

Fig. 581   Characters of the life plant. a) Habit (with appearance of miniature palm). b) Inflorescence. c) Dehiscing capsule. *Biophytum* sp.

Malpighiaceae - *Barbados Cherry Family*

Habit: Lianas, shrubs, small trees.
Leaves: Typically opposite, simple, entire. Pinnate venation. Frequently with glands on petiole or blade. Stipules present.
Inflorescences: Cyme, raceme, panicle.
Flowers: Showy, bilateral, bisexual.
Calyx: 5 sepals, typically with oil glands. Persistent.
Corolla: 5 petals, distinct, differentiated. Typically clawed with margin toothed.
Stamens: 10, basally connate, in 2 whorls.
Carpels: 3.
Ovary: Superior.
Placentation: Axile.
Fruit: Capsule, or drupe with 3-seeded pit, or 3-segmented samaroid schizocarp.
Other: T-shaped hairs.
Genera: *Byrsonima, Galphimia, Malpighia, Stigmaphyllon.*

Fig. 582   Inflorescence of butterfly vine. *Mascagnia macroptera*

a   b

Fig. 583   Floral and fruit characters of butterfly vine. a) Flower with distinctively clawed petals. b) Developing schizo-carps, each consisting of three winged mericarps. *Mascagnia macroptera*

a   b   c   d

Fig. 584   Floral and fruit characters of thryallis. a) Inflorescence. b) CU of flowers. c) Immature capsule. d) Capsule after dehiscence. *Galphimia glauca*

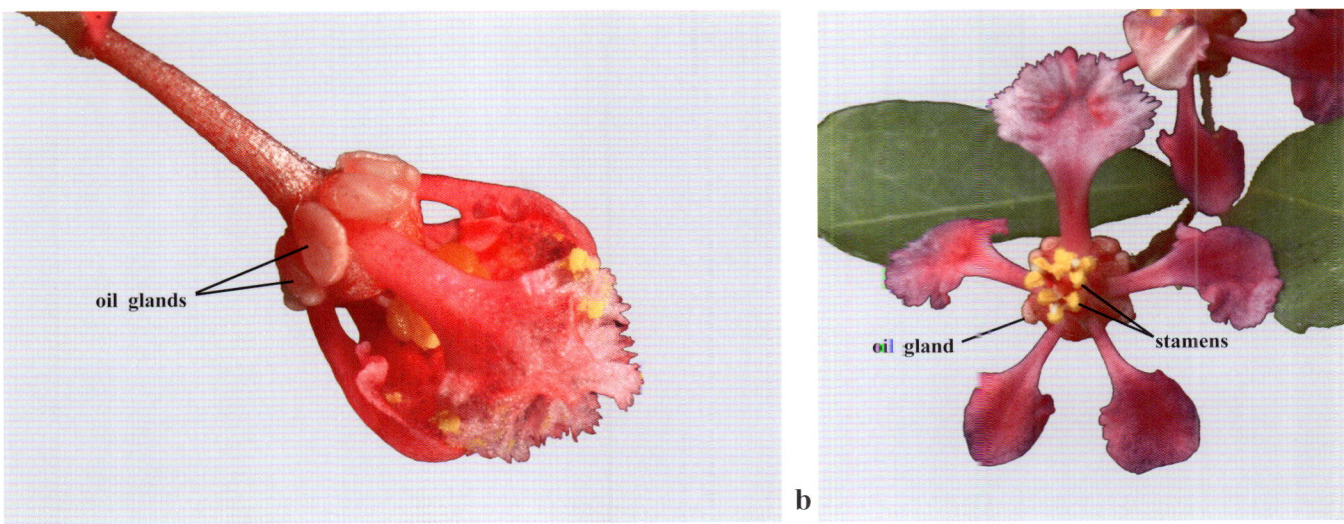

Fig. 585   Floral characters of Barbados cherry. a) Flower bud. b) Flower.
Malpighia glabra

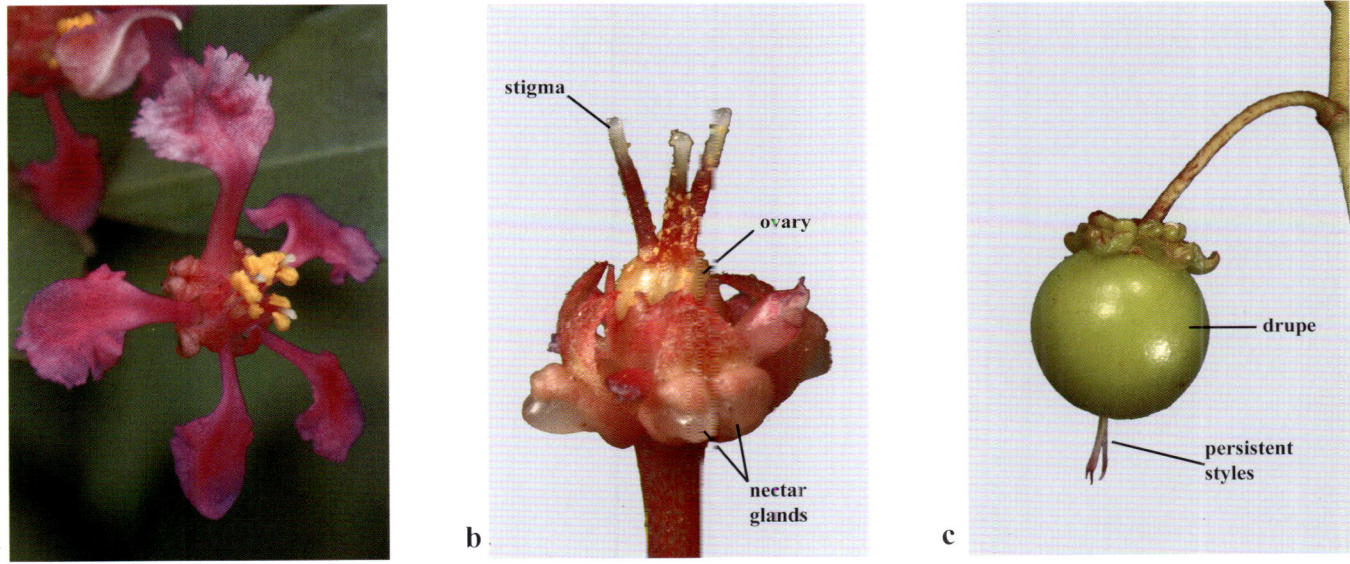

Fig. 586   Floral and fruit characters of Barbados cherry. a) Flower. b) Gynoecium (with petals removed) showing three pistils. c) Immature drupe.   *Malpighia glabra*

Fig. 587   Fruit characters of Barbados cherry. a) Fruiting branch. b) CU of mature cherry-like drupe.
Malpighia glabra

Euphorbiaceae - *Spurge Family*

Note: The genus *Phyllanthus* illustrated here is considered by many botanists to belong in a separate family (Phyllanthaceae) based on having two ovules per locule rather than one. *Phyllanthus* also has dimorphic stems, one erect (orthotropic) and many lateral (plagiotropic) ones with leaves.

Note: The spurges are an extremely diverse family ecologically, vegetatively, and chemically.

Major Subfamilies: Euphorbioideae, Crotonoideae, Acalyphoideae.

Habit: Herbs, shrubs, trees, or vines. Sometimes succulent or cactus-like.

Leaves: Alternate or opposite or whorled, simple or compound, entire or lobed or serrate. Palmate to pinnate venation. Sometimes with stipules. Sometimes with nectar glands.

Inflorescences: Cyme, spike, head, or solitary flower. Sometimes with showy bracts. Some modified into **cyathia** appearing as 'false flowers' (see Euphorbioideae).

Flowers: Showy to inconspicuous, radial, unisexual. Frequently reduced. Often with nectar disk.

Calyx: Typically 5 sepals, distinct.

Corolla: Often apetalous, typically 5 petals when present.

Stamens: 1-numerous, distinct to completely connate.

Carpels: 3, with styles often branched.

Ovary: Superior.

Placentation: Axile.

Fruit: Typically 3-segmented schizocarp. Dehiscence is often ballistic. Sometimes extra-floral nectaries present.

Seeds: Sometime arillate.

Other: Sometimes with milky or colored sap.

Genera: (See subfamilies.)

Fig. 588   Fruiting branch of phyllanthus.
Phyllanthus juglandifolius

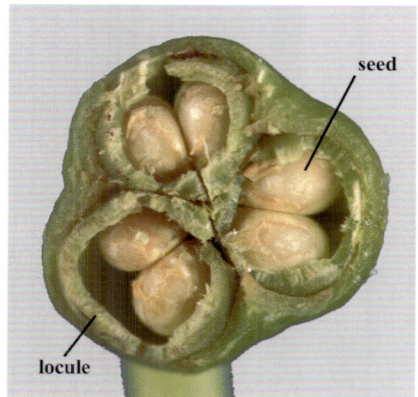

a   b   c

Fig. 589   Floral and fruit characters of phyllanthus. a) Flowers. b) Developing schizocarps. c) XS of fruit showing the two seeds per locule inside.
Phyllanthus juglandifolius

Subfamily Euphorbioideae

Latex: Sap or latex white, caustic.   **Hairs:** Simple in form.

Specialized Inflorescence: Some genera of this subfamily contain an inflorescence that is a 'false flower' (**pseudanthium**), collectively formed by numerous flowers and structures.

Cyathium = one carpellate flower in association with several to numerous staminate flowers, each of which has been reduced to a single stamen. All are located within a cup-like structure or **involucre** of bracts. These bracts often bear extrafloral nectaries, sometimes with petaloid appendages.

Genera: *Chamaesyce, Euphorbia, Poinsettia, Sapium.*

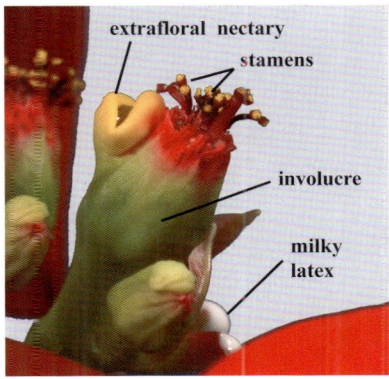

Fig. 590   Leaf and floral characters of the Christmas poinsettia. a) Habit. b) CU of inflorescences or cyathia with distinctive labiate extrafloral nectaries. c) CU of single cyathium.   *Poinsettia* sp.

Fig. 591   Leaf and floral characters of wild poinsettia. a) Habit. b) Cyathia and associated developing schizocarps. c) CU of schizocarps and of cyathia bearing round extrafloral nectaries.   *Euphorbia cyathophora*

Fig. 592   Floral and fruit characters of the popcorn tree.
a) Male inflorescence. b) Schizocarps both intact (dark) and after covering has dehisced to reveal white mericarps.
Sapium sebiferum

Fig. 593   Inflorescence of crown-of-thorns.
Euphorbia mili.

Subfamily Crotonoideae

Latex: Colored or white, non-caustic.
Hairs: Not simple (branched, stellate, peltate).
Genera: *Aleurites, Croton, Jatropha, Manihot.*

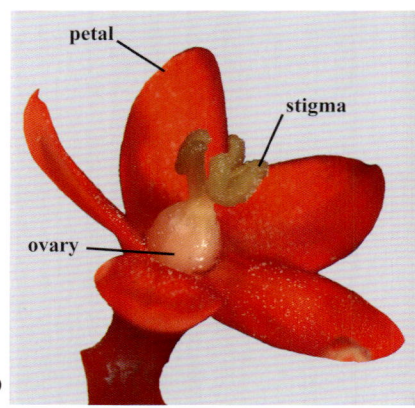

Fig. 594   Floral characters of jatropha. a) Inflorescence. b) Carpellate flower. c) Staminate flower.
Jatropha podagrica

Fig. 595   Fruit characters of jatropha. a) Immature schizocarp.
b) XS of developing schizocarp.   *Jatropha* sp.

Fig. 596   Flower of manioc. (Note bright orange nectar disk.)
Manihot esculenta

Fig. 597   Floral and fruit characters of the tung tree.
a) Flower. b) Immature schizocarp.   *Aleurites fordii*

Fig. 598   Flower of croton.
Croton sp.

Subfamily Acalyphoideae

Latex: No sap or latex present.
Genera: *Acalypha, Ricinus*.

Fig. 599 Floral characters of castor bean.
a) Inflorescence showing red stigmas of carpellate flowers on top and yellow staminate flowers on bottom.
b) CU of staminate flower.
c) CU of carpellate flower.
Ricinus communis

Fig. 600   Fruit characters of castor bean.
a) Infructescence.
b) CU of immature fruit with spiny covering intact.
c) CU of the dehiscing schizocarp.
d) Seed (castor bean) with white aril on top.
Ricinus communis

Clusiaceae - *St. John's - Wort Family*

Note: Two correct Latin names for family:
Clusiaceae and Guttiferae.
Habit: Herbs, shrubs, trees, lianas.
Leaves: Opposite or whorled, simple, entire.
Pinnate venation. Some with pellucid or black dots.
Inflorescences: Cyme, umbel, or single flower.
Flowers: Showy, radial, bisexual or unisexual.
Calyx: 2-5 sepals, distinct, frequently differentiated.
Corolla: 4-5 petals, distinct, frequently asymmetrical.
Stamens: Numerous, often in clusters or tufts.
Carpels: 2-5.
Ovary: Superior.
Placentation: Axile.
Fruit: Dehiscent capsule, berry or drupe.
Seeds: Sometimes arillate.
Other: Clear or colored sap. Paired glands sometimes at nodes.
Genera: *Clusia, Hypericum, Triadenum.*

Photo by Stephen Timme

Fig. 601   Habit of clusia.
Clusia rosea

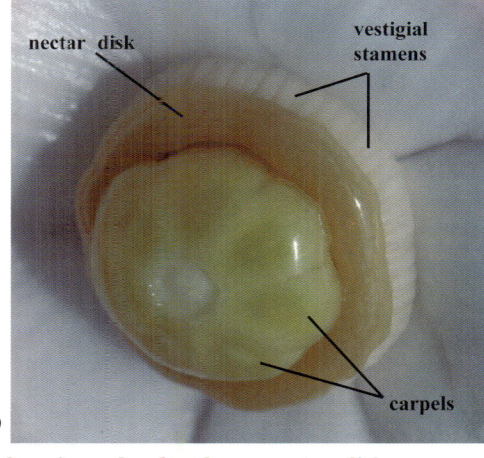

Fig. 602   Floral characters of clusia. a) Flower. b) CU of gynoecium and androecium showing large nectar disk.
Clusia rosea

Fig. 603   Fruit and stem characters of clusia. a) Developing capsule. b) Mature dehisced woody capsule. c) Yellow sap at site of freshly removed leaf.   *Clusia rosea*

Fig. 604   Floral and fruit characters of St. John's-wort. a) Inflorescence. b) Capsules.
Hypericum sp.

Fig. 605   Floral characters of St. John's-wort. a) Flowering branch. b) CU of flower showing multiple stamens.
c) CU of gynoecium with three pistils (stamens have been cut with bases remaining).
Hypericum tetrapetalum

Fig. 606   Fruit characters of St. John's-wort. a) Immature capsule almost covered by bracts. b) Profile of same capsule.
c) LS of capsule showing developing seeds.   *Hypericum hypericoides*

Rhizophoraceae - *Red Mangrove Family*

Habit: Shrubs or trees. Often grow in flooded habitats. Typically with prop roots and pneumatophores.

Leaves: Opposite, simple, entire. Pinnate venation. Large interpetiolar stipules.

Inflorescences: Cyme, raceme.

Flowers: Radial, bisexual, frequently with hypanthium.

Calyx: 4-5 sepals, fleshy.

Corolla: 4-5 petals, typically hairy.

Stamens: 8-10 to numerous. Frequently in pairs and opposite petals.

Carpels: 2-6.

Ovary: Superior to inferior.

Placentation: Axile.

Fruit: Berry or septicidal capsule.

Seeds: Sometimes arillate.

Other: Seeds sometime viviparous, germinating while still in the fruit while still on tree.

Genera: *Bruguiera, Cassipourea, Rhizophora.*

Fig. 607   Fruiting branch of red mangrove showing seedling (vivipary).   *Rhizophora mangle*

Fig. 608   Root characters of red mangrove. a) Pneumatophores protruding from water near mangrove tree. These specialized roots function in gas exchange. b) Prop roots of red mangrove.   *Rhizophora mangle*

Fig. 609   Floral and fruit characters of red mangrove. a) Flower. b) Developing fruit. c) Fruiting branch with seedlings. *Rhizophora mangle*

Violaceae - *Violet Family*

Habit: Herbs. Mostly shrubs, trees, and lianas in tropics.
Leaves: Alternate or basal rosette, simple to lobed, entire to serrate. Palmate to pinnate venation. Stipules present.
Inflorescences: Raceme, sometimes a solitary flower.
Flowers: Showy, typically bilateral and bisexual. Sometimes cleistogamous.
Calyx: 5 sepals, distinct, typically differentiated.
Corolla: 5 petals, distinct, typically with nectar spur.
Stamens: 5, forming ring around the gynoecium. Anthers with dorsal nectar glands.
Carpels: 3.
Ovary: Superior.
Placentation: Parietal.
Fruit: 3-segmented loculicidal capsule.
Seeds: Frequently arillate, explosively dispersed.
Genera: *Hybanthus, Leonia, Rinorea, Viola.*

Fig. 610   Flower of hybanthus. *Hybanthus communis*

  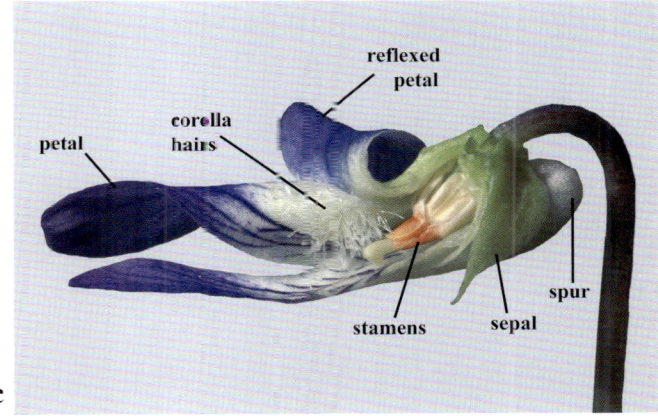

Fig. 611   Floral characters of violet. a) Flower. b) Flower in profile. c) LS of flower. *Viola sororia*

Fig. 612   Fruit characters of violet. a) Dehisced and immature capsules. b) Immature capsule. c) Capsule during dehiscence before ejection of seeds. *Viola sororia*

Fig. 613   Pansy flower. *Viola tricolor*

Passifloraceae - *Passion Flower Family*

Habit: Vines to lianas with axillary tendrils.
Sometimes shrubs, but without tendrils.
Leaves: Alternate, simple, entire or serrate. Frequently lobed.
Palmate venation. Typically with nectar glands on petiole.
Inflorescences: Cyme, typically reduced to a single flower.
Flowers: Showy, radial, bisexual. Tubular to cup-like
hypanthium with nectar disk at base. Conspicuous bracts.
Calyx: 5 sepals, petaloid.
Corolla: 5 petals, distinct. Corona of 1-several whorls of
filaments from hypanthium.
Stamens: 5, borne on raised stalk (androgynophore) along
with the gynoecium.
Carpels: 3.
Ovary: Superior, borne on same raised stalk as androecium.
Placentation: Parietal.
Fruit: Berry or loculicidal capsule.
Seeds: Arillate, flattened, with pitted surface.
Genera: *Adenia, Passiflora.*

Fig. 61-   Flower of passion vine.
Passiflora yucatanensis

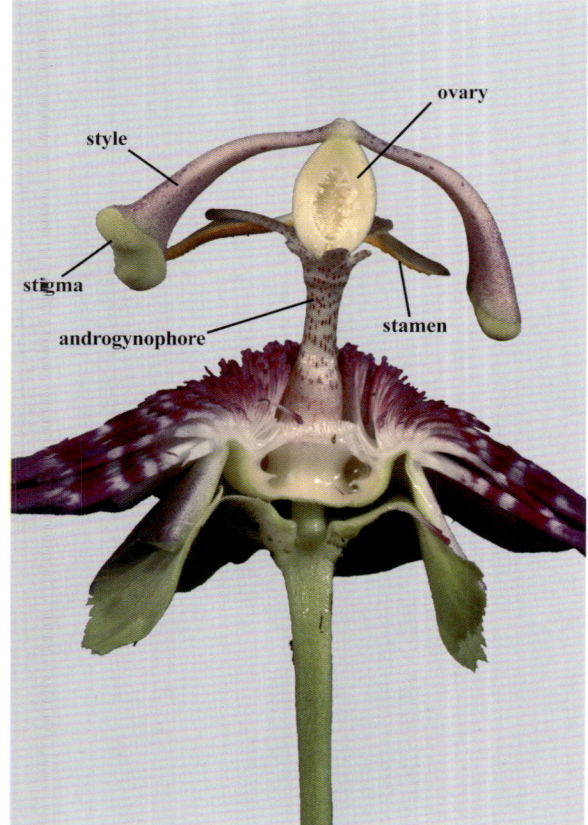

Fig. 615   LS of flower of passion vine.
Passiflora 'Incense'
(*P. incarnata* x *P. circinnata*)

Fig. 616   Flower of passion vine.
Passiflora 'Incense'

Fig. 617   Fruit characters of maypop. a) Berry. b) XS of
berry showing seeds with arils and parietal placentation.
Passiflora incarnata

a   b

Fig. 618   Tendril characters of passion vine. a) Terminal portion of vine showing new leaves and tendrils. **b) CU of axil region showing extrafloral nectary and axillary origin of tendril.** *Passiflora 'Incense'*

Fig. 619   Floral and leaf characters of red passion vine. *Passiflora vitifolia*

Fig. 620   Plant characters of tumbo passion vine. (Note EFN's are extrafloral nectaries.) *Passiflora quadrangularis*

Fig. 621   Floral and leaf characters of maracuya. a) Flower. b) Leaf, flower bud, and tendril. *Passiflora murucuja*

Salicaceae - *Willow Family*

Habit: Shrubs, trees. Often found in wet areas.

Leaves: Alternate, simple, serrate to dentate. Pinnate to palmate venation. Stipules typically obvious. Teeth of leaves with small glandular ball (**salicoid tooth**).

Inflorescences: Erect to pendent catkins.

Flowers: Reduced, radial, unisexual. Subtended by hairy bract.

Perianth: Absent or reduced to small nectar glands (*Salix*) or cup-like disk (*Populus*).

Stamens: 2-numerous.

Carpels: 2-4.

Ovary: Superior.

Placentation: Parietal.

Fruit: Loculicidal capsule.

Seeds: Small, with tuft of hair for aerial dispersal.

Genera: *Populus, Salix*.

Note: Many botanists now treat this family in a much broader sense, including much of the traditional Flacourtiaceae.

Fig. 622   Leaf of poplar tree.
Populus deltoides

a   b   c

Fig. 623   Twig and floral characters of poplar. a) Terminal and axillary buds. b) Early female inflorescence.
c) CU of carpellate flowers on female inflorescence.
Populus deltoides

a   b   c

Fig. 624   Floral and fruit characters of poplar. a) Mature female catkin. b) Infructescence. c) CU of developing fruits
on infructescence.   *Populus deltoides*

Fig. 625 Staminate floral characters of willow. a) Flowering branch with male catkins. b) CU of male catkin prior to full dehiscence. c) Male catkin fully dehisced. *Salix caroliniana*

Fig. 626 Carpellate floral characters of willow. a) Flowering branch with female catkins b) CU of female catkin. c) Female catkin with immature capsules. *Salix caroliniana*

Fig. 627 Fruit characters of willow. a) CU of developing capsule. b) Female catkin with mature capsules dehiscing. c) CU of capsules with white tufts of hair used for aerial dispersal of seeds. *Salix caroliniana*

Fabaceae - *Legume Family*

Note: Two correct Latin names for family: Fabaceae and Leguminosae. 3rd largest family.

Subfamilies: Mimosoideae, Caesalpinoideae, and Faboideae (also known as the Papilionoideae).

Habit: Herbs, shrubs, trees, vines, lianas.

Leaves: Typically alternate and entire. Bipinnately compound, pinnately compound, trifoliate, and sometimes one leaf. Pulvinus at base of petiole and petiolules. Stipules. Leaflets fold down at night and sometimes are sensitive to the touch. Pinnate venation. Sometimes with tendrils.

Inflorescences: Various, sometimes a single flower.

Flowers: Showy, radial to bilateral, bisexual. Short cup-like hypanthium. Usually with nectar disk. (See individual subfamilies for further floral characters.)

Carpels: 1.

Ovary: Superior.

Placentation: Parietal.

Fruit: Typically legume, occasionally loment. Other fruit types possible. Sometimes modified into other fruit types such as samaras, drupes, or indehiscent pods. (See pages 52-53.)

Seeds: Typically flat, sometimes arillate. Commonly referred to as beans.

Other: Nodules on roots containing nitrogen-fixing bacteria. (See page 4, Figs. 11-13.)

Fig. 628   Bean diversity in legumes.

Fig. 629   Fruit of English pea dissected to show seeds within. *Pisum sativum*

a

c

d

Fig. 630   Leaf diversity in legumes. a) Garden bean   *Phaseolus vulgaris.* b) Coral bean   *Erythrina herbacea.* c) Cassia or candle tree   *Cassia alata.* d) Black locust   *Robinia pseudoacacia.*

Subfamily Mimosoideae

Leaves: Typically bipinnately compound, occasionally reduced to **phyllodes**.
Flowers: Showy, radial, frequently in dense heads.
Calyx: 5 sepals, connate, greatly reduced.
Corolla: 5 petals, equal, connate forming tube.
Stamens: 10-many, showy, extending far beyond corolla.
Genera: *Acacia, Albizia, Calliandra, Mimosa.*

Fig. 631   Inflorescence of golden mimosa.   *Acacia baileyana*

a

b

Fig. 632   Floral characters of powderpuff. a) Inflorescence in bud. b) Inflorescence in bloom.   *Calliandra emarginata*

a

b

c

Fig. 633   Floral characters of powderpuff. a) LS of powderpuff inflorescence. b) CU of LS of inflorescence showing individual flowers. c) Single flower with multiple long stamens.   *Calliandra emarginata*

Fig. 634   Fruit and foliage of mimosa.
Albizia julibrissin

a

b

Fig. 635   Foliage and fruit of the sensitive plant. a) Leaf with one leaflet showing 'touch response'. b) Cluster of loments.   *Mimosa pudica*

Subfamily Caesalpinoideae

Leaves: Typically pinnately or bipinnately compound.
Flowers: Showy, typically bilateral.
Calyx: 5 sepals, distinct to connate.
Corolla: 5 petals. Differentiated into 1 **banner** (uppermost),
2 **wings** (laterals), and 2 **keels** (petals not fused).
Banner petal frequently smaller than laterals and at least one side
develops on the inside of the wings.
Stamens: 10 or less, typically distinct.
Genera: *Bauhinia, Caesalpinia, Cassia, Cercis, Senna.*

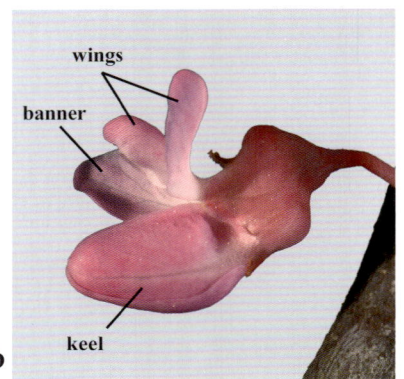

Fig. 636   Floral characters of redbud. a) Inflorescence. b) Flower. c) Flowering branch.
Cercis canadensis

Fig. 637   Floral characters of senna. a) Inflorescence in bud. b) Flower.
Senna obtusifolia

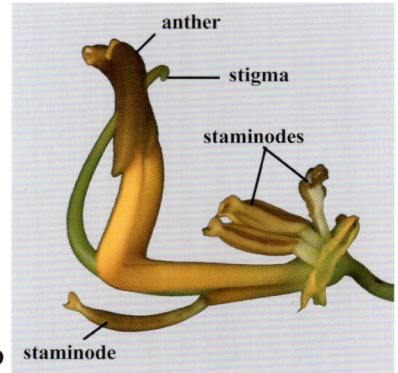

Fig. 638   Floral characters of senna with petals removed. a) & b) Androecium and gynoecium. c) CU of staminodes.
Senna obtusifolia

Subfamily Faboideae (Papilionoideae)

Leaves: Typically pinnately compound or trifoliate.
Sometimes 1 leaflet (unifoliate).
Sometimes leaflets modified into tendrils.
Flowers: Showy, bilateral. Termed papilionaceous due
to their resemblance to a butterfly.
Calyx: 5 sepals, connate.
Corolla: 5 petals. Differentiated into 1 **banner** (uppermost),
2 **wings** (laterals), and a **keel** (fusion of 2 bottom petals).
Both edges of banner petal develop on outside of wings.
Stamens: 10, usually 9 connate and 1 distinct.
Other: Nectar disk at ovary base.
Genera: *Crotalaria, Desmodium, Phaseolus, Pisum, Wisteria.*

Fig. 639 Floral characters of rattlebox. a) Flower. b) Flower with one wing petal removed to show fused keel.
c) Partial inflorescence. *Crotalaria spectabilis*

Fig. 640 Fruit of rattlebox. a) Immature legume.
b) LS of legume showing seeds. *Crotalaria sp.*

Fig. 641 Inflorescence and flowers of wisteria.
Wisteria sinensis

Fig. 642 Loment of beggarweed.
Desmodium incanum

Fig. 643 Legume of coral bean.
Erythrina herbacea

Fig. 644 Dehisced legume of galactia.
Galactia volubilis

Polygalaceae - *Milkwort Family*

Habit: Herbs, shrubs, trees, lianas.

Leaves: Typically alternate, simple, entire. Pinnate venation.

Inflorescences: Raceme, panicle, or a single flower.

Flowers: Showy, bilateral, bisexual. Subtended by bract and bracteoles. Sometimes with nectar gland or disk.

Calyx: 5 sepals, differentiated with 2 lateral and petaloid.

Corolla: 3 petals (2 upper/1 lower), distinct. Lower petal (**keel**) boat-shaped.

Stamens: 8, adnate to petals.

Carpels: 2.

Placentation: Axile.

Fruit: Typically loculicidal capsule with one seed per locule. Sometimes drupe or samara.

Seeds: Arillate, frequently covered with hairs.

Genera: *Monnina, Polygala, Securidaca.*

Fig. 645   Milkwort in flower. *Polygala rugelii*

a

b

c

Fig. 646   Floral characters of milkwort. a) Inflorescence. b) LS of inflorescence. c) CU of flower. *Polygala lutea*

Fig. 647   Flower of milkwort. *Polygala grandiflora*

Photo by Richard Abbott

Fig. 648   Samaras. *Securidaca elliptica*

Photo by Richard Abbott

Fig. 649   Seeds. (Note hairy coat and orange aril.) *Polygala virgata*

Rhamnaceae - *Buckthorn Family*

Habit: Shrubs, trees. Sometimes with thorns or spines.
Sometimes lianas with twining stems or tendrils.
Leaves: Alternate, simple, entire to serrate.
Pinnate or palmate venation. Stipules small, sometimes
modified into spines.
Inflorescences: Various, sometimes 1 flower.
Flowers: Inconspicuous, small, radial, typically bisexual.
Hypanthium with nectar disk.
Calyx: 4-5 sepals, distinct.
Corolla: 4-5 petals, distinct, often clawed.
Typically concave and enclosing stamens.
Stamens: 4-5, distinct, opposite to and enclosed by petals.
Carpels: 2-3.
Ovary: Superior to inferior, but usually in-between.
Fruit: Dehiscent or indehiscent drupe with 1-several pits.
Genera: *Ceanothus, Colubrina, Gouania, Rhamnus.*

Fig. 650   Flowers and fruit of
Asian nakedwood.
Colubrina asiatica

Fig. 651   Floral and fruit characters of Asian nakedwood. a) Flowers. Note stamen and concave petal in 12 o'clock
position of flower on right. b) Immature drupes.   *Colubrina asiatica*

Fig. 652   Mature drupe.
Colubrina asiatica

Fig. 653   Fruit and foliage characters of Carolina buckthorn.
a) Developing drupes. b) Leaf.   *Rhamnus caroliniana*

Rosaceae - *Rose Family*

Subfamilies: Rosoideae, Amygdaloideae, Maloideae and Spiroideae.
Note: Current cladistic research shows that the traditional subfamilies above are not distinct.
Habit: Herbs, shrubs, trees. Frequently with thorns or prickles.
Leaves: Alternate, serrate, simple to pinnately-palmately compound. Stipules.
Inflorescences: Raceme, cyme, or a single flower.
Flowers: Showy, radial, bisexual. **Epicalyx** sometimes present.
Bearing a cup-like hypanthium with nectar disk.
Calyx: 5 sepals, basally connate.
Corolla: 5 petals, distinct, frequently clawed. Ephemeral.
Stamens: Typically numerous.
Carpels: 1-many.
Ovary: Superior to inferior.
Placentation: Axile.
Fruit: Drupe, aggregate of drupelets, achene, aggregate of achenes, pome, follicle.
Genera: (See subfamily descriptions.)

Fig. 654 Cultivated ornamental rose.
Rosa sp.

Subfamily Rosoideae

Habit: Herbs, shrubs.
Leaves: Pinnately or palmately compound. Stipules.
Carpels: Typically numerous, distinct.
Ovary: Superior.
Fruit: Aggregate of achenes, aggregate of drupelets.
Sometimes an accessory fruit.
Genera: *Fragaria, Potentilla, Rosa, Rubus.*

Fig. 655   Floral and fruit characters of blackberry. a) Flower. b) Branch with young fruit.
Rubus argutus

Fig. 656   Fruit development of blackberry. a) Immature aggregate of drupelets. b) Fruit as it nears maturity.
c) Mature blackberry.   *Rubus argutus*

Fig. 657   Flower of swamp rose.
Rosa palustris

Fig. 658   Fruit characters of rose. a) Fruit (aggregate of achenes surrounded by a fleshy hypanthium). b) LS of fruit showing achenes and hypanthium.
Rosa sp.

Subfamily Amygdaloideae

Habit: Shrubs, trees. Smooth bark with horizontal lenticels.
Leaves: Simple, stipules present.
Carpels: Typically 1.
Ovary: Superior.
Fruit: Drupe.
Genera: *Prunus*.

Fig. 659   Inflorescence of hog plum.
Prunus umbellata

Fig. 660   Early foliage of black cherry.
Prunus serotina

Fig. 661   Floral characters of black cherry. a) Inflorescence in bloom.
b) Intact and LS of two individual flowers.   *Prunus serotina*

Fig. 662   Drupes of laurel cherry.
Prunus caroliniana

Fig. 663   Flower of peach. a) Flower. b) LS of flower. c) CU of LS of flower base and hypanthium.
Prunus persicus

Subfamily Maloideae

Habit: Shrubs, trees. Some with thorns.
Leaves: Simple or pinnately compound, stipules present.
Carpels: 2-5, connate, fused to hypanthium.
Ovary: Inferior.
Fruit: Fleshy with cartilaginous core (**pome**).
Other: Some fruits (ex. pear) gritty due to stone cells.
Genera: *Crataegus, Malus, Pyracantha, Pyrus.*

Fig. 664   **Floral and fruit characters of firethorn. a) Flowering branch. b) Fruiting branch.**
Pyracantha coccinea

Fig. 665   **Floral and fruit characters of one-flower hawthorn. a) Flower. b) LS of flower. c) Fruit.**
Crataegus uniflora

Fig. 666   **Fruit and seed characters of apple. a) LS of apple fruit or pome. b) Apple seeds.**
Malus sylvestris

Ulmaceae - *Elm Family*

Habit: Trees, shrubs.

Leaves: Alternate, simple, entire to doubly serrate.
2-ranked. Pinnate to pinnate-palmate intermediate venation.
In *Ulmus*, 2ndary veins reach tooth; in *Celtis* 2ndary veins
form loops. Bases asymmetrical. Stipules.

Inflorescences: Cyme, raceme. Flowers in clusters (fascicles)
or solitary.

Flowers: Inconspicuous, small, radial, bisexual or unisexual.
Hypanthium present (*Ulmus*) or absent (*Celtis*).

Perianth: 4-9 tepals.

Stamens: 4-9, distinct, opposite tepals.

Carpels: 2.

Ovary: Superior.

Placentation: Apical.

Fruit: Samara or nutlet. In *Celtis*, drupe with a pit.

Seeds: Flat, winged, or spherical.

Genera: *Celtis, Planera, Trema, Ulmus*.

Note: Many botanists place *Celtis* in a distinct family (Celtidaceae).

Fig. 667   Hackberry leaf.
Celtis laevigata

Fig. 668   Floral characters of hackberry. a) Flowering branch. b) Staminate flowers. c) Carpellate flowers.
Celtis laevigata

**Fig. 669   Fruit and foliage characters of hackberry. a) Immature drupe. b) Mature drupe. c) Distinctive gall formed by
an insect on the leaf petiole.   *Celtis laevigata***

Fig. 670   Characters of elms. a) Leaf of Chinese elm.   *Ulmus parvifolia*
b) Woody growths on twig of winged elm.   *Ulmus alata*   c) Flowering branch of winged elm.   *Ulmus alata*

Fig. 671   Floral characters of winged elm. a) Flowers with stamens extruded. b) Flowers with stigmas extruded.
c) CU of stigma.   *Ulmus alata*

Fig. 672   Fruit characters of elms. a) Developing fruits of winged elm   *Ulmus alata*
b) Developing fruits of Chinese elm   *Ulmus parvifolia*   c) Samara of Chinese elm   *Ulmus parvifolia*

Moraceae - *Fig Family*

Habit: Shrubs, trees, lianas. Some start as epiphytes. Rarely herbs.
Leaves: Usually alternate, simple, entire to serrate.
Sometimes lobed. 2-ranked. Venation palmate to pinnate.
Large sheathing stipules.
Inflorescences: Various. Flowers often clustered with axis enlarged,
or embedded in fleshy receptacle.
Flowers: Inconspicuous, small, radial, unisexual.
Perianth: 4-5 tepals. May change form with fruit development.
Stamens: 1-5, distinct, opposite tepals.
Carpels: 2.
Ovary: Superior or inferior.
Placentation: Apical.
Fruit: Achene, drupe, or multiples of each. Syconium.
(Syconium = Fleshy multiple fruit with hollow receptacle
containing numerous minute achenes within).
Other: Milky sap throughout plant. 2 style branches.
Genera: *Artocarpus, Brosimum, Dorstenia, Ficus, Maclura, Morus.*

Fig. 673   Fruiting branch of red mullberry.   *Morus rubra*

Fig. 674   Floral characters of red mullberry. a) Branch with staminate inflorescences. b) Male catkin as stamens emerge.
c) Male catkin with stamens fully extruded.   *Morus rubra*

Fig. 675   Floral characters of red mullberry. a) Branch with carpellate inflorescences. b) Female catkin with receptive
stigmas. c) Developing fruit (multiple of drupes).   *Morus rubra*

a   **b**   **c**

Fig. 676   Fruit characters of edible fig. a) Fruiting branch. b) Syconium. c) LS of syconium showing internal florets. *Ficus carica*

a   **b**   **c**

Fig. 677   Stem characters of figs. a) Milky latex produced with removal of leaf. b) Stem showing large sheathing stipules. c) New leaf and protective red stipule that protects leaf bud.   *Ficus* sp.

Fig. 678   Fruiting branch of creeping fig showing leaves and syconium. *Ficus pumila*

Fig. 679   Strangler fig on host. *Ficus* sp.

Urticaceae - *Nettle Family*

Habit: Herbs, shrubs, vines, occasionally trees.
Leaves: Alternate or opposite, simple, entire to serrate.
Pinnate to palmate venation. Bases sometimes asymmetrical.
Inflorescences: Variously clustered, or sometimes 1 flower.
Flowers: Inconspicuous, small, radial, unisexual.
Perianth: 4-5 tepals.
Stamens: 4-5, distinct, opposite tepals.
Carpels: 1 full size, 1 very reduced.
Ovary: Superior.
Placentation: Basal.
Fruit: Achene.
Other: Milky or clear sap. 1 style and 1 stigma (pseudomonomerous).
Sometimes plants with stinging hairs.
Genera: *Boehmeria, Laportea, Parietaria, Pilea, Urtica.*

Fig. 680   Habit of false nettle.
Boehmeria cylindrica

a

b

Fig. 681   Floral character of false nettle. a) Inflorescence. b) Portion of
inflorescence with CU of flowers.   *Boehmeria cylindrica*

Fig. 682   Flowering branch of the
aluminum plant.
Pilea caderoi

a

b

Fig. 683   Inflorescence of aluminum
plant.   *Pilea caderoi*

Fig. 684   a) Flowering branch of moon valley plant. b) Inflorescence.
Pilea involucrata

Begoniaceae - *Begonia Family*

Habit: Succulent herbs or shrubs. Grow from rhizomes or tubers.
Leaves: Alternate, simple, entire or serrate. Palmate venation.
Fleshy. Commonly with oblique bases.
Inflorescences: Cyme.
Flowers: Typically showy, bilaterally symmetrical, imperfect.
Perianth: 2-5 tepals, petaloid. Male and female flowers differ.
Stamens: Numerous, in multiple whorls. Distinct to basally connate.
Carpels: 3.
Ovary: Inferior.
Placentation: Axile or parietal.
Fruit: Winged loculicidal capsule.
Genera: *Begonia* and *Hillebrandia*.

Fig. 685   Habit of wax begonia.
Begonia sp.

Fig. 686   Floral and fruit characters of wax begonia. a) Plant in flower and with immature fruits. b) Mature capsule.
Begonia sp.

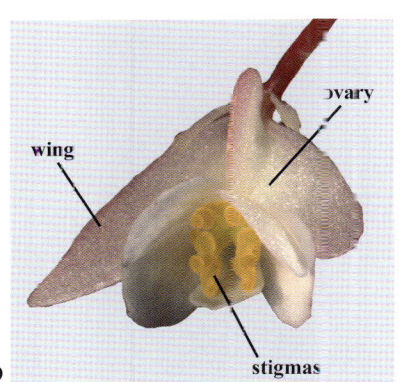

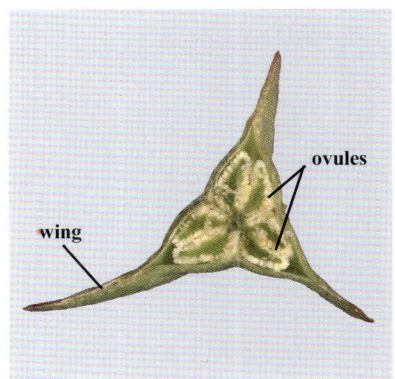

Fig. 687   Floral characters of wax begonia. a) Staminate flower. b) Carpellate flower. c) XS of ovary.
Begonia sp.

Cucurbitaceae - *Cucurbit Family*

Habit: Vines, herbaceous to succulent to slightly woody. Coiled axillary tendrils from petiole base arising at a right angle.

Leaves: Alternate, simple, palmately lobed. Palmate venation. Sometimes rough-textured (**scabrous**).

Inflorescences: Cyme, sometimes 1 flower.

Flowers: Showy, radial, unisexual, ephemeral. Hypanthium. Male and female flowers similar in appearance.

Calyx: 5 sepals, connate.

Corolla: 5 petals, connate.

Stamens: 3-5, adnate to hypanthium. Fused and twisted to resemble stigmas.

Carpels: 3.

Ovary: Inferior.

Placentation: Parietal.

Fruit: Berry, dehiscent capsule. Frequently a pepo. (**pepo** = modified berry with hard or leathery rind. See page 41.)

Seeds: Flat, large, numerous. Sometimes arillate.

Genera: *Citrullus, Cucumis, Cucurbita, Luffa, Momordica.*

Fig. 688　Cucurbit flower showing large inferior ovary. *Cucurbita* sp.

a

b

flower bud

tendril

Fig. 689　Cucurbit tendrils. a) Squash plant showing coiled nature of tendrils. b) Loofah stem showing origin of tendril and the manner in which it projects at a right angle from the stem. *Luffa aegyptica*

a

ovary

b

c

Fig. 690　Floral characters of a cucurbit. a) Female flower bud with inferior ovary. b) Male flower with twisted stigma-like stamens. c) Male flower with petals removed to show CU of stigma-appearing stamens.

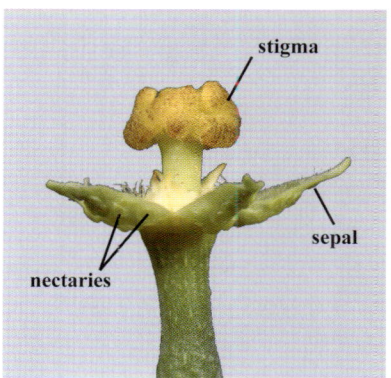

Fig. 691   Floral characters of loofah. a) Carpellate flower. b) Sepals bearing nectaries. (View from behind flower.) c) Gynoecium after petals have been removed.
Luffa aegyptica

Fig. 692   Characters of loofah. a) Leaf (Note: Typical in shape of many squash plants.) b) Inflorescence in bud stage. c) Flower showing clearly the inferior ovary.
Luffa aegyptica

Fig. 693   Fruit and leaf characters of balsampear. a) Immature capsule b) Maturing capsule and leaf on vine. c) Dehisced capsule showing seeds.
Momordica charantia

Fagaceae - *Oak & Beech Family*

Habit: Trees, shrubs.
Leaves: Alternate, simple, frequently lobed. Entire to serrate. Pinnate venation. Leaf shape may vary greatly on same tree (**heterophylly**).
Inflorescences: Catkins/spike (male), clusters or single flower (female).
Flowers: Inconspicuous, small, radial, unisexual. Males with bracts, females 1-3 together with cupule.
(**cupule** = cup-like involucre of spiny to scaly bracts)
Stamens: 4-numerous, distinct.
Carpels: 3.
Ovary: Inferior.
Placentation: Axile.
Fruit: Nut, acorn. (**acorn** = nut partially enclosed by a cupule).
Fagus with 2 triangular nuts in spiny cupule.
Other: White oak group has leaves with rounded lobes and female flowers with short style branches. Red oak group has leaves with sharp lobes with bristles and female flowers with long style branches.
Genera: *Castanea, Fagus, Quercus.*

Fig. 694   Ridged bark of mature oak tree.   *Quercus* sp.

Leaf Diversity

Fig. 695   Shumard oak.
Quercus shumardii

Fig. 696   Bluff oak.
Quercus austrina

Fig. 697   Water oak.
Quercus nigra

Fig. 698   Live oak.
Quercus virginiana

Acorn Diversity

Fig. 699   Immature S. Red Oak
Quercus falcata

nut

cupule

Fig. 700   Shumard oak.
Quercus shumardii

Fig. 701   Live oak
Quercus virginiana

Fig. 702 Staminate inflorescence characters of oaks. a) & b) Male catkins of swamp chestnut oak. *Quercus michauxii* c) CU of stamens on male catkin of Shumard oak. *Quercus shumardii*

Fig. 703 Carpellate floral characters of oaks. a) Female flower of water oak. *Quercus nigra* b) Female flowers of Shumard oak. *Quercus shumardii* c) Female flowers of bluff oak. *Quercus austrina*

Fig. 704 Floral and fruit characters of chinquapin. a) Flowering branch with multiple staminate inflorescences. b) CU of portion of male inflorescence showing stamens. c) Immature spiny capsule. *Castanea pumila*

Betulaceae - *Birch Family*

Habit: Trees, shrubs. Bark sometimes with horizontal lenticels or peeling.

Leaves: Alternate, simple, doubly serrate. Pinnate venation. Stipules.

Inflorescences: Separate male and female catkins, pendent or erect, with conspicuous bracts.

Flowers: Inconspicuous, small, radial, unisexual. Associated with bracts and bracteoles.

Perianth: 0-4 tepals, minute, scale-like.

Stamens: 1-4.

Carpels: 2.

Ovary: Inferior.

Placentation: Axile.

Fruit: Nut, nutlet, achene, 2-winged samara. Subtended to enclosed by bracts and bracteoles.

Genera: *Alnus, Betula, Carpinus, Corylus, Ostrya.*

Fig. 705   Leaf of river birch. *Betula nigra*

a

b

c

Fig. 706   Staminate floral characters of river birch. a) Male catkins at tip of winter branch. b) CU of male catkin. c) Male catkins during dehiscence and release of pollen. *Betula nigra*

a

b

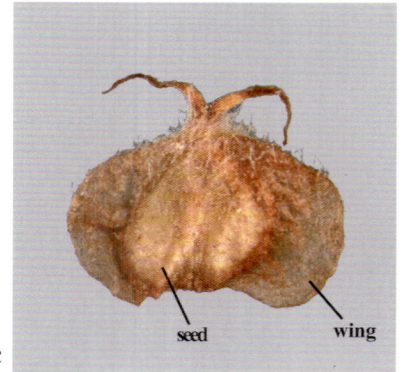

c

Fig. 707   Carpellate floral characters of river birch. a) Flowering branch with female catkins. b) CU of female catkin. c) Samara.   *Betula nigra*

Fig. 708   Floral characters of American hornbeam. a) Flowering branch with male and female inflorescences.
b) CU of male catkin.   *Carpinus caroliniana*

Fig. 709   Floral and fruit characters of American hornbeam. a) Carpellate inflorescence. b) Infructescence with two fruits.
c) CU of nutlet with wing-like subtending bracts.   *Carpinus caroliniana*

Fig. 710   Floral and fruit characters of hop hornbeam. a) Male catkins. b) Infructescence with papery hop-like bracts.
c) Fruit with bract intact. d) Fruit with half of bract removed to show nutlet within.   *Ostrya virginiana*

Casuarinaceae - *She Oak Family*

Habit: Trees, shrubs. Branches drooping, slender, green, grooved, jointed.

Leaves: Reduced, scale-like, triangular. Simple. In whorls of 4-20 at nodes.

Inflorescences: Males clustered in whorls at tips of lateral branches. Females in dense heads with leafy-woody bract.

Flowers: Inconspicuous, small, radial, unisexual. One per inflorescence bract-bracteole structure. Carpellate flowers with red feathery stigma.

Perianth: Absent.

Stamens: 1.

Carpels: 2.

Ovary: Superior.

Placentation: Axile.

Fruit: Cone-like with samaras.

Seeds: 1 per samara.

Other: Roots with nitrogen-fixing nodules.

Genera: *Casuarina* (monotypic family).

Fig. 711   Needle-like lateral branches of Australian pine. *Casuarina equisetifolia*

Fig. 712   Plant characters of Australian pine. a) CU of lateral branches. b) Dehisced male inflorescences. *Casuarina equisetifolia*

a   b   c

Fig. 713   Fruiting characters of Australian pine. a) Immature female 'cone'. b) Mature female 'cone' after dehiscence. c) Samaras.   *Casuarina equisetifolia*

Myricaceae - *Bayberry Family*

Habit: Shrubs, trees. Aromatic wood and leaves.

Leaves: Alternate, simple, entire to serrate.
Pinnate venation. Underside with golden glandular hairs.

Inflorescences: Catkin-like spikes. Separate male and female.

Flowers: Inconspicuous, small, radial, unisexual.
Subtended by scale-like bract.

Perianth: Absent.

Stamens: 2-9.

Carpels: 2.

Ovary: Superior to inferior.

Placentation: Basal.

Fruit: Drupe with waxy, bumpy surface, or an achene.

Other: Roots typically with nitrogen-fixing nodules.

Genera: *Comptonia, Myrica.*

Fig. 714 Underside of wax myrtle leaf. Yellow 'dots' are glandular hairs. *Myrica cerifera*

Fig. 715   Floral and fruit characters of wax myrtle. a) Flowering branch with staminate inflorescences. b) Flowering branch with carpellate inflorescences. c) Fruiting branch.   *Myrica cerifera*

Fig. 716   Floral and fruit characters of wax myrtle. a) Male catkin. b) Female inflorescence. c) CU of waxy drupes. *Myrica cerifera*

Juglandaceae - *Walnut Family*

Habit: Trees, with aromatic wood and leaves.
Leaves: Typically alternate, pinnately compound, entire to serrate. Underside dotted with glandular hairs. Pinnate venation.
Inflorescences: Males with numerous flowers in catkins. Females with 1-several flowers in clusters.
Flowers: Inconspicuous, small, radial, unisexual. Typically with bracts/bracteoles.
Perianth: 0-4 tepals. Sometimes reduced or absent.
Stamens: 3-numerous, distinct, short.
Carpels: 2.
Ovary: Inferior.
Placentation: Basal.
Fruit: Drupe with dehiscent or indehiscent leathery husk (bracts & perianth), or nut/nutlet with winglike bract/bracteoles.
Seed: 1, large, with large wrinkled cotyledons.
Other: Twigs often with chambered pith.
Genera: *Carya, Juglans*.

Fig. 717   Leaf of pignut hickory. *Carya glabra*

a

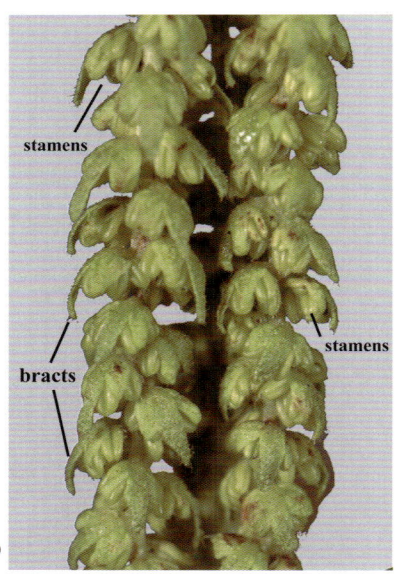

b

c

Fig. 718   Floral and fruit characters of black walnut. a) Staminate inflorescences. b) CU of portion of male catkins. c) Female flower.   *Juglans nigra*

a

b

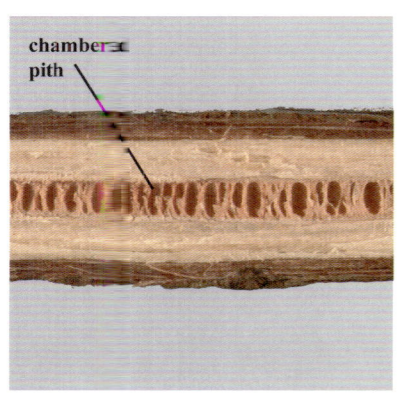

Fig. 719   Fruit characters of pecan. a) Infructescence. b) Drupe with portion of husk removed to show nut-like pit.   *Carya illinoiensis*

Fig. 720   LS of twig of black walnut with chambered pith. *Juglans nigra*

Lythraceae - *Loosestrife Family*

Habit: Herbs typically, also shrubs and trees.
Leaves: Opposite or whorled, simple, entire. Pinnate venation.
Inflorescences: Clusters in various arrangements or 1 flower.
Flowers: Showy, radial to bilateral, bisexual.
Hypanthium with nectar disk. Frequently an epicalyx.
Calyx: 4-8 sepals, distinct. Often thick and triangular.
Corolla: 4-8 petals, distinct, often wrinkled. Sometimes absent.
Stamens: 8-16 (2X # of petals), distinct, unequal in length.
Attached to lower part of hypanthium.
Carpels: 2-numerous.
Ovary: Superior.
Placentation: Axile.
Fruit: Dehiscent capsule, sometimes enclosed by
persistent hypanthium. Sometimes a berry.
Seeds: Flat and/or winged.
Genera: *Ammaria, Cuphea, Decodon, Didiplis, Lagerstroemia,
Lythrum, Rotala.*

Fig. 721　**Flowering branch of lythrum.**　*Lythrum flagellare*

a　　　　　　　　　　　　　b

Fig. 722　a) & b) Flowering branch of cuphea.
Cuphea hyssopifolia

Fig. 723　Flower of crape myrtle.
Lagerstroemia indica

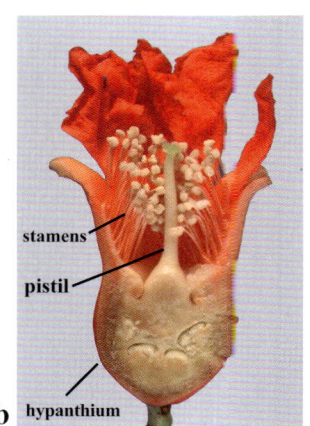

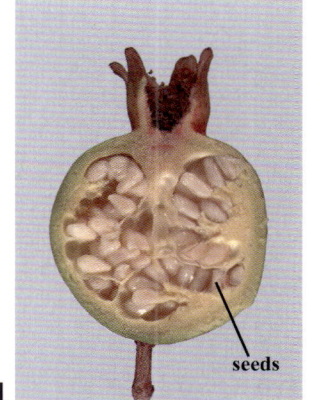

a　　　　　　　b　　　　　　　c　　　　　　　d

Fig. 724　Floral and fruit characters of pomegranate. a) Flower. b) LS of flower. c) Fruit. d) LS of fruit.
Punica granatum

Onagraceae - *Evening Primrose Family*

Habit: Mostly herbs, some shrubs and trees.
Leaves: Alternate or opposite or whorled. Simple.
Entire, lobed, or serrate. Pinnate venation.
Inflorescences: Spike, raceme, panicle, or 1 flower.
Flowers: Showy, radial or bilateral, typically bisexual.
Well developed hypanthium (typically long and tubular).
Calyx: 4 sepals, distinct.
Corolla: 4 petals, distinct, sometimes clawed.
Stamens: 4 or 8, attached near rim of hypanthium.
Pollen clumped in viscin threads.
Carpels: 4.
Ovary: Inferior.
Placentation: Axile.
Fruit: Loculicidal capsule, berry, nutlet.
Seeds: Typically numerous, sometimes winged.
Other: Stigma with 4 lobes.
Genera: *Clarkia, Fuchsia, Ludwigia, Oenothera.*

Fig. 725   Flower of fuchsia.
Fuchsia sp.
(See Fig. 172)

Fig. 726   Flower of pink ladies.
Oenothera speciosa

a

b

Fig. 727   Floral and fruit characters of cutleaf primrose. a) Flowering branch. b) Fruiting branch with immature capsules   *Oenothera laciniata*

a

b

c

Fig. 728   Floral and fruit characters of cutleaf primrose. a) Flower. b) CU of androecium showing viscin threads of pollen.
c) XS of capsule showing four distinct locules.   *Oenothera laciniata*

Fig. 729 Floral characters of Peruvian primrosewillow. a) Flower. b) Flower from behind showing sepals.
Ludwigia peruviana

Fig. 730 Floral diversity of primrosewillows. a) CU of gynoecium and androecium. *Ludwigia peruviana*
b) Inflorescence of an apetalous species. *Ludwigia suffruticosa* c) Flower. *Ludwigia* sp.

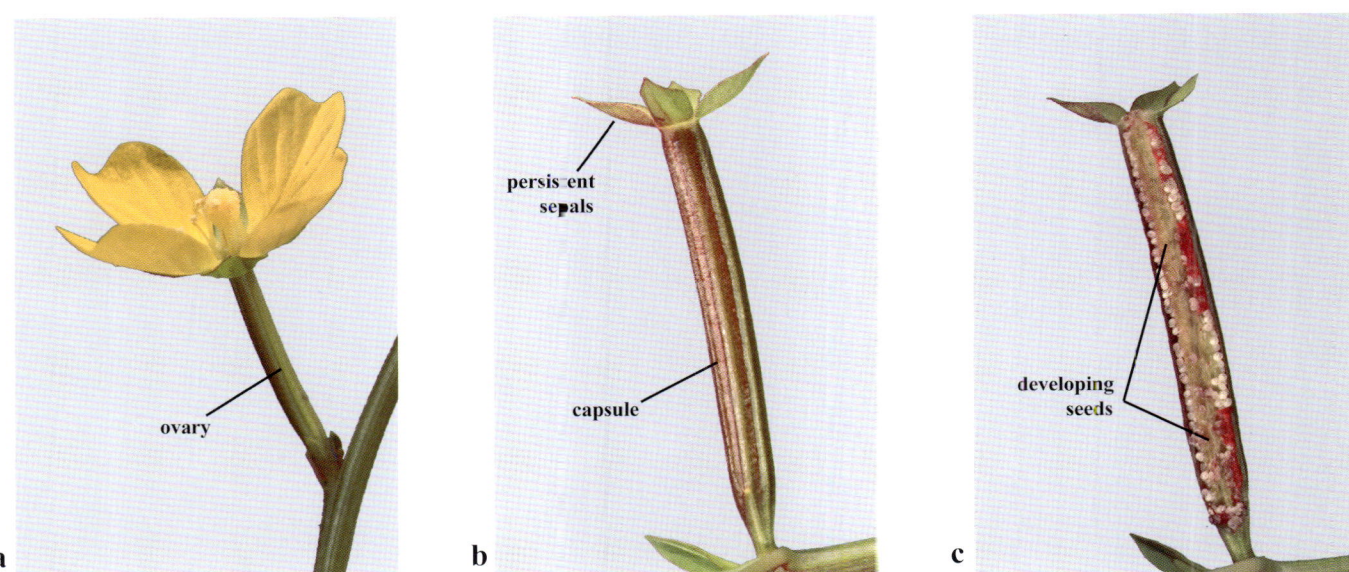

Fig. 731 Floral and fruit characters of primrose-willow. a) Flower. b) Capsule. c) LS of capsule.
Ludwigia sp.

Myrtaceae - *Myrtle Family*

Habit: Trees, shrubs.
Leaves: Opposite or alternate, simple, entire.
Pinnate venation. Pellucid dots. Aromatic.
Inflorescences: Various, sometimes 1 flower.
Flowers: Showy, radial, bisexual. Well developed hypanthium
(sometimes hard and cup-like).
Stamens: Numerous, in tufts.
Carpels: 2-5.
Ovary: Inferior to half inferior.
Placentation: Axile.
Fruit: Berry, loculicidal capsule.
Other: Some species are commercial spices.
Genera: *Eucalyptus, Eugenia, Myrcia, Psidium, Syzgium.*

Fig. 732 Budding branch of bottlebrush.   *Callistemon speciosus*

Fig. 733   Floral and fruit characters of bottlebrush. a) Flowering branch. b) Developing capsules surrounded by a cup-like hypanthium.   *Callistemon speciosus*

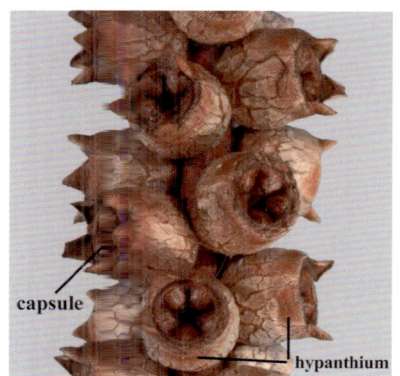

Fig. 734   Floral and fruit characters of bottlebrush. a) Opening flower bud. b) Flowers with multiple showy stamens. c) Dehisced capsules surrounded by a woody hypanthium.   *Callistemon speciosus*

Fig. 735   Floral characters of pineapple-guava. a) Flowering branch. b) Flower showing multiple showy stamens.
Feijoa sellowiana

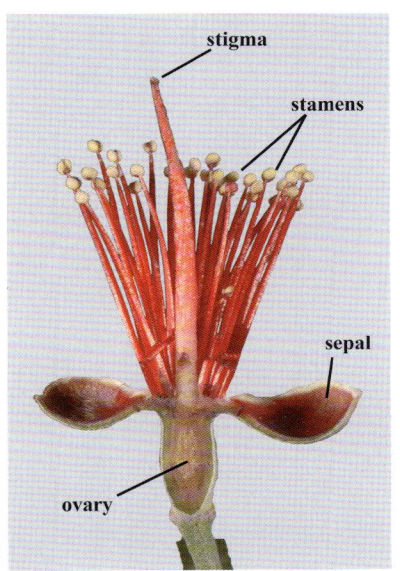

Fig. 736   Floral and fruit characters of pineapple-guava. a) LS of flower. b) Immature fruit. c) Nearly mature fruit.
Feijoa sellowiana

Fig. 737   Floral and fruit characters of guava. a) Flowering branch. b) Nearly mature berries.
Psidium guajava

Melastomataceae - *Melastome Family*

Habit: Herbs, shrubs, vines, trees.
Leaves: Opposite, decussate, simple, entire to serrate. 3-5 subparallel secondary veins diverge from base and converge towards tip. Tertiary veins perpendicular to secondaries.
Inflorescences: Cymes.
Flowers: Bilateral, bisexual. Well developed hypanthium. Sometimes with showy bracts.
Calyx: 3-5 sepals.
Corolla: 3-5 petals, distinct.
Stamens: 6-10, distinct. Often brightly colored. Anthers open by terminal pores.
Carpels: 2-10.
Ovary: Superior to inferior.
Placentation: Axile.
Fruit: Loculicidal capsule, berry.
Seeds: Small, numerous.
Genera: *Medinilla, Miconia, Rhexia, Tetrazygia.*

Fig. 733   Characteristic melastome venation.   *Tibouchina* sp.

a

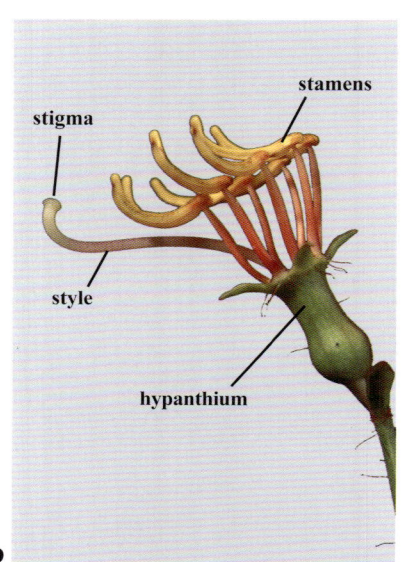

b

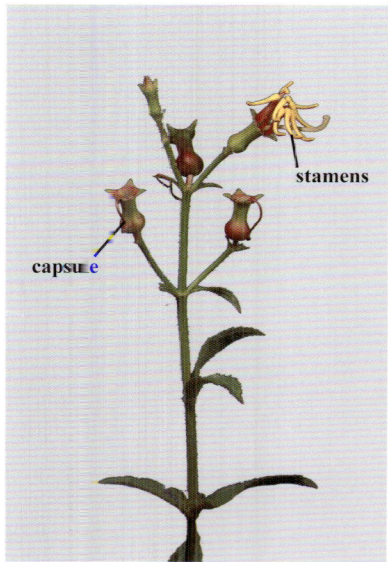

c

Fig. 739   Floral and fruit characters of meadow beauty. a) Flower. b) Flower with corolla removed. c) Fruiting branch showing cymose form of inflorescence.   *Rhexia mariana*

a

b

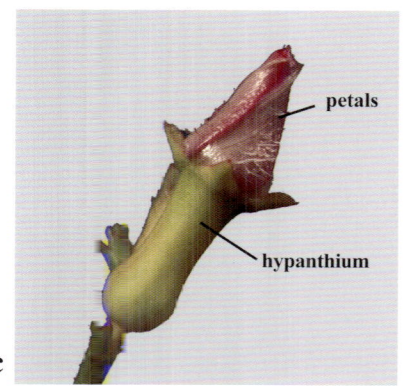

c

Fig. 740   Fruit and floral characters of meadow beauty. a) Immature fruit surrounded by persistent hypanthium. b) LS of capsule showing hypanthium. c) Flower bud.   *Rhexia mariana*

Combretaceae - *White Mangrove Family*

Habit: Trees, shrubs, lianas. Frequently in coastal swamps.

Leaves: Alternate or opposite, simple, entire. Pinnate venation. Blade frequently with paired glands (nectaries) at base and several pits (domatia) on undersurface.

Inflorescences: Raceme, spike, head.

Flowers: Small, typically radial, bisexual or unisexual. Hypanthium prolonged beyond ovary.

Calyx: 4-5 sepals.

Corolla: 4-5 petals, distinct. Sometimes absent.

Stamens: 4-10, distinct.

Carpels: 2-5.

Ovary: Inferior.

Placentation: Apical.

Fruit: Drupe, frequently flat, ribbed, or winged.

Seeds: 1, large.

Genera: *Bucida, Combretum, Conocarpus, Laguncularia, Terminalia.*

Fig. 741   Inflorescence of monkey's brush.   *Combretum aubletii.*

a   b   c

Fig. 742   Characters of button mangrove.   a) Flowering branch. b) Staminate inflorescence. c) Infructescence.   *Conocarpus erectus*

Fig. 743   Leaf of white mangrove with insect domatia. *Laguncularia racemosa*

a   b   c   d

Fig. 744   Floral and fruit characters of white mangrove. a) Flowering branch. b) Inflorescence. c) Flowers. d) Infructescence with drupes.   *Laguncularia racemosa*

Brassicaceae - *Mustard Family*

Note: Two correct Latin names for family: Brassicaceae and Cruciferae.
Habit: Herbs, shrubs.
Leaves: Alternate or in basal rosettes, simple. Frequently pinnately lobed. Pinnate or palmate venation. Aromatic mustard oils.
Inflorescences: Raceme or corymb.
Flowers: Radial or bilateral, bisexual.
Calyx: 4 sepals, distinct.
Corolla: 4 petals, distinct. Arranged in form of a cross (hence the name Cruciferae). Petals with narrow basal claw and wide apical limb.
Stamens: 6, 2 outer shorter than 4 inner (tetradynamous).
Carpels: 2.
Ovary: Superior.
Placentation: Parietal.
Fruit: Silique (= 2-valved capsule with persistent septum).
replum = thick hard placental rim to which seeds are attached.
(See also page 51)
Other: Aromatic mustard oils.
Genera: *Brassica, Cardamine, Descuraenia, Lepidium, Raphanus*.

Fig. 745 Cruciform flower of the wild radish.
Raphanus raphanistrum

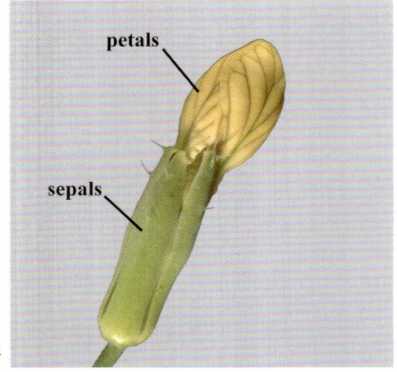

Fig. 746 Floral characters of wild radish. a) Flower bud. b) Cruciform flower. c) Flower with perianth removed.
Raphanus raphanistrum

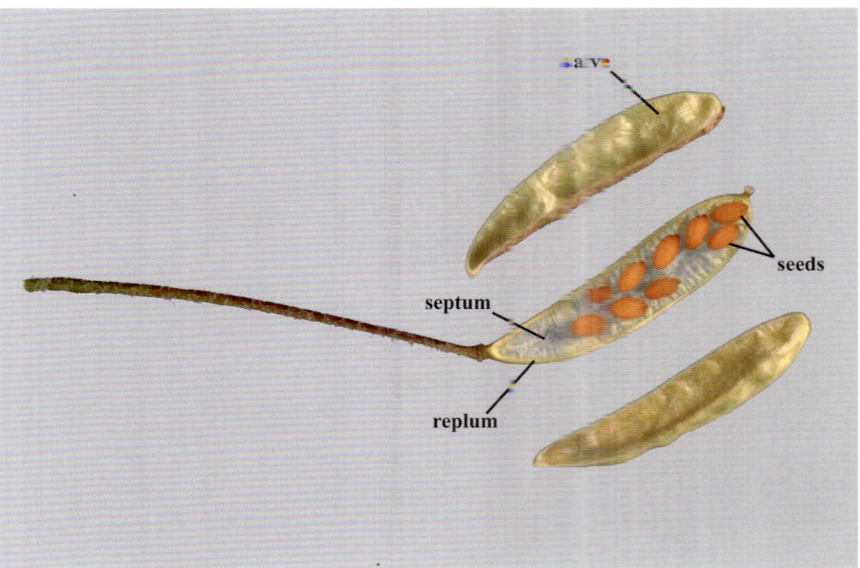

Fig. 747 Fruit of wild radish.
Raphanus raphanistrum

Fig. 748 Silique of tansy mustard.
Descurainia pinnata

Fig. 749   Floral and fruit characters of peppergrass. a) Habit of plant with flowers / fruit. b) Flowering / fruiting branch. c) Sequence of siliques.   *Lepidium virginicum*

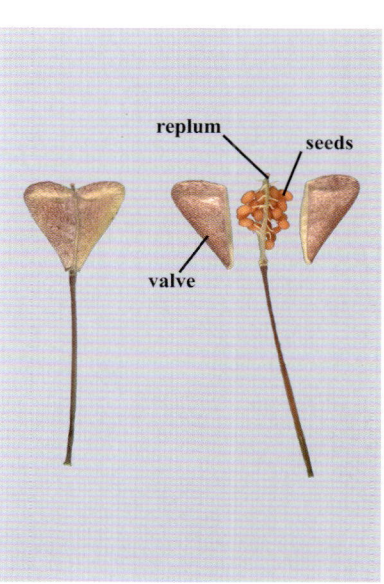

Fig. 750   Floral and fruit characters of shepherd's purse. a) Habit of plant with flowers / fruit. b) Inflorescence. c) Siliques.   *Capsella bursa-pastoris*

Fig. 751   Floral and fruit characters of tansy mustard. a) Habit of plant with flowers / fruit. b) Flowering / fruiting branch.   *Descurainia pinnata*

Fig. 752   Broccoli inflorescence with edible florets.   *Brassica oleracea*

Capparaceae - *Caper Family*

Habit: Herbs, shrubs.

Leaves: Alternate, simple or palmately compound. Pinnate to palmate venation.

Inflorescences: Raceme, sometimes 1 flower.

Flowers: Radial to bilateral, bisexual.

Calyx: 2-6 sepals.

Corolla: 2-6 petals. Frequently with narrow basal claw and wide apical limb.

Stamens: 4-many, distinct. Sometimes extremely long.

Carpels: 2-12.

Ovary: Superior.

Placentation: Parietal.

Fruit: Dehiscent or indehiscent berry, drupe, capsule, samara, nut, silique.

Other: Aromatic mustard oils present.

Genera: *Capparis, Cleome, Polanisia.*

Note: Some botanists include this family within the Brassicaceae.

Fig. 753   Flowering spider flower. *Cleome hassleriana*

a   b   c   d

Fig. 754   Floral characters of spider flower. a) Flowering / fruiting plant. b) Flower. c) Flower bud. d) Clawed petal. *Cleome hassleriana*

a   b   c   d

Fig. 755   Fruit characters of spider flower. a) Developing siliques. b) Dehiscing silique. c) Dehisced silique. d) Seeds. *Cleome hassleriana*

Tiliaceae - *Linden Family*

Habit: Trees, shrubs, and a few herbs.
Leaves: Alternate, simple, serrate. Distichous.
Palmate venation. Bases asymmetrical.
Inflorescences: Cymes, sometimes 1 flower.
Flowers: Small, radial, bisexual. Sometimes with epicalyx.
Nectar glands or nectar disk present.
Calyx: 5 sepals.
Corolla: 5 petals, distinct. Sometimes absent.
Stamens: 10-numerous, in groups (fascicles).
Adnate to base of petals.
Carpels: 2.
Ovary: Superior.
Placentation: Axile.
Fruit: Loculicidal capsule, drupe, schizocarp, nut.
Other: Flower/fruit sometimes borne on large leafy bract.
Genera: *Corchorus, Grewia, Tilia, Triumfetta.*
Note: Many botanists include this family within the Malvaceae.

Fig. 756   Foliage, bracts, and budding inflorescence of basswood.   *Tilia americana*

Fig. 757   Floral characters of basswood. a) Flowering branch. b) Flower.
c) Flower with several petals removed to show stamens. d) Flower with
petals and stamens removed to show pistil.   *Tilia americana*

Fig. 758   Flower of basswood.
Tilia americana

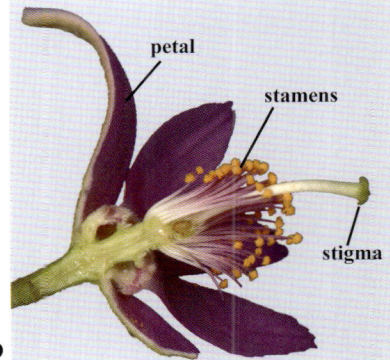

Fig. 759   Floral characters of star flower. a) Flowering branch.
b) LS of flower.   *Grewia occidentalis*

Malvaceae - *Mallow Family*

Habit: Herbs, shrubs, trees, lianas.
Leaves: Alternate, simple, frequently palmately lobed.
Entire to serrate. Palmate venation. Stipules.
Inflorescences: Cymes, sometimes 1 flower.
Flowers: Showy, radial, bisexual.
Calyx: 5 sepals, frequently subtended by an **epicalyx**
(involucre of bracts).
Corolla: 5 petals, distinct. Adnate to base of filament tube.
Petal bases with numerous nectar-producing hairs.
Stamens: Numerous, connate forming a tube (**monadelphous**).
Carpels: 2-numerous.
Ovary: Superior.
Placentation: Axile.
Fruit: Loculicidal capsule, schizocarp.
Seeds: Often hairy.
Other: Mucilaginous sap. Stellate hairs.
Genera: *Abutilon, Hibiscus, Malvaviscus, Pavonia, Sida.*

Fig. 760   Flower of rose of Sharon.
Hibiscus syriacus

Fig. 761   Flower and fruit of Indian hemp.
a) Flower. b) Schizocarp.
Sida rhombifolia

Fig. 762   Flower of
flowering maple.
Abutilon sp.

Fig. 763   Flower of cotton.
Gossypium sp.

Fig. 764   Flower of Turk's
cap mallow.
Malvaviscus penduliflorus

Fig. 765   Floral and fruit characters of Caesarweed. a) Flower. b) Mature schizocarp.
c) Three isolated mericarps from fruit.
Urena lobata

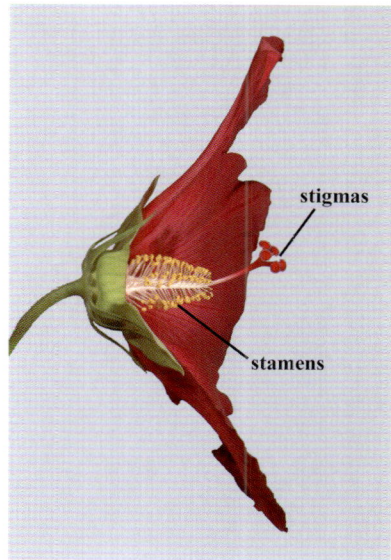

Fig. 766   Floral characters of swamp hibiscus. a) Flower bud. b) Flower. c) LS of flower.
Hibiscus   moscheutos

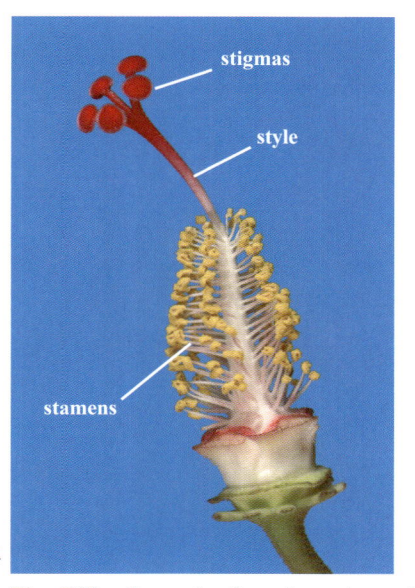

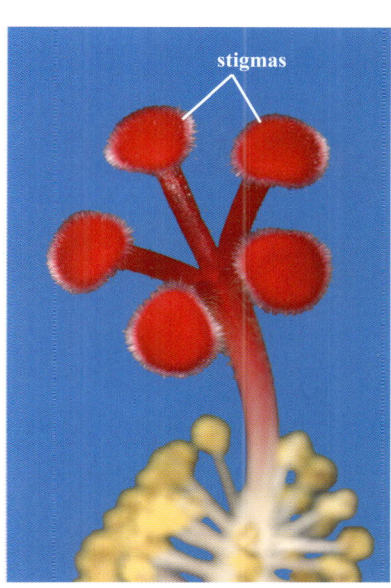

Fig. 767   Reproductive characters of swamp hibiscus. a) Androecium and gynoecium. b) Androecium. c) Multiple stigmas.   *Hibiscus moscheutos*

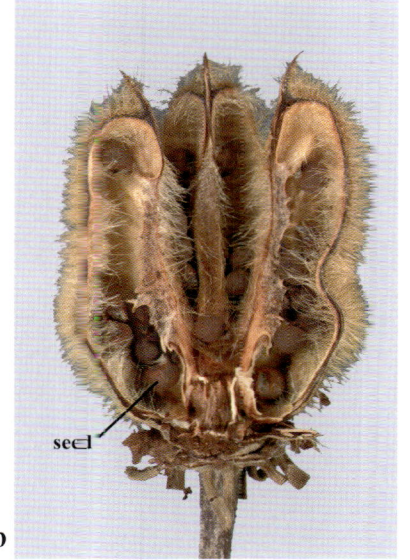

Fig. 768   Fruit characters of swamp hibiscus. a) Dehisced capsule. b) LS of capsule. c) Seeds.
Hibiscus moscheutos

Sapindaceae - *Soapberry Family*

Habit: Trees, shrubs, lianas (Some with tendrils).
Leaves: Alternate or opposite. Palmately compound or pinnately compound or trifoliate or unifoliate. Entire or serrate.
Inflorescences: Panicles, racemes.
Flowers: Inconspicuous to showy, radial to bilateral, typically unisexual. Nectar disk present.
Calyx: 4-5 sepals, typically differentiated.
Corolla: 4-5 petals, frequently clawed. Often with hairy-scaly basal appendages. Sometimes petals absent.
Stamens: 4-10 in 2 whorls, distinct, filaments hairy.
Carpels: 2-3.
Ovary: Superior.
Placentation: Typically axile.
Fruit: Loculicidal or septifragal capsule, schizocarp, berry, nut, and others.
Seeds: Frequently arillate.
Genera: *Koelreuteria, Paullinia, Sapindus.*

Fig. 769   Partial inflorescence of golden rain tree.
Koelreuteria elegans

a

b

c

Fig. 770   Floral and fruit characters of golden rain tree. a) Staminate flower. b) Carpellate flower. c) Immature capsule.
Koelreuteria elegans

a

b

Fig. 771   Fruit of golden rain tree. a) Mature capsule. b) Capsule with one valve removed exposing seeds. *Koelreuteria elegans*

Fig. 772   Dehisced capsule of harpullia   *Harpullia arborea*

Hippocastanaceae - *Horse Chestnut Family*

Habit: Shrubs, trees.
Leaves: Opposite, palmately compound, entire to serrate.
Leaflets with pinnate venation.
Inflorescences: Raceme or panicle.
Flowers: Showy, large, bilateral, bisexual or staminate.
Nectar disk present.
Calyx: 4-5 sepals, distinct or basally connate.
Corolla: 4-5 petals, distinct, clawed, differentiated.
Stamens: 5-8, distinct.
Carpels: Typically 3.
Ovary: Superior.
Placentation: Axile.
Fruit: Loculicidal capsule, leathery to spiny.
Seeds: Very large, hard. Large obvious hilum scar.
Genera: *Aesculus.*
Note: Many botanists place this family within the Sapindaceae.

Fig. 773   Habit of red buckeye with fruit.   *Aesculus pavia*

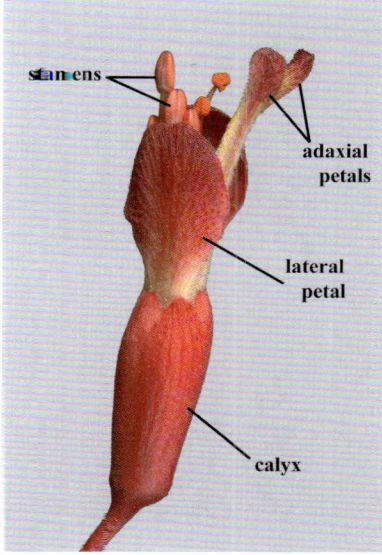

a   b   c

Fig. 774   Foliage and fruit characters of red buckeye. a) Palmately compound leaf. b) Inflorescence. c) Staminate flower.
Aesculus pavia

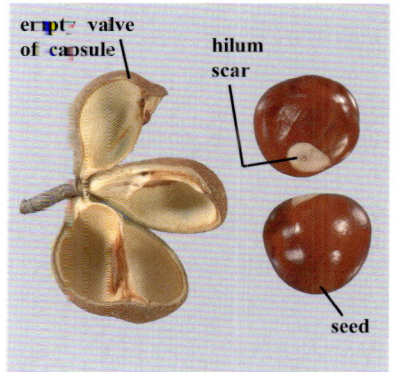

a   b   c

Fig. 775   Fruit characters of red buckeye. a) Capsule dehiscing. b) Capsule with one valve removed exposing seeds.
c) Empty valves of capsule and two seeds.   *Aesculus pavia*

Aceraceae - *Maple Family*

Habit: Shrubs, trees.

Leaves: Opposite, simple/palmately lobed or pinnately compound. Palmate venation.

Inflorescences: Raceme, panicle, corymb, umbel.

Flowers: Showy to inconspicuous, small, radial, bisexual or unisexual. Nectar disk.

Calyx: Typically 5 sepals, distinct.

Corolla: Typically 5 petals, distinct.

Stamens: Typically 8 (4-12 possible), distinct.

Carpels: 2.

Ovary: Superior.

Placentation: Axile.

Fruit: Winged schizocarp with two samaroid segments.

Seeds: 1 per locule.

Genera: *Acer.*

Note: Many botanists treat this family as a subgroup within the Sapindaceae.

Fig. 776   Compound leaf of boxelder.   *Acer negundo*

a

b

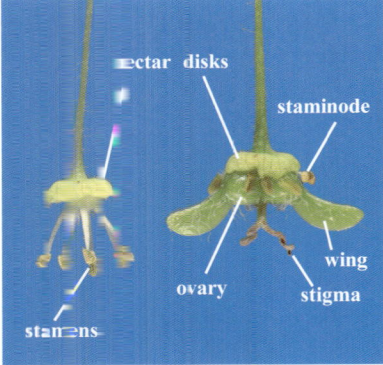

c

Fig. 777   Floral characters of boxelder. a) Extruded stamens of staminate inflorescence. b) Staminate inflorescence and new leaves. c) Carpellate flowers: left with well developed stamens and right with reduced staminodes.   *Acer negundo*

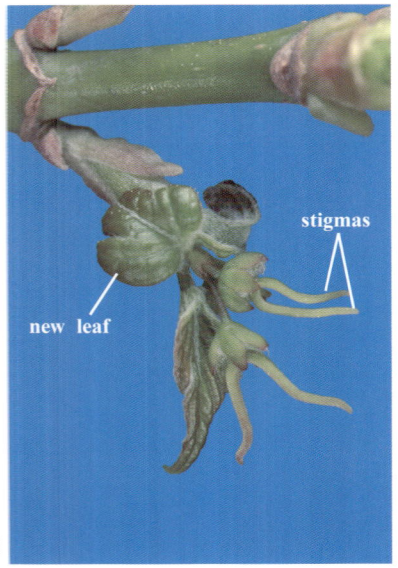

a

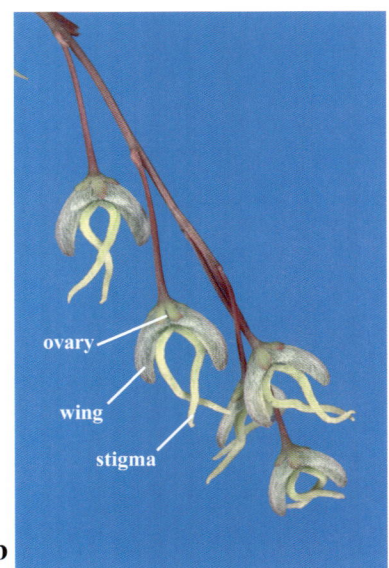

b

c

Fig. 778   Floral and fruit characters of boxelder. a) Young carpellate inflorescence. b) Mature carpellate inflorescence. c) Developing infructescence of several schizocarps.   *Acer negundo*

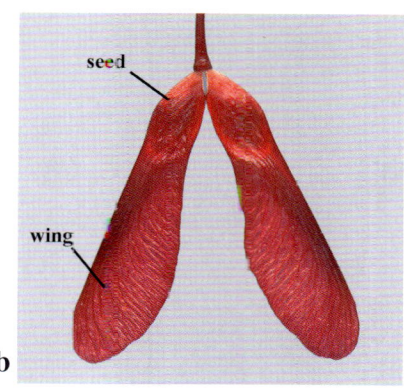

Fig. 779   Leaf and fruit characters of maples.
a) Leaf of red maple.   *Acer rubrum*
b) Schizocarp of red maple.   *Acer rubrum*
c) Schizocarp of Florida sugar maple.   *Acer saccharum* ssp. *floridanum*

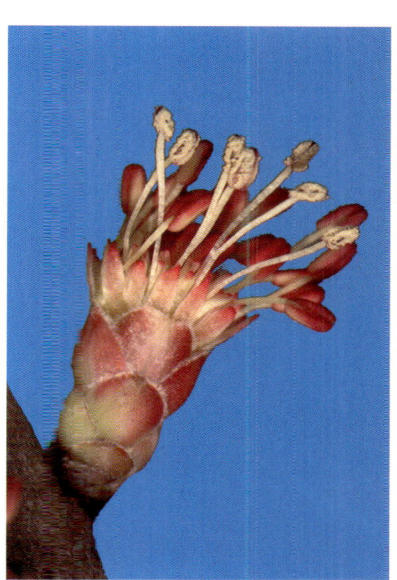

Fig. 780   Staminate floral characters of red maple. a) Flowering branch. b) Staminate inflorescences. c) Staminate flowers.
Acer rubrum

Fig. 781   Carpellate floral characters of red maple. a) Flowering branch. b) Carpellate inflorescence. c) Two carpellate
flowers.   *Acer rubrum*

Rutaceae - *Citrus Family*

Habit: Shrubs, trees. Often with prickles, spines, thorns.
Leaves: Alternate, entire to crenate. Aromatic. Unifoliate, trifoliate, pinnately or palmately compound. Pinnate venation. Pellucid dots.
Inflorescences: Terminal or axillary, determinate, sometimes 1 flower.
Flowers: Showy, radial, bisexual or unisexual. Nectar disk.
Calyx: 4-5 sepals.
Corolla: 4-5 petals, distinct.
Stamens: 8-10 to numerous.
Carpels: 4-5.
Ovary: Superior.
Placentation: Axile.
Fruit: Hesperidium (= berry with thick leathery rind), samara, drupe, capsule, schizocarp, aggregate of follicles.
Seeds: Large.
Genera: *Citrus, Poncirus, Ptelea, Zanthoxylum.*

Fig. 782 Leaf of valencia orange. *Citrus aurantifolium*

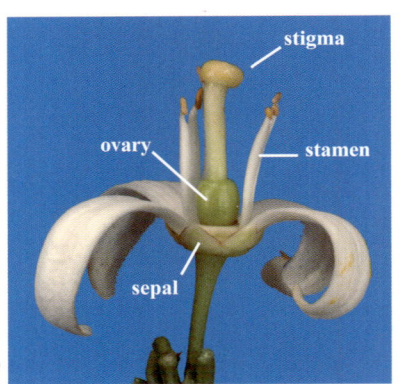

Fig. 783   **Floral and fruit characters of orange tree. a) Flower. b) Flower with one petal and numerous stamens removed to show pistil. c) Developing orange fruit (hesperidium).**   *Citrus* sp.

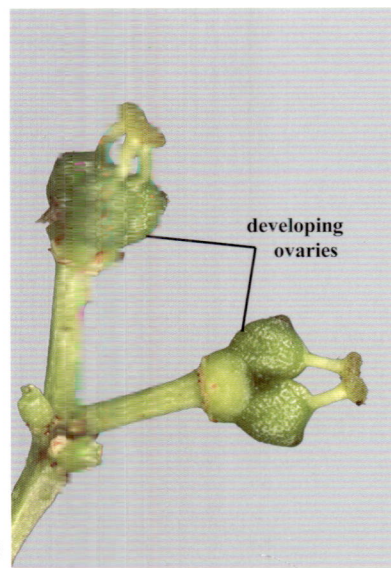

Fig. 784   **Floral and fruit characters of Hercules' club. a) Flowering branch. b) Carpellate flowers. c) Very young fruits.**   *Zanthoxylum clava-herculis*

Meliaceae - *Mahogany Family*

Habit: Shrubs, trees.
Leaves: Typically alternate, pinnately or bipinnately compound, entire or serrate. Pinnate venation.
Inflorescences: Cyme, panicle.
Flowers: Small, radial, unisexual. Nectar disk.
Calyx: 4-5 sepals.
Corolla: 4-5 petals.
Stamens: 4-10, typically connate forming a tube (**monadelphous**). Often with apical teeth or appendages.
Carpels: 2-5.
Ovary: Superior.
Placentation: Axile.
Fruit: Loculicidal or septicidal capsule, berry, drupe.
Seeds: Papery and winged, or fleshy and arillate.
Genera: *Cedrela, Guarea, Melia, Swietenia.*

Fig. 785   Infructescence of Chinaberry.   *Melia azedarach*

Fig. 786   Floral and fruit characters of Chinaberry. a) Flowers showing long stamen tube. b) LS of flower. c) Fruit. *Melia azedarach*

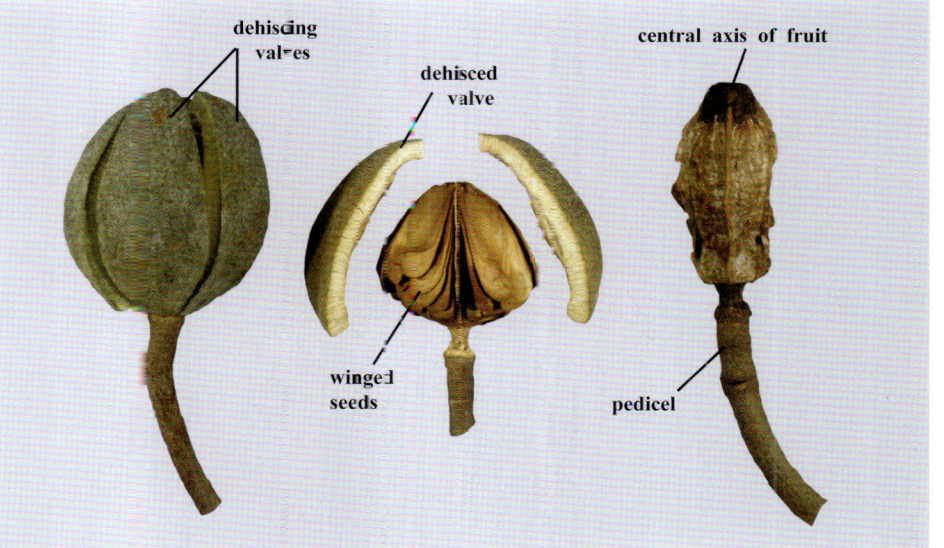

Fig. 787   Floral and fruit characters of West Indian mahogany a) Flower. b) Sequence of fruits. *Swietenia mahagoni*

Photo by Roblin Giblin - Davis

Anacardiaceae - *Sumac Family*

Habit: Shrubs, trees, lianas.
Leaves: Alternate, pinnately compound or trifoliate or unifoliate.
Entire to serrate. Pinnate venation.
Inflorescences: Cyme, panicle.
Flowers: Small, radial, unisexual. Nectar disk.
Calyx: 5 sepals.
Corolla: 5 petals.
Stamens: 5-10.
Carpels: Usually 3.
Ovary: Usually superior.
Placentation: Various.
Fruit: Drupe, frequently flat and asymmetrical.
Other: Resinous allergenic compounds.
Genera: *Anacardium, Mangifera, Pistacia, Rhus, Schinus, Spondias, Toxicodendron.*

Fig. 788 Leaf of winged sumac.
Rhus copallina

Fig. 789 Floral and fruit characters of winged sumac. a) Part of staminate inflorescence. b) Part of carpellate inflorescence. c) Infructescence.   *Rhus copallina*

Fig. 790 Floral and fruit characters of winged sumac. a) Staminate flower. b) Carpellate flower. c) Drupes.
Rhus copallina

a   b   c

Fig. 791   Leaf and floral characters of poison ivy. a) Leaf. b) Carpellate inflorescence. c) Carpellate flowers.
Toxicodendron radicans

stigma

a   b   c

Fig. 792   Floral and fruit characters of Chinese pistachio. a) Carpellate inflorescence. b) Carpellate flowers.
c) Infructescence with developing drupes.   *Pistacia chinensis*

Fig. 793   Fruiting branch of
mango tree.   *Mangifera indica*

Fig. 794   Fruit of cashew tree. The red cashew 'apple' is the fleshy expanded
receptacle, while the edible 'nut' (actually the seed) grows in a drupe from the
bottom of the fruit (shown upside-down here).   *Anacardium occidentale*

Cornaceae - *Dogwood Family*

Habit: Shrubs, trees.
Leaves: Opposite, occasionally alternate, simple. Entire to serrate.
Pinnate venation with arching secondary veins.
Inflorescence: Various, sometimes 1 flower.
Sometimes with large showy petaloid bracts.
Flowers: Small, radial, bisexual or unisexual. Nectar disk.
Calyx: 4-5 small sepals, distinct, frequently reduced to teeth.
Corolla: 4-5 petals, distinct.
Stamens: 4-10, distinct.
Carpels: 2-3.
Ovary: Inferior.
Placentation: Axile.
Fruit: Drupe with 1 ridged pit (**pyrene**).
Other: A torn leaf often remains connected by
thread-like tracheids (the 'dogwood test').
Genera: *Cornus, Davidia, Nyssa.*

Fig. 795 Leaf and venation of flowering dogwood. *Cornus florida*

Fig. 796 Floral characters of flowering dogwood. a) Inflorescence in bud. b) Inflorescence in flower. c) CU of one flower after all other flowers have been removed.   *Cornus florida*

Fig. 797 Floral and fruit characters of flowering dogwood. a) Inflorescence without bracts. b) Immature infructescence.
c) Infructescence of mature drupes.   *Cornus florida*

Fig. 798   Floral characters of Florida dogwood. a) Flowering branch. b) CU of part of inflorescence.
Cornus foemina

Fig. 799   Staminate floral characters of tupelo. a) Staminate flowering branch. b) Staminate inflorescence.
Nyssa ogeche

Fig. 800   Carpellate floral characters of tupelo. a) Carpellate flowering branch b) Carpellate flower.
Nyssa ogeche

Ericaceae - *Heath Family*

Habit: Herbs, shrubs, trees, lianas, epiphytes.

Leaves: Typically alternate, simple, entire to serrate. Often leathery, sometimes needle-like. Pinnate or palmate venation.

Inflorescences: Variable.

Flowers: Showy, radial to slightly bilateral, bisexual. Typically pendent.

Calyx: 4-5 sepals, connate.

Corolla: 4-5 petals, connate. Typically shaped urceolate, funnelform, or campanulate.

Stamens: Typically 8-10 (2X # of petals) and distinct. Sometimes adnate to petals. Open by terminal pores. Sometimes with spurs at/near anthers.

Carpels: 2-10.

Ovary: Superior to inferior.

Placentation: Axile or parietal.

Fruit: Loculicidal or septicidal capsule, berry, drupe.

Seeds: Small, sometimes winged.

Other: Pollen shaken out by pollinators. Occurs in clumps.

Genera: *Agarista, Allotropa, Arbutus, Azalea, Calluna, Erica, Gaultheria, Kalmia, Leucothoe, Lyonia, Monotropa, Oxydendrum, Pieris, Pterospora, Pyrola, Rhododendron, Vaccinium.*

Fig. 801 Flowering branch of azalea. *Rhododendron simsii*

Fig. 802 Floral and fruit characters of blueberry. a) Flowering branch with developing fruits. b) Infructescence of developing berries. *Vaccinium* sp.

Fig. 803 Floral and fruit characters of blueberry. a) Two urceolate flowers. b) LS of flower. c) Two mature berries. *Vaccinium* sp.

Fig. 804   **Floral characters of fetterbush. a) Flowering branch. b) LS of flower.**
Lyonia lucida

Fig. 805   **Fruit characters of fetterbush. a) Developing undehisced capsules. b) Young capsules.**
Lyonia lucida

Fig. 806   **Dehisced capsules of fetterbush. a) Portion of fruiting branch.**
b) CU of several capsules.   *Lyonia lucida*

Fig. 807   **Immature capsules of coastal-plain staggerbush.**   *Lyonia fruticosa*

a   b   c

Fig. 808   Floral characters of Florida rosemary. a) Staminate flowering branch. b) CU of staminate flowering branch. c) CU of carpellate flowering / fruiting branch.   *Ceratiola ericoides*   Note: This species and its relatives are traditionally placed in the family Empetraceae.

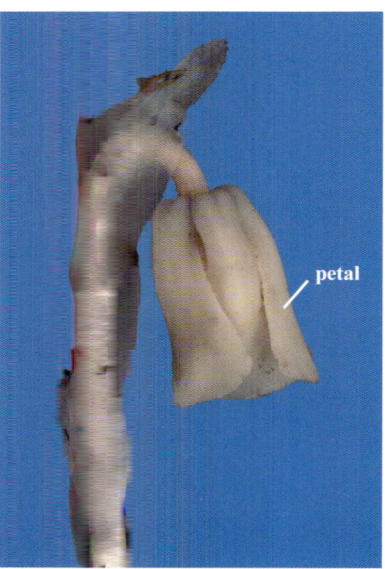

petal

Fig. 809   Indian pipes are fleshy herbs lacking chlorophyll. They are parasitic on mycorrhizal fungi associated with the roots of certain trees. a) In-situ on forest floor. b) Flower.
Monotropa uniflora

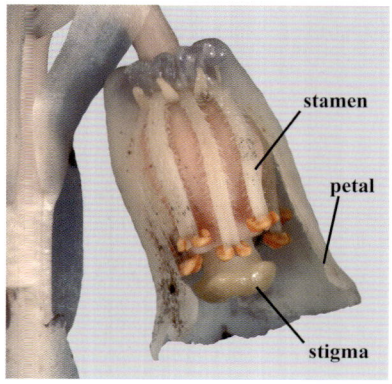
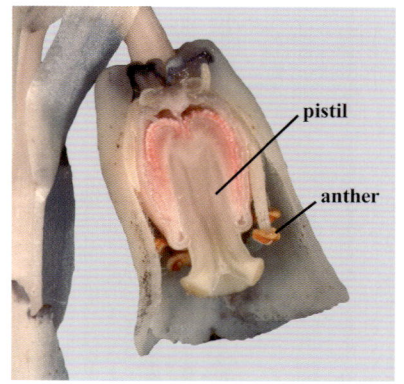
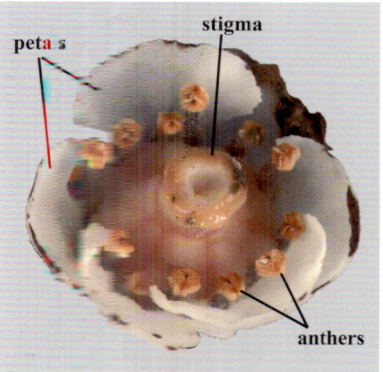

Fig. 810   Floral characters of Indian pipes. a) Flower with several petals removed. b) LS of flower. c) Flower head-on.
Monotropa uniflora

Sarraceniaceae - *Pitcher Plant Family*

Habit: Insectivorous herbs. Grow from rhizomes.

Leaves: Extremely modified. Alternate, simple, in basal rosettes. Petiole hollow, tubular, filled with liquid. Blade small, flat, lid-like.

Inflorescences: Solitary, terminal and nodding.

Flowers: Large, radial, bisexual. Frequently with large bracts.

Calyx: Typically 5 sepals, distinct, persistent. Petaloid and showy.

Corolla: Typically 5, distinct. Frequently asymmetrical. Strap-like.

Style: Umbrella-like with stigmas at tips. Sexual parts hidden.

Stamens: Numerous.

Carpels: 3-5.

Ovary: Superior.

Placentation: Axile.

Fruit: Loculicidal capsule.

Seeds: Small, numerous, sometimes arillate or winged.

Other: Specialized retrorse and glandular hairs contribute to the capture and digestion of insects. Typically found in bogs and marshes.

Genera: *Darlingtonia, Heliamphora, Sarracenia.*

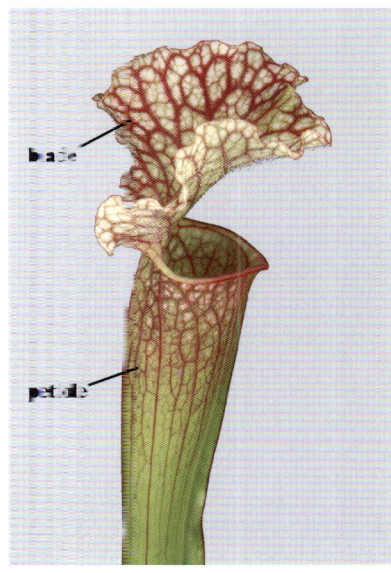

Fig. 811   Insect-trapping modified leaf of the white-top pitcher plant. *Sarracenia leucophylla*

Fig. 812   Leaves of parrot pitcher plant.   *Sarracenia psittacina*

Fig. 813   Floral and leaf characters of the hooded pitcher plant. a) Flower and leaf. b) LS of flower.   *Sarracenia minor*

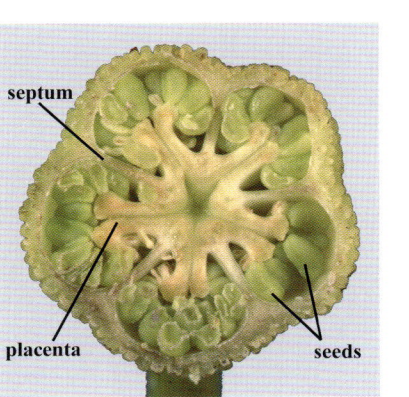

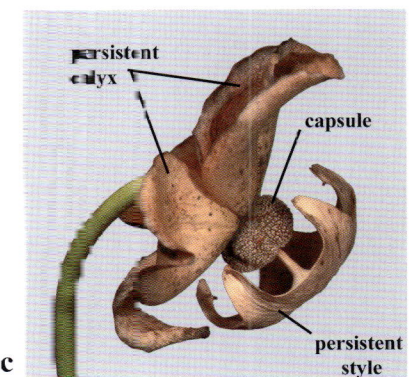

Fig. 814   Fruit characters of the hooded pitcher plant. a) XS of immature fruit. b) Developing capsule. c) Mature capsule with persistent calyx.   *Sarracenia minor*

Sapotaceae - *Sapodilla Family*

Habit: Shrubs, trees.

Leaves: Alternate, simple, entire, leathery. Spirally arranged and sometimes clustered at tips of branches. Pinnate venation.

Inflorescences: Cyme, umbel, sometimes 1 flower. Sometimes with cauliflory.

Flowers: Small, radial, bisexual.

Calyx: 4-8 sepals, sometimes differentiated.

Corolla: 4-8 petals, connate. Sometimes with appendages.

Stamens: 8-16, distinct, opposite and adnate to petals. Often in whorls and alternating with staminodes.

Carpels: 2-numerous.

Ovary: Superior.

Placentation: Axile.

Fruit: Berry.

Seeds: 1-several, large. Obvious hilum scar.

Other: Milky sap. Brownish T-shaped hairs. Source of *chicle*.

Genera: *Chrysophyllum, Manilkara, Pouteria, Sideroxylon*.

Fig. 315   Leaf of chicle tree. *Manilkara zapota*

Fig. 816   Floral characters of chicle tree. a) Flowering branch. b) Flower. c) LS of flower showing stamens adnate to petals.   *Manilkara zapota*

Fig. 817   Floral and fruit characters of chicle tree. a) Flower, top view. b) Fruiting branch. c) Seed showing large hilum scar.   *Manilkara zapota*

Fig. 818   Flowering branch of gum bully.
Sideroxylon lanuginosum

Fig. 819   Immature berry of gum
bully.   *Sideroxylon lanuginosum*

Fig. 820   Inflorescence of Spanish
cherry.   *Mimusops elengi*

a

b

Fig. 821   Foliage characters of caimito. a) Adaxial surface of leaves.
b) Abaxial surface of leaves with coppery sheen.   *Chrysophyllum cainito*

a

b

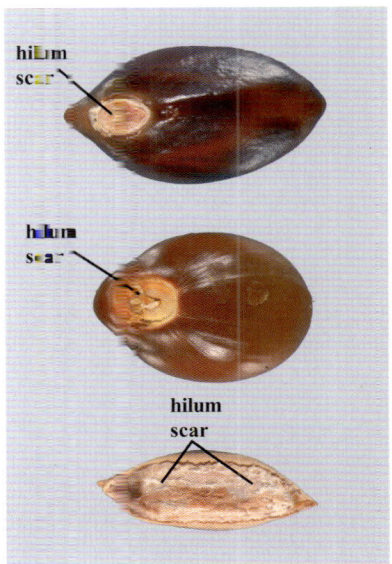

hilum
scar

hilum
scar

hilum
scar

c

Fig. 822   Fruit and seed characters of the Sapotaceae. a) Fruiting branch. *Mimusops balata* b) LS of fruit. *Mimusops balata* c) Seeds: top = *Mimusops balata*, middle = *Manilkara zapota*, bottom = *Chrysophyllum* sp.

Myrsinaceae - *Myrsine Family*

Habit: Shrubs, trees, occasionally lianas.
Leaves: Alternate, simple, entire to crenate, leathery.
Pinnate venation. Often marked with dots or lines.
Inflorescences: Cyme, corymb, panicle.
Flowers: Small, radial, bisexual or unisexual.
Often marked with dots or lines.
Calyx: 4-5 sepals.
Corolla: 4-5 petals, connate at bases with apical lobes
expanding outwards (**rotate**).
Stamens: 4-5, opposite and adnate to petals.
Carpels: 3-5.
Ovary: Superior.
Placentation: Free-central.
Fruit: Drupe with 1 pit (pyrene). Often with dots or lines.
Seeds: Small, dark, 1-several.
Other: Usually with translucent or reddish to blackish lines and
dots on the leaves and sometimes on other structures.
Genera: *Ardisia, Maesa, Myrsine.*

Fig. 823 Flowering and fruiting branch of ardisia. *Ardisia crenata*

Fig. 824   Plant characters of ardisia. a) Habit. b) Flower.
Ardisia crenata

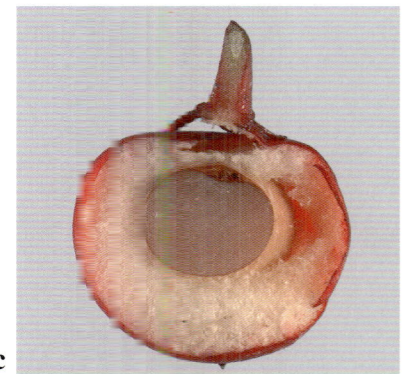

Fig. 825   Floral and fruit characters of ardisia. a) Flower. b) LS of flower. (Note lines and dots on peduncle and petals.) c) LS of drupe.

Primulaceae - *Primrose Family*

Note: This family is closely related to the Myrsinaceae and to the Theophrastaceae, with some genera more closely related to them.

Habit: Herbs.

Leaves: Alternate or opposite or whorled or basal rosette. Pinnate venation.

Inflorescences: Various to highly modified. Sometime 1 flower.

Flowers: Radial, bisexual. Sometimes heterostylous.

Calyx: 5 sepals, connate, lobed or toothed.

Corolla: 5 petals, connate, sometimes reflexed.

Stamens: 5, opposite and adnate to petals.

Carpels: 5.

Ovary: Superior to half-inferrior.

Placentation: Free-central.

Fruit: Capsule, sometimes circumscissile.

Seeds: Sometimes arillate.

Other: Frequently with resinous hairs.

Genera: *Cyclamen, Dodecatheon, Lysimachia, Primula.*

Fig. 826   Leaves of cyclamen. *Cyclamen* sp.

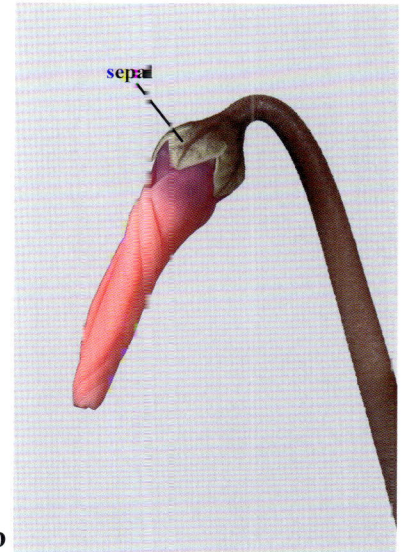

a   b

Fig. 827   Floral characters of cyclamen. a) Flower viewed from behind showing calyx. b) Bud. *Cyclamen* sp.

a   b   c   d

Fig. 828   Floral characters of cyclamen. a) Flower. b) LS of flower. c) LS of flower with with petals removed. d) LS of flower with petals and one stamen removed.   *Cyclamen* sp.

Theaceae - *Tea Family*

Habit: Shrubs, trees.
Leaves: Alternate, simple, serrate. Leathery. Pinnate venation.
Inflorescences: Axillary, solitary flowers.
Flowers: Large, showy, radial, bisexual. Subtended by 1-2 bracts.
Calyx: Typically 5 sepals.
Corolla: Typically 5 petals, distinct, edges frequently wrinkled.
Stamens: Numerous, distinct to connate forming a ring.
Adnate to base of corolla.
Carpels: 3-5.
Ovary: Superior.
Placentation: Axile.
Fruit: Woody loculicidal capsule.
Seeds: Flat or spherical, sometimes winged.
Genera: *Camellia, Franklinia, Gordonia, Stewartia, Ternstroemia.*
Note: Many botanists now place the genus *Ternstroemia* in
its own family, the Ternstroemiaceae.

Fig. 829   Flowering branch of camellia. *Camellia japonica*

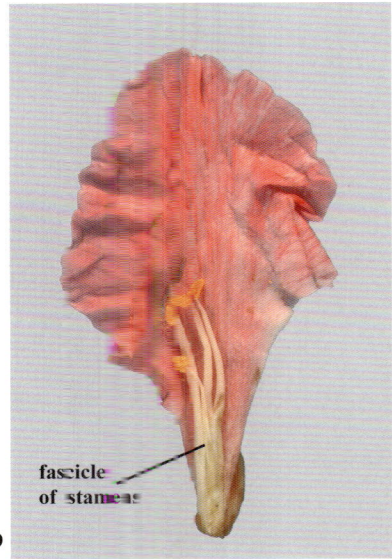

fascicle
of stamens

Fig. 830   Foliage and floral characters of camellia. a) Leaves and buds. b) Petal with adnate fascicle of stamens.
Camellia japonica

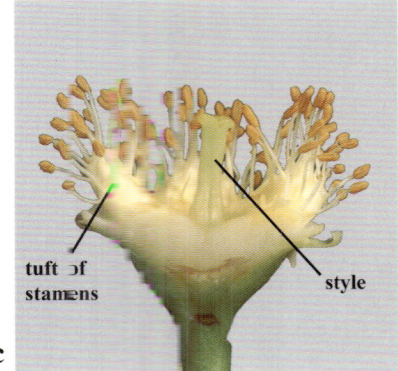

tuft of
stamens

style

Fig. 831   Floral characters of loblolly bay. a) Flower. b) CU of androecium and gynoecium. c) Profile of androecium
and gynoecium, with two tufts of stamens removed.   *Gordonia lasianthus*

Polemoniaceae - *Phlox Family*

Habit: Typically herbs, shrubs. Rarely trees, lianas.
Leaves: Alternate or opposite. Simple or compound or dissected. Pinnate venation.
Inflorescences: Various.
Flowers: Showy, radial, bisexual. Nectar disk.
Calyx: 5 sepals, connate.
Corolla: 5 petals, connate forming tube with limbs then flaring out. Convolute in bud, fold lines on petals (**plicate**).
Stamens: 5, distinct. Adnate to corolla and often attached at different levels.
Carpels: 3.
Ovary: Superior.
Placentation: Axile.
Fruit: Loculicidal capsule.
Seeds: Sticky when wet.
Other: Frequently with glandular hairs. Sometimes with bad odor
Genera: *Gilia, Ipomopsis, Linanthus, Phlox, Polemonium.*

Fig. 832   Habit of phlox.
Phlox drummondii

a   b   c

Fig. 833   Floral characters of phlox. a) Flowering branch. b) Flowers head-on. c) Bud and flower in profile.
Phlox drummondii

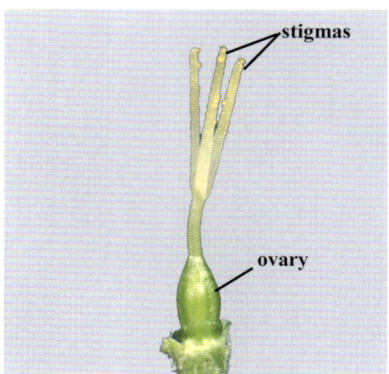

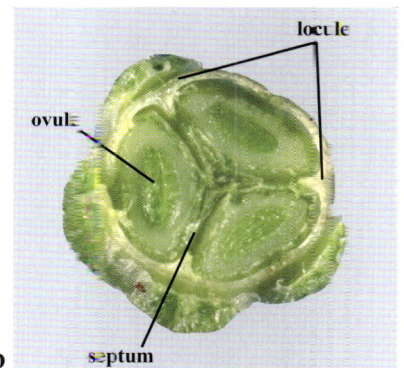

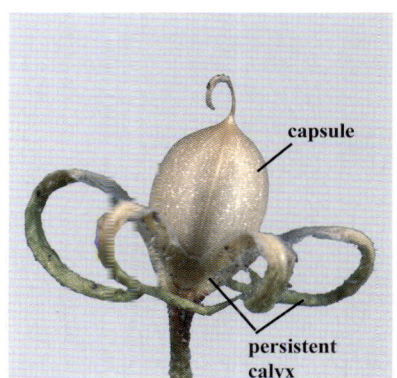

a   b   c

Fig. 834   Fruit characters of phlox. a) Gynoecium with calyx and corolla removed. b) XS of ovary. c) Capsule.
Phlox drummondii

Solanaceae - *Nightshade Family*

Habit: Herbs, shrubs, trees, vines. Frequently with prickles.
Leaves: Alternate, simple or pinnately compound or lobed, entire to serrate. Pinnate venation.
Inflorescences: Cyme, sometimes a single flower.
Flowers: Showy, radial, bisexual.
Calyx: 5 sepals, connate. Persistent and frequently enlarging with fruit development.
Corolla: 5 petals, connate and forming a corolla tube that may be funnel-shaped, bell-shaped, or rotate. Convolute in bud, fold lines on petal (**plicate**).
Stamens: 5, distinct. Adnate to corolla and alternate with petals. Open by terminal pores or slits.
Carpels: 2.
Ovary: Superior.
Placentation: Axile.
Fruit: Berry (sometimes enclosed by calyx), capsule.
Seeds: Small, numerous, flat.
Other: Many species poisonous, many species edible.
Genera: *Brugmansia, Capsicum, Datura, Lycium, Nicotiana, Petunia, Physalis, Solanum.*

Fig. 835   Flower of petunia.
Petunia cultivar

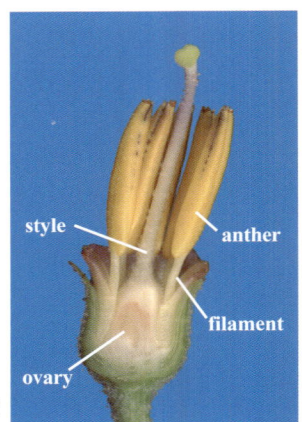

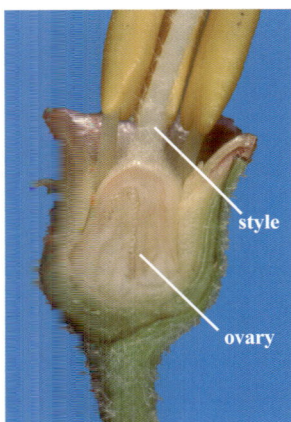

Fig. 836   Floral characters of eggplant. a) Flower. b) Flower with corolla removed. c) LS of flower with corolla removed. d) LS of ovary.   *Solanum melongena*

Fig. 837   Fruit characters of tomatillo. a) Inflated calyx surrounding berry. b) Half of calyx removed to show berry.
c) LS of fruit showing inflated calyx and berry with developing seeds.   *Physalis ixocarpa*

Fig. 838   Floral and fruit characters of tomato. a) Flower. b) Immature fruits. *Solanum lycopersicon*

Fig. 839   Floral and fruit characters of jimsonweed. a) Bud. b) Flower. c) Immature capsule. *Datura* sp.

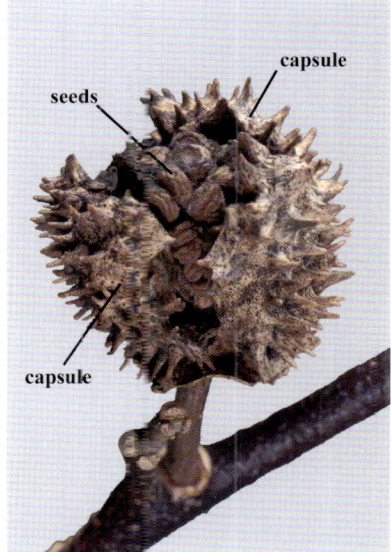

Fig. 840   Fruit characters of jimsonweed. a) LS of capsule. b) XS of capsule. c) Mature capsule dehsicing. *Datura* sp.

Convolvulaceae - *Morning Glory Family*

Habit: Typically twining vines, sometimes herbs, shrubs. Grow from rhizomes or tubers.
Leaves: Alternate, simple or compound, entire or lobed. Pinnate venation.
Inflorescences: Cyme, sometimes a single flower.
Flowers: Showy, small to large, radial, bisexual. Typically with pair of bracts or involucre.
Nectar disk.
Calyx: 5 sepals, distinct, persistent.
Corolla: 5 petals, connate with funnel-shaped tube. Convolute in bud, fold lines on petals (**plicate**).
Stamens: 5, distinct, unequal lengths. Adnate to corolla.
Carpels: 2.
Ovary: Superior.
Placentation: Axile, basal, or basal-axile.
Fruit: Typically 4-valved septifragal capsule. Also circumscissile or irregularly dehiscing capsule.
Seeds: Large, typically 4. Frequently hairy.
Other: Milky sap.
Genera: *Calystegia, Convolvulus, Cuscuta, Dichondra, Evolvulus, Ipomoea, Merremia.*

  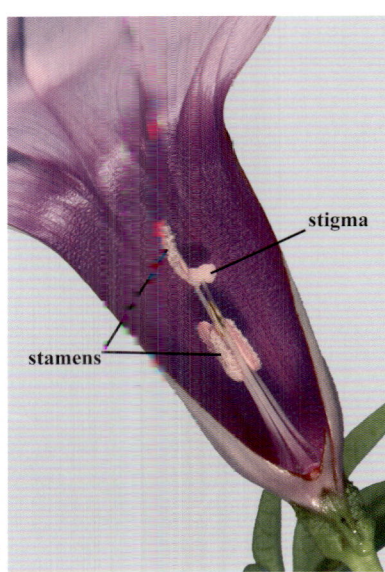

a   b   c

Fig. 841   Foliage and floral characters of morning glory. a) Leaf and bud, and twining growth habit shown. b) Flower.
c) LS of flower.   *Ipomoea cairica*

a   b   c

Fig. 842   Unfurling of morning glory flower. a) Tightly furled petals in bud. b) Partially opened flower. c) Nearly fully
opened flower.   *Ipomoea cairica*

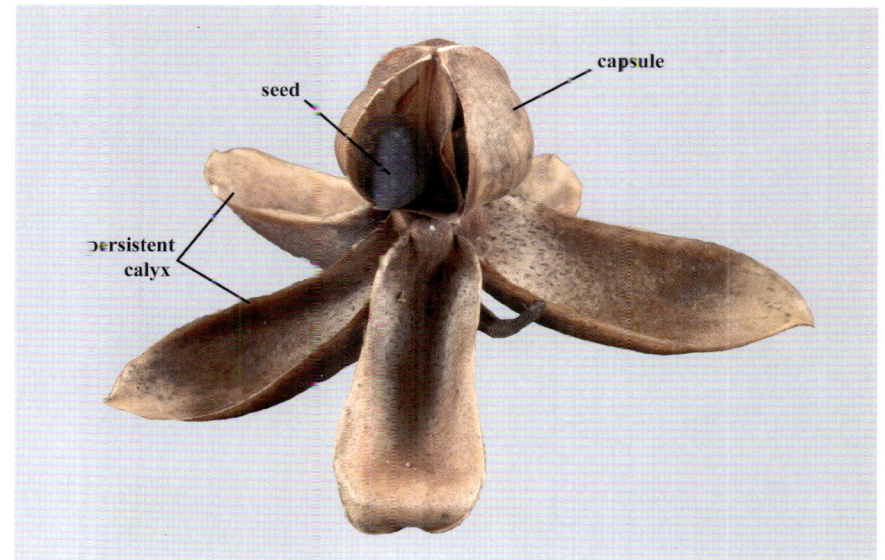

Fig. 843   Flowers and foliage of morning glory.
Ipomoea pandurata

Fig. 844   Capsule of cut-leaf morning glory with one valve removed to show seed inside.
Merremia dissecta

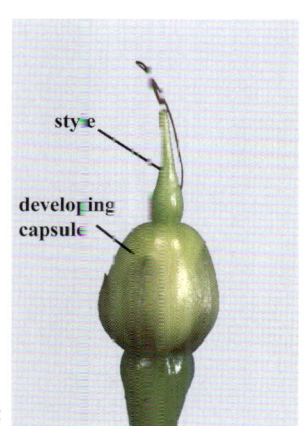

a   b   c   d

Fig. 845   Floral and fruit characters of cypress vine. a) Flower and foliage. b) Flower. c) Immature capsule. d) Mature capsule.
Ipomoea quamoclit

a   b   a   b

Fig. 846   Plant characters of pony-foot. a) Habit on ground surface. b) Flower.
Dichondra caroliniensis

Fig. 847   Plant characters of dodder, a parasitic vine. a) Habit on host plant. b) Flowers.
Cuscuta gronovii

Boraginaceae - *Borage Family*

Habit: Herbs, shrubs, trees.
Leaves: Alternate, simple, entire. Often rough.
Inflorescences: Helicoid or scorpioid cymes.
Flowers: Typically showy, radial, bisexual.
Calyx: 5 sepals, frequently persistent.
Corolla: 5 petals, connate forming corolla tube from saucer-shaped to bell-shaped. Sometimes with hairy scales or appendages in throat of corolla. Fold lines (**plicate**).
Stamens: 5, distinct. Adnate to corolla, alternate to lobes.
Carpels: 2. Sometimes style branched as in *Cordia*.
Ovary: Superior.
Placentation: Axile.
Fruit: Schizocarp yielding 4 nutlets, or drupe with 1-4 seeds. Frequently bumpy, wrinkled, or with stiff hairs.
Other: Plants rough to touch (**scabrous**).
Genera: *Borago, Cordia, Cryptantha, Cynoglossum, Heliotropium, Lithospermum, Mertensia, Myosotis, Tournefortia.*

Fig. 848   Habit of heliotrope.
Heliotropium amplexicaule

a

b

c

Fig. 849   **Floral and fruit characters of heliotrope. a) Scorpioid inflorescence. b) Flowers on partial inflorescence. c) Young developing schizocarps.**   *Heliotropium amplexicaule*

a

branched
style

b

c

Fig. 850   **Floral and fruit characters of cordia. a) Inflorescence. b) Flower. c) Drupe.**   *Cordia globosa*
Note: In *Cordia* and relatives the style is 4-cleft. In the photos above one lobe has broken off.

Rubiaceae - *Madder Family*

Habit: Herbs, lianas, shrubs, trees.

Leaves: Opposite or whorled, simple, entire. Often decussate. Pinnate venation. Interpetiolar stipules at nodes between petioles (appearing as tiny fused leaves). Adaxial surface of stipules with glands (**colleters**).

Inflorescences: Cyme, panicle, head. Sometimes 1 flower.

Flowers: Small, radial, bisexual. Frequently heterostylous. Nectar disk.

Calyx: 4-5 sepals, connate. Sometimes differentiated or with colleters.

Corolla: 4-5 petals, connate, rotate to funnel-form. Throat of corolla tube often hairy.

Stamens: 4-5 (# stamens = # petals), distinct, adnate to corolla and alternate with petals.

Carpels: 2.

Ovary: Inferior.

Placentation: Axile.

Fruit: Loculicidal or septicidal capsule, drupe, berry, or 2-segmented schizocarp.

Genera: *Cephalanthus, Coffea, Diodia, Galium, Gardenia, Hamelia, Hedyotis, Ixora, Psychotria.*

Fig. 851   Flower and foliage of innocence.   *Hedyotis procumbens*

  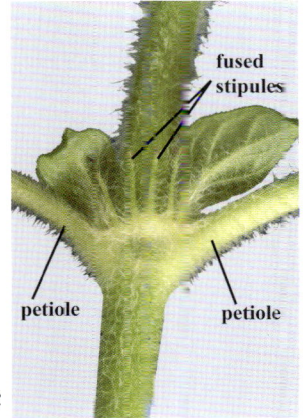 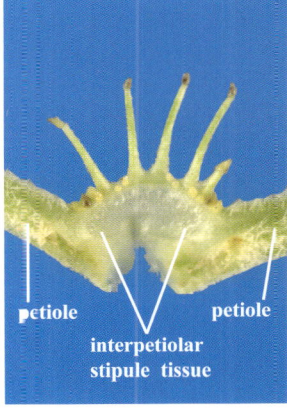

Fig. 852   Floral and stem characters of pentas. a) Inflorescence. b) Flower. c) Interpetiolar stipules. d) Interpetiolar stipules and adjoining tissue isolated.   *Pentas lanceolata*

Fig. 853   Interpetiolar stipules.
a) Buttonbush   *Cephalanthus occidentalis*
b) Firebush   *Hamelia patens*

Fig. 854   Coffee beans (drupes) in clusters on fruiting branch.   *Coffea arabica*

Apocynaceae - *Dogbane Family*

Note: Many botanists now include the Asclepiadaceae within this family.
Habit: Herbs, shrubs, trees, lianas.
Leaves: Usually opposite or whorled, simple, entire. Pinnate venation.
Inflorescences: Various, sometimes 1 flower.
Flowers: Showy, large, radial, bisexual. Sometimes with nectar glands.
Calyx: 5 sepals, connate, sometimes with appendages.
Corolla: 5 petals, connate, funnel-like or campanulate.
Frequently with appendages in corolla tube. Petals convolute in bud.
Stamens: 5, distinct, adnate to corolla & alternate with lobes.
Carpels: 2.
Ovary: Superior or half-inferior.
Placentation: Axile or parietal.
Fruit: Drupe, berry, schizocarp, pair of follicles.
Seeds: 2-numerous, sometimes with tuft of hairs.
Other: Usually milky sap. Plunger-shaped stylar head.
Genera: *Allamanda, Amsonia, Apocynum, Catharanthus,
Mandevilla, Nerium, Plumeria, Vinca.*

corollar appendages

Fig. 855   **Floral characters of oleander. a) Flower. b) CU of flower with one petal removed showing corollar appendages at base of petals. c) Flowering branch.**   *Nerium oleander*

seed

Fig. 856   **Fruit characters of oleander. a) Immature follicles. b) Mature follicle after dehiscence. c) Seed with tuft of hairs for aerial dispersal.**   *Nerium oleander*

Fig. 857 Plant characters of confederate jasmine. a) Flowers. b) Leaves and fruit. *Trachelospermum jasminoides*

Fig. 858 Madagascar periwinkle flower. *Catharanthus roseus*

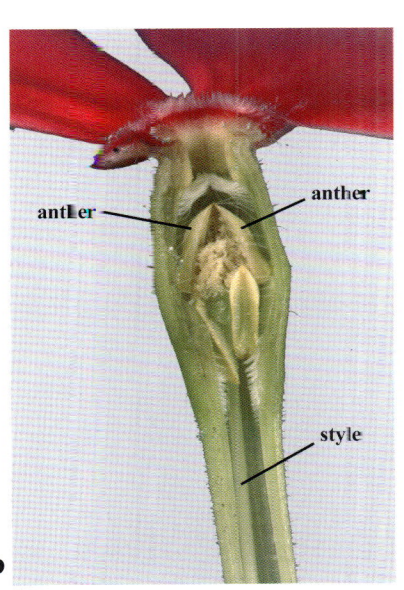

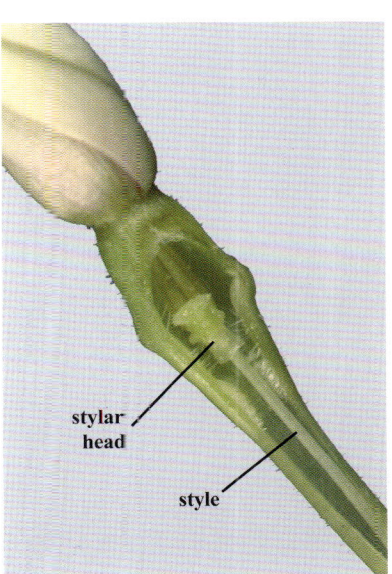

Fig. 859 Floral characters of Madagascar periwinkle. a) Bud. b) LS of upper part of flower. c) LS of upper part of flower with anthers removed to show stylar head that functions in 'plunger pollination'. *Catharanthus roseus*

Fig. 860 Fruit characters of Madagascar periwinkle. a) Fruiting branch. b) Pair of immature follicles. c) Mature follicle after dehiscence showing seeds. *Catharanthus roseus*

Asclepiadaceae - *Milkweed Family*

Note: Many botanists now include this family as part of the Apocynaceae.
Habit: Herbs, succulents, shrubs, trees, lianas.
Leaves: Opposite or whorled, simple, entire. Pinnate venation.
Inflorescences: Cyme, raceme, umbel, or sometimes 1 flower.
Flowers: Highly modified. Showy, radial, bisexual.
Calyx: 5 sepals, basally connate. Sometimes reflexed.
Corolla: 5 petals, connate. Sometimes reflexed. Typically with appendages.
Stamens: 5, connate. Adnate to stigma (collectively forming **gynostegium**), with additional 5 append-ages each with a cup-like **hood** enclosing a spur-like **horn**. Appendages collectively form the corona and are adnate to corolla and alternate with petals. Each anther produces two waxy packs of pollen (**pollinia**). Each **pollinium** is attached to the closest pollinium of neighboring anther by a connective (**translator arm**). Connective of each pollinium attached to a gland. The 2 pollinia (which originate from each of 2 adjacent anthers), 2 connectives, and 1 gland are removed as a single structure by pollinators.
Carpels: 2.
Ovary: Superior to partially inferior.
Placentation: Parietal.
Fruit: Follicle.
Seeds: Flat, with tuft of silky hair.
Other: Milky sap.
Genera: *Asclepias, Ceropegia, Cynanchum, Hoya, Matelea, Stapelia.*

 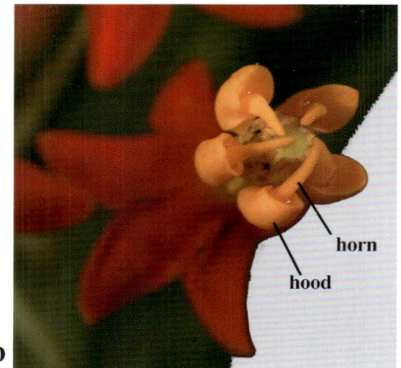 

Fig. 861 Floral characters of milkweed. a) Inflorescence. b) Flower. c) Flowering plant. *Asclepias curassavica*

   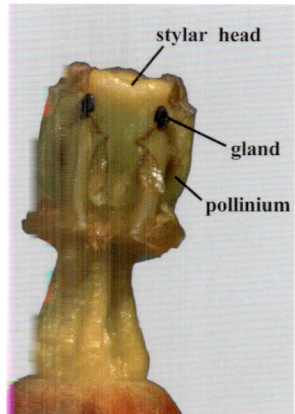

Fig. 862 Floral characters of milkweed. a) Flower. b) Flower with one hood and one horn removed. c) Flower with all hoods and horns removed. d) CU of gynoecium. *Asclepias curassavica*

Fig. 863 Fruit characters of milkweed. a) Immature follicles. b) Follicles at beginning of dehiscence. c) Follicle at full dehiscence releasing seeds with tuft of hairs for aerial dispersal.   *Asclepias curassavica*

Fig. 864 Sandhill milkweed. a) Seeds. b) Habit. *Asclepias humistrata*

Fig. 865 Partial inflorescence of butterfly weed.   *Asclepias tuberosa*

Fig. 866 Flower of giant milkweed. *Calotropis gigantea*

Fig. 867 Plant characters of hoya. a) Foliage and inflorescence. b) Flowers.   *Hoya carnosa*

Lamiaceae - *Mint Family*

Note: Two correct Latin names for family: Lamiaceae and Labiatae. Recent studies have shown that many genera traditionally placed in the Verbenaceae should be placed in the Lamiaceae.

Habit: Herbs, shrubs, trees. Stems square in x-section.

Leaves: Opposite, sometimes whorled. Simple to lobed or pinnately/palmately compound. Entire to serrate. Aromatic.

Inflorescences: Cymes clustered in pseudowhorls. Sometimes a single flower.

Flowers: Showy, bilateral, bisexual. Frequently with nectar disk.

Calyx: 5 sepals, connate, radial to bilateral. Tubular to bell-shaped.

Corolla: 5 petals, connate. Typically bilabiate with an upper lip of 2 lobes and a lower lip of 3 lobes.

Style: Slender style branches with small dot-like stigmas. Sometimes style attached near base of ovary.

Stamens: 2 or 4, distinct, adnate to corolla. Sometimes with 2 short & 2 long stamens (**didynamous**).

Carpels: 2.

Ovary: Superior.

Placentation: Axile.

Fruit: Schizocarp yielding 4 nutlets or drupe with 1-4 pits. Persistent calyx present.

Other: Many species used as spices.

Genera: *Coleus, Hyptis, Monarda, Plectranthus, Salvia, Scutellaria, Stachys, Thymus.*

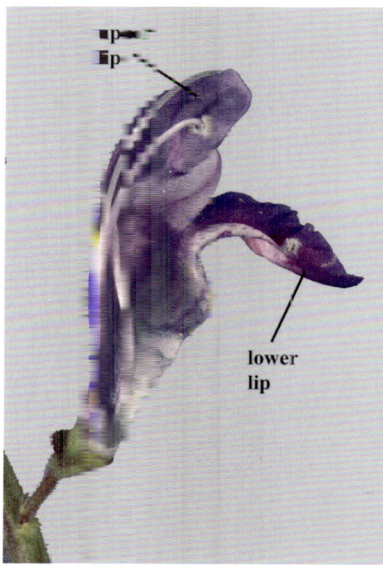

Fig. 868   **Floral characters of skullcap. a) Flowering branch. b) Flower. c) LS of flower.**
Scutellaria incana

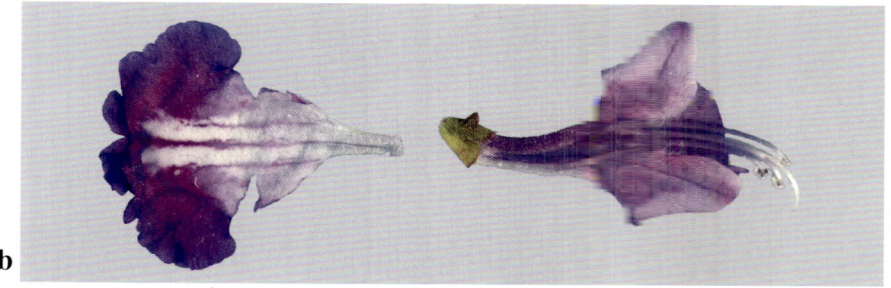

Fig. 869   **Floral characters of skullcap. a) Flower. b) Flower in expanded view. Lower lip (left) and upper lip (right).**
Scutellaria incana

Fig. 870 Floral characters of horsemint. a) Flowering branch. b) Inflorescences in pseudowhorls. c) Flower showing pistil with slender style branches. *Monarda punctata*

Fig. 871 Floral and fruit characters of bluecurls. a) Flower. b) Fruiting branch. c) Schizocarp with four nutlets that are slightly fused at the base. *Trichostema dichotomum*

Fig. 872 Fruit characters of salvia. a) Fruiting branch. b) Infructescences in pseudowhorl arrangement. c) Schizocarp. d) LS of schizocarp. *Salvia coccinea*

Verbenaceae - *Verbena Family*

Habit: Herbs, shrubs; sometimes trees, lianas. Often with prickles & thorns. Stem square in x-section.

Leaves: Opposite, decussate, sometimes whorled. Simple, entire or lobed or serrate. Pinnate venation. Aromatic.

Inflorescences: Raceme, spike, head, sometimes with involucre.

Flowers: Showy, bilateral, bisexual. Frequently with nectar disk.

Calyx: 5 sepals, connate, lobed or toothed. Tubular or bell-shaped, persistent. Sometimes enlarging on fruit.

Corolla: 5 petals, connate, with narrow tube and spreading lobes. Typically bilabiate with an upper lip of 2 lobes and a lower lip of 3 lobes.

Style: Stigma thick, expanded.

Stamens: 4, distinct, adnate to corolla. Didynamous with 2 stamens long and 2 short.

Carpels: 2.

Ovary: Superior.

Placentation: Axile.

Fruit: Schizocarp yielding 2 or 4 nutlets, or drupe with 2 or 4 pits. Persistent calyx present.

Genera: *Duranta, Glandularia, Lantana, Lippia, Stachytarpheta, Verbena.*

Note: Several genera traditionally placed here are now in the Lamiaceae, such as *Callicarpa, Clerodendrum, Holmskioldia, Vitex.*

a   b   c

Fig. 873   Floral and fruit characters of lantana. a) Flowering branch. b) Flowering branch from above showing decussate leaves. c) Infructescence of developing drupes.   *Lantana camara*

a   b   c

Fig. 874   Floral characters of lantana. a) Inflorescence or head. b) LS of inflorescence or head. c) CU lateral view of flower.   *Lantana camara*

Fig. 875　Floral characters of glandularia. a) Inflorescence. b) Flower c) Flowers in profile.
Glandularia pulchella

Fig. 876　Floral characters of match heads. a) Inflorescence with leaves. b) Inflorescence. c) Inflorescence.
Lippia nodiflora

  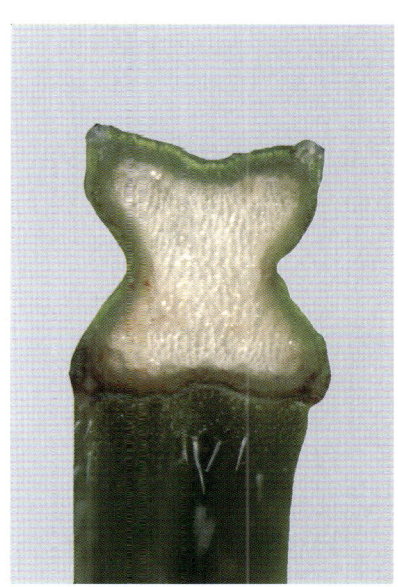
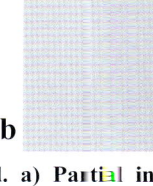

Fig. 877　Floral characters of porterweed. a) Partial inflorescence.
b) Flower in profile.　　*Stachytarpheta mutabilis*

Fig. 878　XS of stem of bitter mint.
Hyptis mutabilis

Oleaceae - *Olive Family*

Habit: Shrubs, trees, lianas.
Leaves: Opposite, simple or trifoliate or pinnately compound.
Entire to serrate. Pinnate venation.
Inflorescences: Cyme, panicle, raceme.
Flowers: Usually small, sometimes showy.
Radial, bisexual or unisexual. Commonly with nectar disk.
Calyx: 4 sepals, connate.
Corolla: 4 petals, connate.
Stamens: 2, distinct, short. Adnate to corolla.
Carpels: 2.
Ovary: Superior.
Placentation: Axile.
Fruit: Berry, drupe, samara, loculicidal or circumscissile capsule.
Other: Peltate scales sometimes present.
Genera: *Chionanthus, Forestiera, Forsythia, Fraxinus, Jasminum,
Ligustrum, Olea, Osmanthus*.

Fig. 879   New leaves of ash.
Fraxinus sp.

a

b

c

Fig. 880   Floral and fruit characters of ash. a) Staminate inflorescences. b) Carpellate inflorescences.
c) Infructescences of samaras.   *Fraxinus* sp.

a

b

c

Fig. 881   Floral and fruit characters of ash. a) Staminate flowers. b) Carpellate flowers. c) Samaras.
Fraxinus sp.

Fig. 882   Floral and fruit characters of glossy privet. a) Flowering branch. b) Flowers. c) Partial infructescence of drupes.   *Ligustrum lucidum*

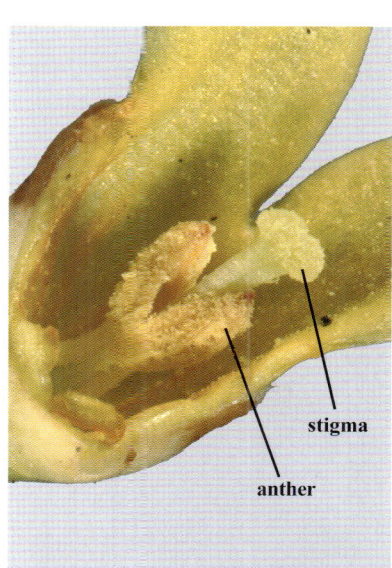

Fig. 883   Floral characters of forsythia. a) Flowers. b) LS of flower. c) CU of gynoecium and androecium.
***Forsythia* sp.**

Fig. 884   Flower and fruit characters of fringe tree. a) Flowers. b) Drupes.
Chionanthus retusus

Fig. 885   Flowers of wild olive.
Osmanthus americanus

Scrophulariaceae - *Snapdragon Family*

Habit: Herbs, shrubs. Alternate or opposite, simple to pinnately compound, entire or lobed.

Inflorescences: Various, sometimes 1 flower.

Flowers: Showy, radial, bisexual. Commonly with nectar disk.

Calyx: 4-5 sepals, connate, persistent.

Corolla: 4-5 petals, connate, sometimes nectar spur at base. Rotate or bell-shaped or bilabiate with 2 upper lobes and 3 lower lobes.

Stamens: 4, distinct, adnate to corolla. Didynamous (2 long / 2 short).

Carpels: 2.

Ovary: Superior.

Placentation: Axile or parietal.

Fruit: Septicidal capsule with persistent calyx.

Seeds: Small, numerous, angular.

Genera: *Antirrhinum, Digitalis, Linaria, Penstemon, Scrophularia, Veronica.*

Note: The genera illustrated here are traditional Scrophulariaceae, but are now placed by some in the Veronicaceae and the Plantaginaceae.

Fig. 886   Flowering snapdragon. *Antirrhinum majus*

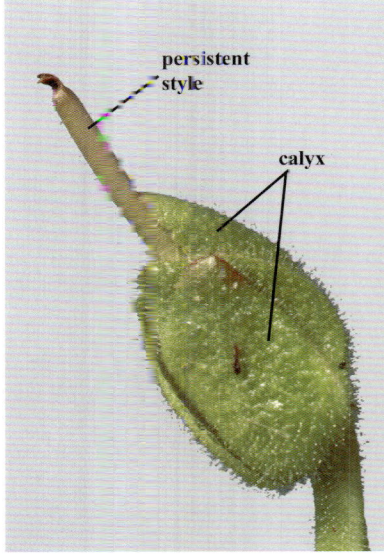

Fig. 887   **Floral characters of snapdragon. a) Flower. b) Flower in profile. c) Developing fruit showing persistent calyx.** *Antirrhinum majus*

Fig. 888   **Fruit characters of snapdragon. a) XS of immature capsule. b) LS of immature capsule. c) Mature capsule. d) LS of mature capsule.**   *Antirrhinum majus*

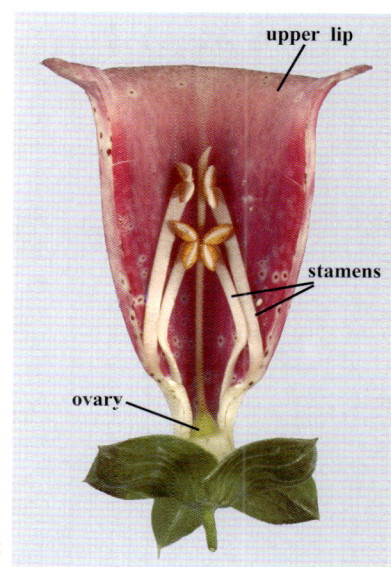

Fig. 889   Floral characters of foxglove. a) Flowering branch. b) Flower and bud. c) Upper lip of flower showing four adnate didynamous stamens.   *Digitalis purpurea*

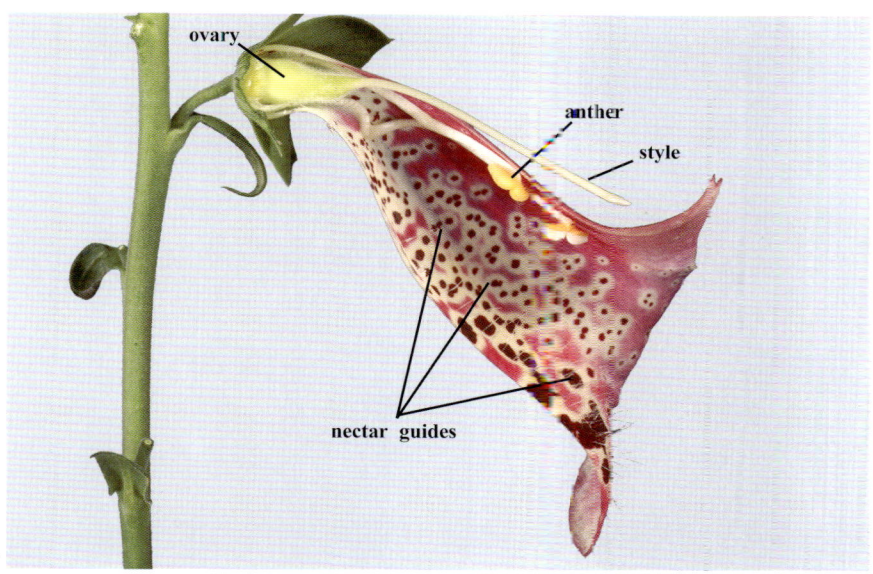

Fig. 890   Floral and fruit characters of foxglove. a) LS of flower. b) Immature capsules (lower one with part of calyx removed).   *Digitalis purpurea*

Fig. 891   Floral characters of bacopa. a) Flowering branch. b) Flower.   *Bacopa careliniana*
c) Flower.   *Bacopa innominata*

Orobanchaceae - *Broomrape Family*

Note: Traditionally included within the Scrophulariaceae or restricted to the non-photosynthetic holoparasites.

Habit: Fleshy parasitic herbs, attach to roots of hosts. Completely lacking chlorophyll or green hemiparasites.

Leaves: Alternate, simple, frequently lobed. Sometimes reduced and scale-like.

Inflorescences: Spike, raceme, corymb, sometimes 1 flower.

Flowers: Bilateral, bisexual. Nectar disk present.

Calyx: 5 sepals, connate.

Corolla: 5 petals, connate, bilabiate. Corolla tube typically curved.

Stamens: 4, distinct, adnate to corolla. 2 long / 2 short (**didynamous**). Sometimes 1 staminode.

Carpels: 2.

Ovary: Superior.

Placentation: Axile or parietal.

Fruit: Loculicidal capsule.

Seeds: Small, numerous, angular.

Other: Roots modified into a haustorium to invade host plant's roots.

Genera: *Castilleja, Conopholis, Orobanche, Pedicularis*.

Fig. 892   Squawroot inflorescences. *Conopholis americana*

flower

Fig. 893   Plant characters of squawroot. a) Inflorescences as they appear on the forest floor. b) Partial inflorescence showing buds and flowers.   *Conopholis americana*

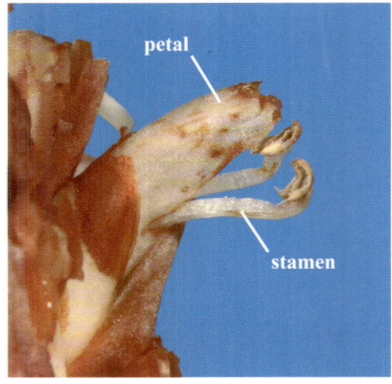

petal

stamen

host root

Fig. 894   Plant characters of squawroot. a) Flower. b) Young inflorescence recently emerged from ground. c) Plant still attached to host root.   *Conopholis americana*

Plantaginaceae - *Plantain Family*

Note: Many botanists now include in this family most genera of autotrophic plants that traditionally were in the Scrophulariaceae.
Habit: Herbs.
Leaves: Basal rosettes, simple, entire to serrate. Oval to lanceolate. Parallel venation.
Inflorescences: Terminal spike or head on long slender scape.
Flowers: Inconspicuous, small, radial, bisexual. Each with a bract.
Calyx: 4 sepals, distinct, persistent.
Corolla: 4 petals, connate. Dry and papery.
Stamens: 4, distinct, adnate to corolla.
Carpels: 2.
Ovary: Superior.
Placentation: Axile.
Fruit: Membranous circumscissile capsule.
Seeds: Sticky when wet.
Genera: *Littorella, Plantago.*

Fig. 895   Habit of plantain in flower.
Plantago virginica

Fig. 896   Plant characters of English plantain. a) Leaves. b) Habit of flowering plant.
Plantago lanceolata

Fig. 897   Floral characters of English plantain. a) Inflorescence head. b) CU of inflorescence with receptive stigmas. c) CU of inflorescence with stamens.   *Plantago lanceolata*

Bignoniaceae - *Trumpet Vine Family*

Habit: Shrubs, trees, lianas.

Leaves: Opposite or whorled (rarely alternate), simple or compound, entire to serrate. Pinnate or palmate venation. Sometimes leaflets are modified into hooks or tendrils. Sometimes glands at base of petioles.

Inflorescences: Cyme, raceme, panicle, sometimes 1 flower.

Flowers: Showy, large, bilateral, bisexual. Nectar disk.

Calyx: 5 sepals, connate.

Corolla: 5 petals, connate, bell-shaped to bilabiate.

Stamens: 4, distinct, adnate to corolla. Didynamous (2 long / 2 short). 1 staminode.

Carpels: 2.

Ovary: Superior.

Placentation: Axile.

Fruit: Loculicidal or septicidal capsule with 2 valves. Typically elongate and woody. Sometimes an indehiscent pod or berry.

Seeds: Numerous, flat, winged.

Genera: *Bignonia, Campsis, Catalpa, Macfadyena, Tabebuia.*

Fig. 898   Inflorescence of cape honeysuckle.   *Tecoma capensis*

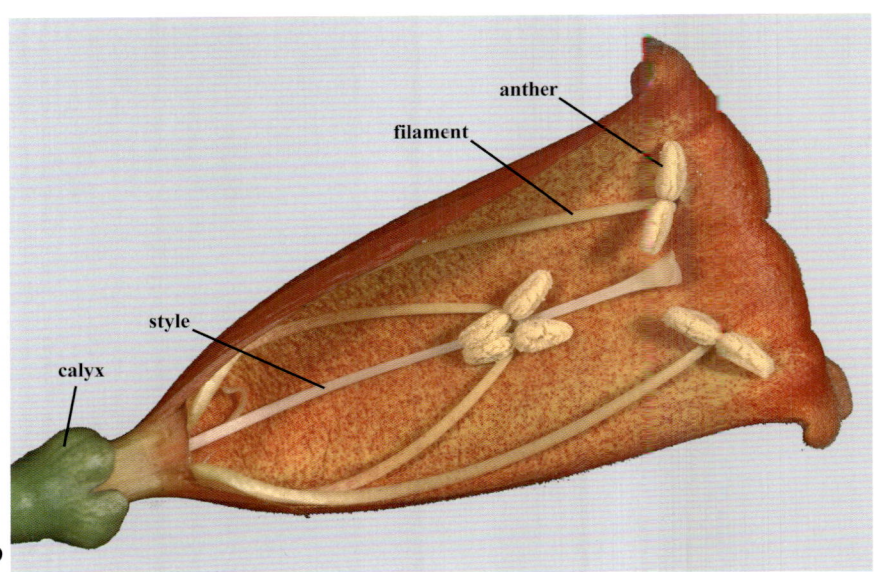

F g. 899   Floral characters of cross vine. a) Inflorescence. b) Upper lip of flower showing four adnate and didynamous stamens.   *Bignonia capreolata*

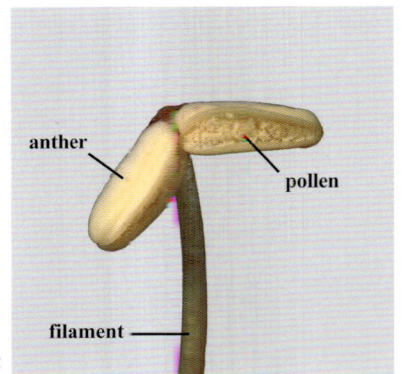

Fig. 900   Floral characters of cross vine. a) Flower. b) Upper lip of flower with adnate stamens attached. c) Tip of stamen showing anther and pollen.   *Bignonia capreolata*

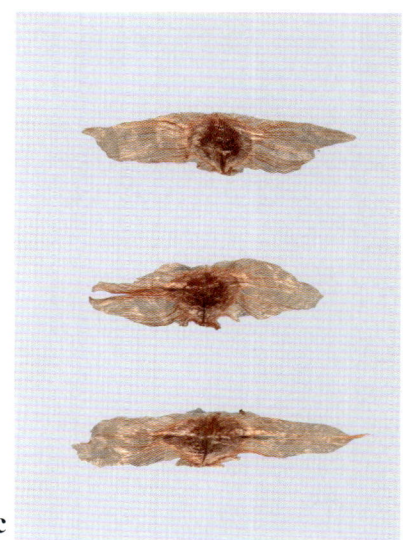

a   b   c

Fig. 901   Fruit characters of trumpet creeper.   a) Immature capsules. b) Mature capsule that has dehisced. c) Winged seeds.   *Campsis radicans*

a   b   c

Fig. 902   Floral and leaf characters of cat's claw vine.   a) Flower head-on. b) Flower in profile. c) Leaves and tendrils. *Macfadyena unguis-cati*

a   b

Fig. 903   Floral and fruit characters of catalpa. a) Inflorescence. b) Mature dehisced fruits. *Catalpa bignonioides*

Acanthaceae - *Acanthus Family*

Habit: Herbs, shrubs, vines.

Leaves: Opposite, simple or lobed, entire to dentate. Decussate. Pinnate venation.

Inflorescences: Various, sometimes 1 flower.

Flowers: Showy, bilateral, bisexual. Frequently with conspicuous or colorful bracts. Nectar disk.

Calyx: 4-5 sepals, connate, persistent.

Corolla: 5 petals, connate, bilabiate.

Stamens: 4 with 1 staminode (**didynamous**), or 2 with 2 staminodes, distinct, adnate to petals. Anthers frequently asymmetrical and/or hairy.

Carpels: 2.

Ovary: Superior.

Placentation: Axile.

Fruit: Loculicidal capsule with 2 valves, explosively dehiscent. Inside of each valve has small hooks or projections (**jaculators** or **retinacula**) supporting seeds. (See Fig. 235 and 905-b).

Genera: *Aphelandra, Justicia, Ruellia, Strobilanthes, Thunbergia.*

Fig. 904   Flowering branch of wild petunia.   *Ruellia caroliniensis*

Fig. 905   Floral and fruit characters of wild petunia. a) Flower. b) Mature capsule after dehiscence. *Ruellia caroliniensis*

Fig. 906   Floral characters of shrimp plant. a) Inflorescences. b) Inflorescence with flower. *Justicia brandegeana*

Lentibulariaceae - *Bladderwort & Butterwort Family*

Habit: Carnivorous herbs. Found in moist habitats or as free-floating aquatics.

Leaves: Highly modified according to prey capture. In floating *Utricularia*, highly divided with numerous small bladders that trap tiny aquatic organisms. In terrestrial *Pinguicula*, a basal rosette of flattened, mucous-covered leaves with edges rolling in.

Inflorescences: Raceme or solitary flower. Borne on a scape.

Flowers: Showy, bilateral, bisexual.

Calyx: 4-5 sepals, 2-lobed in *Utricularia*.

Corolla: 5 petals, connate, 2-lipped. Lower lip with sac or nectar spur, and frequently with rounded 'hump' or projection that closes entrance to corolla throat (corolla type = personate).

Stamens: 2, distinct, adnate to corolla.

Carpels: 2.

Ovary: Superior.

Placentation: Free-central.

Fruit: Capsule (loculicidal, circumscissile, irregularly dehiscent).

Other: Glandular hairs produce mucilage and/or digestive enzymes. Trapped prey often visible in bladders or on leaves.

Genera: *Genlisea, Pinguicula, Utricularia.*

Fig. 907   **Foliage and floral characters of butterwort. a) Sticky insect-trapping leaves. b) Flower.** *Pinguicula primuliflora*

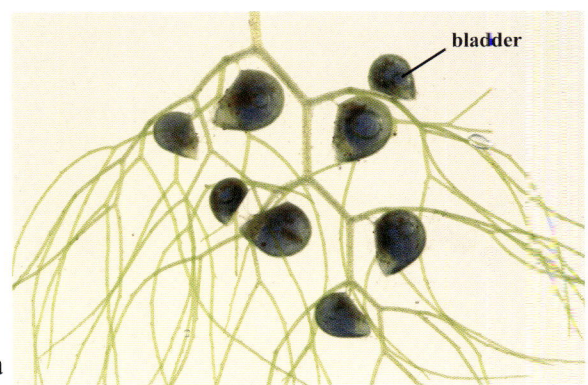

Fig. 908   **Foliage and floral characters of bladderwort. a) Underwater leaves with small dark insect-trapping bladders. b) Flower.**   *Utricularia radiata*

Aquifoliaceae - *Holly Family*

a   *I. vomitoria*

b   *I. cassine*

c   *I. opaca*

Habit: Shrubs, trees.

Leaves: Alternate, simple, entire to serrate. Often spiny and/or leathery. Pinnate venation. Stipules usually small, dark, and triangular.

Inflorescences: Cyme, raceme, panicle, sometimes 1 flower.

Flowers: Small, radial, unisexual.

Calyx: 4-6 sepals, basally connate.

Corolla: 4-6 petals, distinct or basally connate.

Style: Very short, stigma capitate or lobed and more or less sessile.

Stamens: 4-6. Distinct or basally adnate to corolla, alternate with petals.

Carpels: 4-6.

Ovary: Superior.

Placentation: Axile.

Fruit: Drupe (typically brightly colored), with 4-6 pits. Button-like stigma remains on fruit.

Other: Style short, stigma capitate or lobed.

Genera: *Ilex* (monogeneric family).

Fig. 909 Holly leaf diversity. *Ilex* spp. (Not to same scale.)

a

b

drupe
c

Fig. 910 Floral and fruit characters of Burford holly. a) Flowering branch. b) Immature fruits. c) Mature drupes. *Ilex cornuta* 'burfordii'

ovary
stigma
a

staminode
b

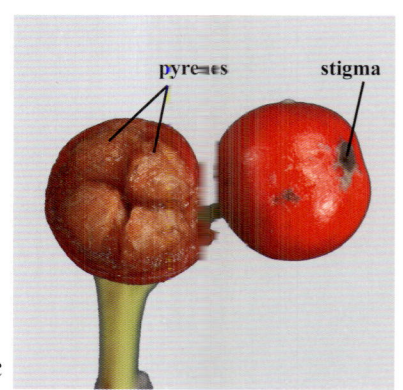
pyrenes   stigma
c

Fig. 911 Floral and fruit characters of Burford holly. a) Flowers. b) Flower and immature drupes. c) Mature drupe dissected to show four pyrenes inside. (Note sessile stigma.)   *Ilex cornuta* 'burfordii'

Pittosporaceae - *Pittosporum Family*

Habit: Shrubs, small trees, lianas.
Leaves: Alternate, simple, entire. Evergreen and leathery.
Inflorescences: Cymes.
Flowers: Radial, perfect.
Sepals: 5.
Petals: 5.
Stamens: 5.
Carpels: 2.
Ovary: Superior.
Placentation: Parietal.
Fruit: Loculicidal capsule or berry.
Genera: *Pittosporum.*

Fig. 912   Flowering branch of pittosporum.   *Pittosporum tobira*

Fig. 913   Floral characters of pittosporum. a) Flowers. b) Flower with petal removed to show pistil.
Pittosporum tobira

Fig. 914   Fruit characters of pittosporum. a) Fruiting branch. b) XS of fruit.
Pittosporum tobira

Apiaceae - *Parsley Family*

Habit: Herbs.
Leaves: Alternate or basal, simple or lobed or compound.
Inflorescences: Umbel (compound or simple).
Flowers: Small, radial, bisexual.
Calyx: 5 sepals, distinct. Often reduced or absent.
Corolla: 5 petals, distinct.
Stamens: 5, distinct, alternate with petals.
Carpels: 2.
Ovary: Inferior.
Placentation: Apical-Axile.
Fruit: Schizocarp with 2 segments (mericarps) connected by a common stalk. Mericarps variously winged, ribbed, or spiny.
Other: Frequently aromatic. Many edible, some poisonous.
Genera: *Apium, Chaerophyllum, Cicuta, Conium, Coriandrum, Daucus, Eryngium, Foeniculum, Hydrocotyle, Thaspium, Zizia.*
Note: Some botanists include the closely related Araliaceae here as part of the Apiaceae.

Fig. 915   Umbel of marsh penny-wort.   *Hydrocotyle umbellata*

Fig. 916   Floral characters of fennel. a) Compound umbel. b) CU of single umbel. c) CU of flowers.
Foeniculum vulgare

Fig. 917   Fruit characters of hairy-fruit chervil. a) Immature schizocarp. b) Nearly mature schizocarp.
c) Mature schizocarp with merocarps separated. d) Umbel of mature schizocarps.   *Chaerophyllum tainturieri*

Araliaceae - *Ginseng Family*

Habit: Herbs, shrubs, trees, lianas.

Leaves: Often large, alternate, simple or lobed or pinnately/palmately compound. Bases sheathing.

Inflorescences: Umbels or heads.

Flowers: Small, radial, bisexual or unisexual. Nectar disk present.

Calyx: 5 sepals, distinct, often reduced to small teeth.

Corolla: 5 petals, distinct.

Stamens: 5, distinct, alternate with petals.

Carpels: 2-15.

Ovary: Inferior.

Placentation: Apical-Axile.

Fruit: Drupe, berry.

Genera: *Aralia, Fatsia, Hedera, Panax, Schefflera.*

Note: This family is sometimes included in the Apiaceae.

Fig. 918   Inflorescence and foliage of ivy.   *Hedera* sp.

Fig. 919   Floral characters of ivy. a) Portion of umbel. b) CU of flowers. *Hedera* sp.

Fig. 920   Plant characters of the devil's walking stick. a) Single flowering umbel from very large inflorescence. b) Immatue drupes. c) CU of developing drupes. d) Mature drupes.   *Aralia spinosa*

Caprifoliaceae - *Honeysuckle Family*

Habit: Herbs, shrubs, small trees, lianas.

Leaves: Opposite, simple or pinnately compound, entire to serrate. Decussate. Pinnate venation.

Inflorescences: Various, sometimes paired or 1 flower.

Flowers: Showy, bilateral, bisexual.

Calyx: 5 sepals, connate, often small lobes or teeth.

Corolla: 5 petals, connate. Tubular to bell-shaped to bilabiate. Often with basal nectar spur or glandular hairs in base of corolla.

Stamens: 4-5, distinct, adnate to corolla.

Carpels: 2-5.

Ovary: Inferior.

Placentation: Axile.

Fruit: Berry, capsule, drupe, achene.

Other: Style long, stigma capitate.

Genera: *Abelia, Dipsacus, Lonicera, Symphoricarpos, Valeriana.*

Note: The genera *Sambucus* and *Viburnum* were traditionally treated here. Some botanists treat *Dipsacus* and *Valeriana* as separate families.

Fig. 921   Flowering branch of Japanese honeysuckle. *Lonicera japonica*

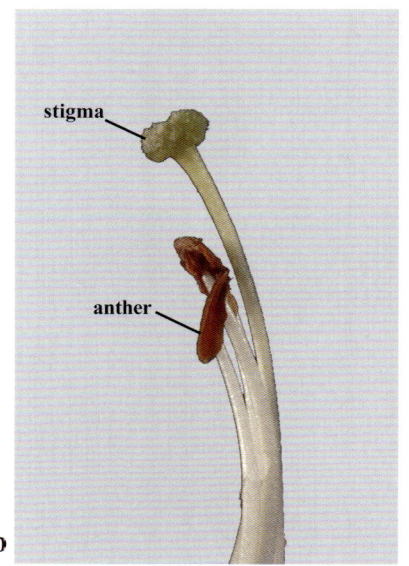

a

b

c

Fig. 922   Floral and fruit characters of Japanese honeysuckle. a) Flower. b) CU of stigma and two stamens. c) Developing fruits.   *Lonicera japonica*

a

b

c

d

Fig. 923   Floral and fruit characters of coral honeysuckle. a) Flowering branch. b) LS of flower. c) Young fruits shortly after petals have fallen. d) Immature berries.   *Lonicera sempervirens*

Adoxaceae - *Elderberry Family*

Habit: Herbs, shrubs, trees.

Leaves: Opposite, simple or pinnately or bipinnately compound, entire to serrate. Pinnate/palmate venation

Inflorescences: Umbel-like or panicle-like cymes.

Flowers: Small, radial, bisexual.

Calyx: 3-5 sepals.

Corolla: 4-5 petals, connate.

Stamens: 4-5, distinct, adnate to corolla.

Carpels: 3-5.

Ovary: Partially inferior.

Placentation: Axile.

Fruit: Drupe with 1-5 pits.

Genera: *Adoxa, Sambucus, Viburnum*

Note: The genera *Sambucus* and *Viburnum* are traditionally treated in the Caprifoliaceae.

Fig. 924   Elderberry flowers. *Sambucus canadensis*

Fig. 925   Floral and fruit characters of elderberry. a) Flowering branch. b) Fruits. *Sambucus canadensis*

stamen

Fig. 926   Floral characters of viburnum. a) Inflorescence. b) Flower. c) LS of flower. *Viburnum* sp.

Campanulaceae - *Bellflower Family*

Habit: Herbs, sometimes shrubs.
Leaves: Alternate, simple to lobed, entire to serrate. Pinnate venation.
Inflorescences: Various, sometimes 1 flower.
Flowers: Showy, radial to bilateral, bisexual.
Hypanthium present. Nectar disk present.
Calyx: 5 sepals, connate.
Corolla: 5 petals, connate. Bell-shaped (**campanulate**)
or tubular or bilabiate.
Stamens: 5, distinct or connate. Anthers form tube through which
the style grows before spreading apart (**plunger pollination**).
Carpels: 2-5.
Ovary: Inferior or half-inferior.
Placentation: Axile.
Fruit: Poricidal or loculicidal capsule, berry.
Seeds: Numerous, sometimes winged.
Genera: *Campanula, Cyanea, Lobelia, Triodanis, Wahlenbergia.*
Other: Milky sap.

Fig. 927   Flowering plant of lobelia.
Lobelia siphilitica

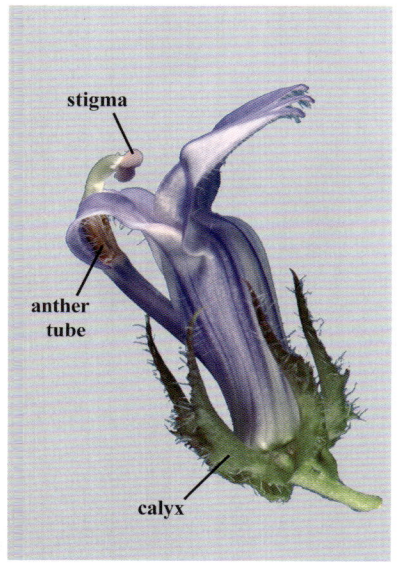

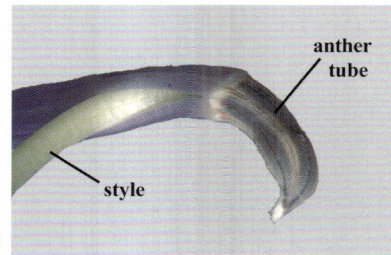

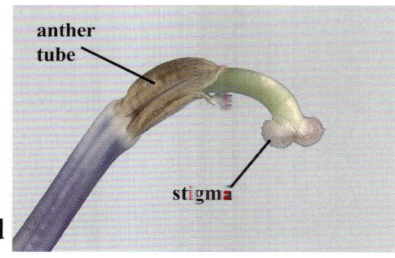

Fig. 928   Floral characters of lobelia. a) Flower. b) Two androecia / gynoecia. c) CU of anther tube formed by connate anthers. d) Expanded stigma protruding through anther tube after dehiscence of anthers.
Lobelia siphilitica

Fig. 929   Floral and fruit characters of Venus' looking-glass. a) Flowering branch. b) Flower. c) Poricidal capsule.
Triodanis perfoliata

Asteraceae - *Sunflower Family*

Note: Two correct Latin names for family: Asteraceae and Compositae.
Subfamilies: Cichorioideae, Asteroideae. (A third subfamily, the Barnadesioideae, is also recognized by some. It consists of primitive woody plants and will not be treated further.) Recent phylogenetic studies are challenging the traditional subfamilial classification and indicate that the taxonomy is much more complex than the three subfamilies mentioned here.
Habit: Herbs, shrubs, trees.
Leaves: Alternate, opposite, whorled, basal rosette. Simple to highly dissected. Pinnate or palmate venation.
Inflorescences: Showy, heads of clustered specialized flowers with an involucre of bracts (**phyllaries**).
Flowers: Showy, small, radial or bilateral, bisexual or unisexual. (See Aster Floral Morphology.)
Calyx: When present, sepals modified into a pappus which may be papery, hairy, bristly, or scaly.
Corolla: 5 connate petals forming a tubular disk flower or a single flat petal with 5 apical teeth forming a ray or ligulate flower.
Stamens: 5, connate at anthers (**syngenesious**) forming tube around style.
Carpels: 2.
Ovary: Inferior.
Placentation: Basal.
Fruit: Achene with modified persistent calyx (**pappus**). Winged or flat or spiny.
Seed: 1.
Other: Resinous or milky sap.

Floral Morphology

The inflorescences of the Asteraceae are densely packed heads of many small flowers. The individual flowers or florets are divided into disk florets, ray florets, and ligulate florets. In association with most florets is a greatly reduced and modified calyx called a pappus. Each inflorescence is subtended by a protective involucre of bracts called phyllaries. The different florets combine to form different types of heads.

disk floret - Small tubular radial bisexual flower consisting of 5 connate petals with short lobes. Stamens are present and connate at the anthers.
ray floret - Bilateral, sterile or carpellate flower with corolla flat and strap-like with 2-3 apical lobes or teeth. No stamens present.
ligulate floret - Bilateral, bisexual flower with corolla flat and strap-like with 5 apical lobes or teeth.
discoid head - Inflorescence consisting entirely of disk florets.
ligulate head - Inflorescence consisting entirely of ligulate florets.
radiate head - Inflorescence consisting of both disk and ray florets.
pappus - Reduced and modified calyx that may take the form of hairs, bristles, spines, scales, or papery or feathery structures used in seed dispersal.
pseudanthium - Many flowers that collectively function as a single flower with all reproductive parts. In a radiate head, the involucre is analagous the the calyx, the ray florets are collectively analagous to the corolla, and the disk florets are collectively analagous to the pistil and stamens.

Aster Floral Morphology

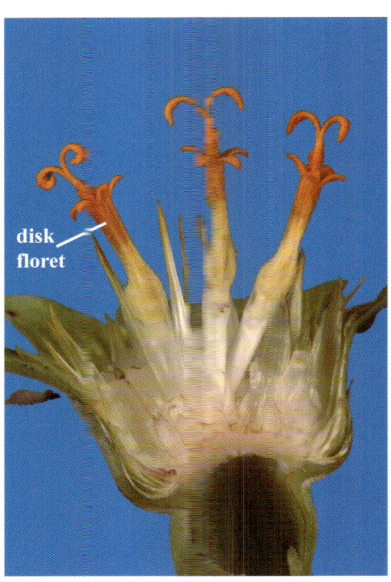

Fig. 930　RADIATE HEAD　The head of the Mexican sunflower is composed of both disk and ray florets. a) Radiate head. b) CU showing centrally located disk florets. c) LS of flower with only three disk florets remaining. *Tithonia* sp.

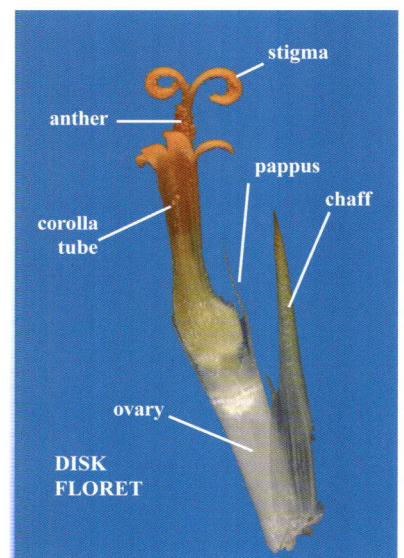

Fig. 931　Disk floret of Mexican sunflower.　*Tithonia* sp.

Fig. 932　Ray floret of Mexican sunflower.　*Tithonia* sp.

Fig. 933　Radiate head of Spanish needles.　*Bidens alba*

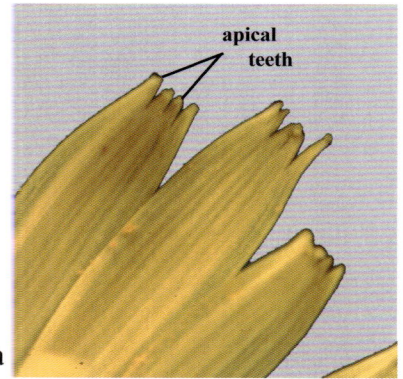

Fig. 934　LIGULATE HEAD　a) Apical teeth on ligulate florets of false dandelion. b) Falsely ligulate head of Stokes' aster. c) Falsely ligulate floret.　*Stokesia laevis*

Pyrrhopappus carolinianus

Aster Floral Morphology

Fig. 935   DISCOID HEAD   The head of lavender paintbrush is composed entirely of disk florets. a) & b) Portion of inflorescence showing disk florets with long slender style branches protruding. c) Buds and flowers.
Carphephorus corymbosus

**Fig. 936   Discoid head of tassel flower. a) Intact inflorescence.
b) Inflorescence with part of calyx removed to show base of disk florets.**
Emilia fosbergii

Fig. 937   Discoid head of thistle.
Cirsium horridulum

Fig. 938   Floral characters of goldenrod. a) Inflorescence of disk florets. b) CU of single head.
Solidago sp.

Subfamily Cichorioideae

Inflorescences: Heads either discoid (composed of all disk florets) or ligulate (composed of all ligulate florets). (Limb or strap of ligulate florets with 5 apical teeth or lobes.) (Note in *Vernonia* and relatives the flowers are falsely ligulate.)

Stigma: Style branches with entire inner surfaces stigmatic or receptive.

Sap: Either resinous or milky.

Genera: *Cichorium, Cirsium, Pyrrhopappus, Sonchus, Stokesia, Taraxacum, Vernonia, Youngia.*

Fig. 939   Plant characters of false hawksbeard. a) Milky sap. b) Inflorescence. c) Ligulate head.
Youngia japonica

Fig. 940   Floral and fruit characters of false dandelion. a) Ligulate head. b) Infructescence of achenes. c) Achene with tuft of hair for aerial dispersal.
Pyrrhopappus carolinianus

  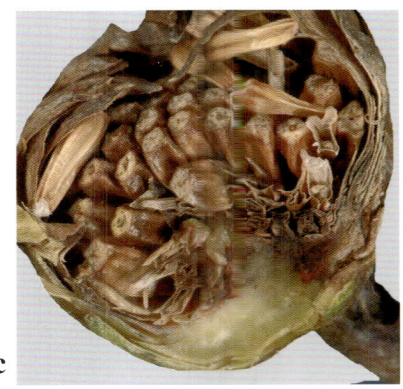

Fig. 941   Floral and fruit characters of Stokes' aster. a) Falsely ligulate head. b) LS of head after petals have dehisced and showing developing achenes. c) Mature achenes on head.
Stokesia laevis

Subfamily Asteroideae

Inflorescences: Heads radiate (with both disk florets and ray florets) or discoid (with only disk florets). (Limb or strap of ray floret with 2-3 apical lobes or teeth.)
Stigma: Style branches each have receptive area restricted to 2 marginal stigmatic lines.
Sap: Only resinous, never milky.
Genera: *Artemisia, Aster, Bidens, Chrysanthemum, Coreopsis, Eupatorium, Helianthus, Liatris, Mikania, Senecio, Solidago, Zinnia.*

 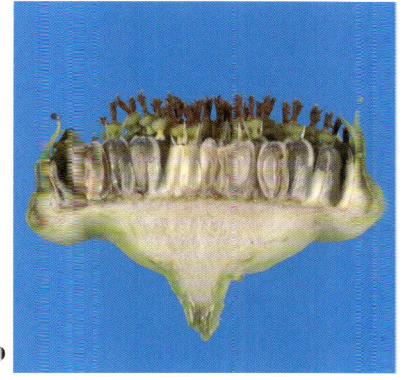 

Fig. 942 Floral and fruit characters of sunflower. a) Radiate head. b) LS of head after petals have dehisced and fruits have begun to develop. c) CU of developing fruits on head.
Helianthus annuus

Fig. 943 Fruit characters of sunflower. a) Single floret in fruit. b) LS of achene showing pericarp and seed.
Helianthus annuus

Fig. 944 Radiate head of wedelia. *Wedelia trilobata*

Fig. 945 Floral and fruit characters of Spanish needles. a) Radiate head. b) Barbed achenes in head. c) Achene.
Bidens alba

Araceae - *Arum Family*

Habit: Herbs, vines, epiphytes. Terrestrial or aquatic. Vines often with adventitious roots. Grow from rhizomes, tubers, or corms.
Leaves: Alternate, simple to pinnately or palmately compound, entire. Pinnate, palmate, or parallel venation. Bases sheathing.
Inflorescences: Spike of many small flowers on a fleshy axis (**spadix**) subtended by a leafy or petal-like bract (**spathe**).
Flowers: Minute, radial, bisexual to unisexual.
Perianth: Absent or reduced. When present, with 4-6 tepals, distinct to connate, inconspicuous.
Stamens: 1-6.
Carpels: 2-3.
Ovary: Superior.
Placentation: Various.
Fruit: Berry, sometimes an utricle or drupe.
Other: Milky or watery aromatic sap. Adventitious roots sometimes present.
Genera: *Anthurium, Caladium, Colocasia, Dieffenbachia, Monstera, Philodendron, Pistia, Pothos, Spathiphyllum.*

Fig. 946   Jack-in-the-pulpit with inflorescence.
Arisaema triphyllum

a

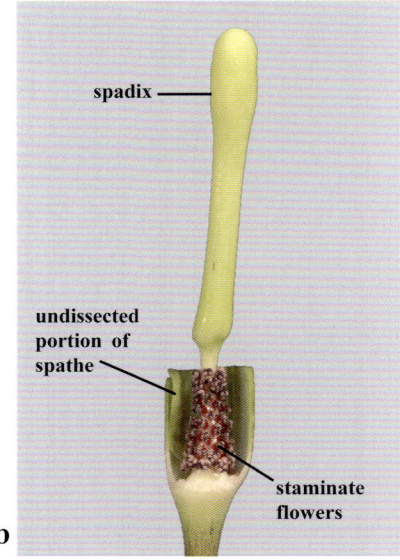

b

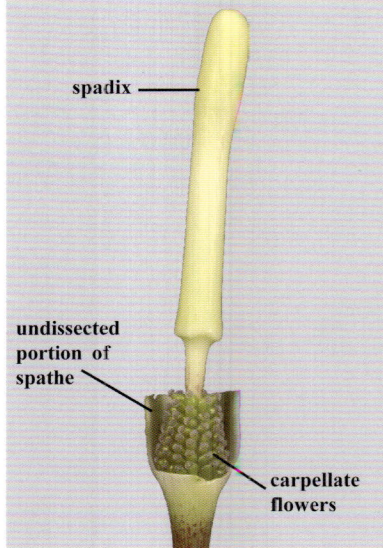

c

Fig. 947   Floral characters of Jack-in-the-pulpit. a) CU of leaf-like spathe. b) Staminate spadix (spathe removed). c) Carpellate spadix (spathe removed).   *Arisaema triphyllum*

a

b

c

Fig. 948   a) Staminate inflorescence (spathe removed). b) Carpellate inflorescence (spathe removed). c) Infructescence. *Arisaema triphyllum*

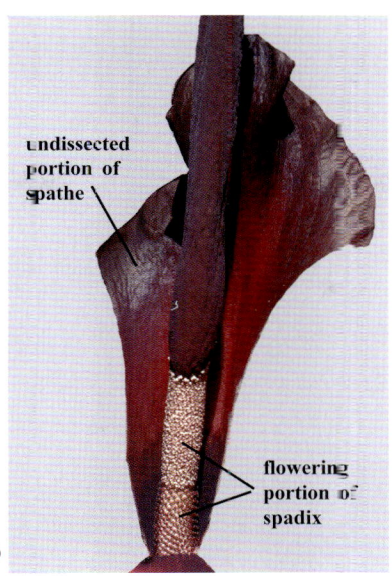

Fig. 949 Floral and fruit characters of the snake arum. a) Two inflorescences. b) CU of flowering portion of spadix with majority of spathe removed. c) Developing infructescence.   *Amorphophallus konjac*

Fig. 950   Floral characters of spathiphyllum. a) Inflorescence showing spathe and spadix. b) CU of spadix. c) CU of flowers on spadix.   *Spathiphyllum* sp.

Fig. 951   Foliage and floral characters of water lettuce. a) Habit of floating plant. b) CU of inflorescence. *Pistia stratiotes*

Lemnaceae - *Duckweed Family*

Habit: Small to minute aquatic herbs. Leaf-like in form.
Floating or submerged. Roots absent or thread-like.
Flowers: Extremely reduced, occur in pouches.
Fruit: Utricle.
Genera: *Landoltia, Lemna, Spirodela, Wolffia,* and *Wolffiella.*
Note: Many botanists treat the duckweed genera as part of the Araceae.

Fig. 952   Duckweed.
Wolffiella gladiata

Fig. 953   Duckweed.
Spirodella sp.

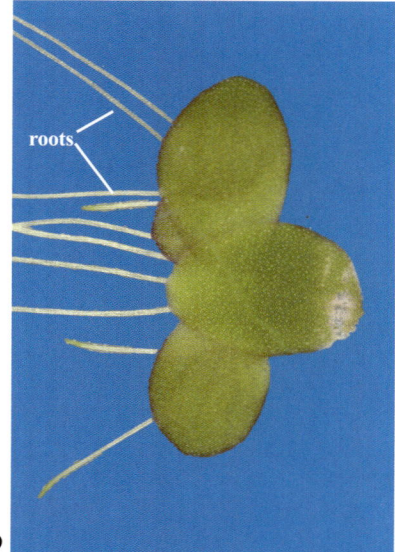

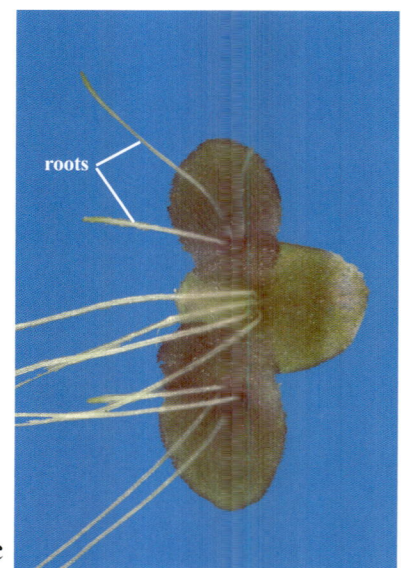

Fig. 954   Characters of duckweed. a) Covering a pond's surface. b) Upper surface of several plants. c) Lower surface
of several plants.   *Landoltia punctata*

Alismataceae - *Water Plantain Family*

Habit: Aquatic herb. Grows from rhizome, often with tubers.
Leaves: Emergent, floating, or submerged. Alternate, simple, entire. Parallel or palmate venation. Emergent leaves with long petioles and sheathing bases.
Inflorescences: Cyme, raceme, panicle at apex of scape.
Flowers: Showy to inconspicuous, radial, bisexual or unisexual.
Calyx: 3 green sepals.
Corolla: 3 white petals.
Stamens: 6-numerous.
Carpels: Numerous, spirally arranged on flat to globose receptacle.
Ovary: Superior.
Placentation: Basal.
Fruit: Aggregate of achenes.
Other: Milky sap.
Genera: *Alisma, Echinodorus, Sagittaria.*

Fig. 955   Leaves and flowers of an ornamental arrowhead. *Sagittaria montevidensis*

a   b   c

Fig. 956   Foliage and floral characters of an arrowhead. a) Leaf. b) Staminate inflorescence. c) Carpellate inflorescence. *Sagittaria lancifolia*

a   b   c

Fig. 957   Floral and fruit characters of an arrowhead. a) Staminate flower. b) Carpellate flower. c) LS of fruit (aggregate of achenes) with milky sap oozing. *Sagittaria lancifolia*

Hydrocharitaceae - *Frog's Bit Family*

Habit: Aquatic herbs. Completely submerged, floating, or partially emergent. Grow from rhizomes.
Leaves: Alternate, opposite or whorled. Entire, simple or serrate. Palmate or parallel venation. Variable in size and shape.
Inflorescences: Staminate in cyme or umbel, carpellate solitary.
Flowers: Showy to inconspicuous, radial, bisexual and unisexual.
Calyx: 3 sepals, distinct.
Corolla: 3 distinct petals or absent. Usually white.
Stamens: 2 or 3-numerous.
Carpels: 3-6.
Ovary: Inferior.
Placentation: Parietal (but may appear axile).
Fruit: Berry or fleshy capsule.
Seeds: Numerous.
Genera: *Elodea, Hydrilla, Hydrocharis, Limnobium.*

Fig. 958   Lower (T) and upper (B) surfaces of frog's bit leaves. *Limnobium spongia*

Fig. 959   Plant characters of frog's bit. a) Flowering and fruiting plant. b) Carpellate flower. *Limnobium spongia*

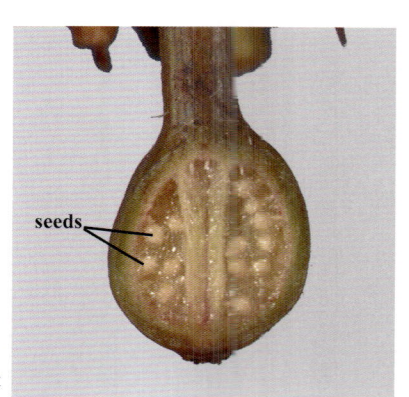

Fig. 960   Floral and fruit characters of frog's bit. a) Staminate flower. b) Fruit. c) LS of fruit. *Limnobium spongia*

Smilacaceae - *Greenbrier Family*

Habit: Herbaceous to woody vines. Grow from thick rhizomes.
Often with prickles on stems and/or leaves.
Leaves: Alternate, simple. Entire to spinose-serrate.
Leathery texture. Often lobed. Palmate to reticulate venation.
A pair of tendrils near the base of the petiole opposite leaf.
Inflorescences: Umbels.
Flowers: Inconspicuous, radial, unisexual.
Perianth: 6 tepals.
Stamens: 6.
Carpels: 3.
Ovary: Superior.
Placentation: Axile.
Fruit: Berry.
Seeds: 1-3, not black (lacking phytomelan).
Other: Nectaries at base of tepals.
Genera: *Smilax*.

Fig. 961   Leaf diversity in greenbrier.
a) *Smilax auriculata*
b) *Smilax bona-nox*

Fig. 962   Plant characters of greenbrier. a) Paired opposite tendrils at nodes. b) Flowering branch. c) Staminate inflorescence. d) Staminate flower.   *Smilax* spp.

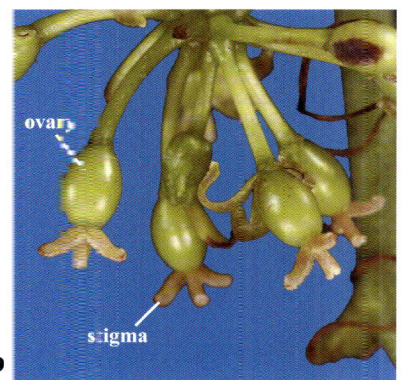

Fig. 963   Floral and fruit characters of greenbrier. a) Carpellate inflorescence. b) Young fruits. c) Infructescence of mature berries.   *Smilax* sp.

Liliaceae - *Lily Family*

Habit: Herbs. Grow from bulbs or rhizomes.
Leaves: Alternate or whorled, simple, entire. Arranged along stem or in a basal rosette. Parallel venation. Sheathing bases.
Inflorescences: Raceme, umbel, or single flower.
Flowers: Showy, radial to slightly bilateral, bisexual.
Perianth: 6 tepals, distinct, petaloid. Often marked with spots or lines.
Stamens: 6.
Carpels: 3.
Ovary: Superior.
Placentation: Axile.
Fruit: Loculicidal capsule, rarely a berry.
Seeds: Flat and disk-shaped. Not black (lack phytomelan).
Other: Nectaries at bases of tepals.
Genera: *Erythronium, Lilium, Medeola.*
Note: Familial limits in the Liliales and Asparagales are not widely agreed upon by botanists.

Fig. 964   Lily plant in flower. *Lilium* sp.

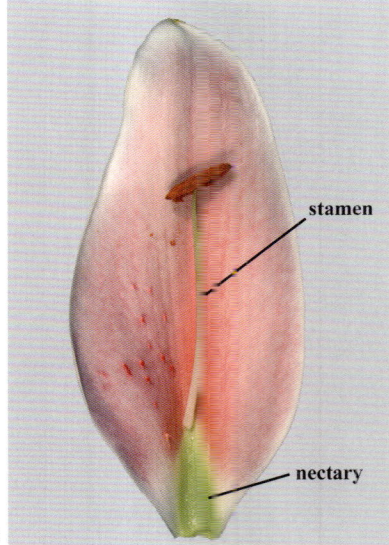

Fig. 965   Floral characters of lily. a) Stigma and anthers. b) CU of bases of tepals showing light green nectar-producing areas. c) One tepal and one stamen.   *Lilium* sp.

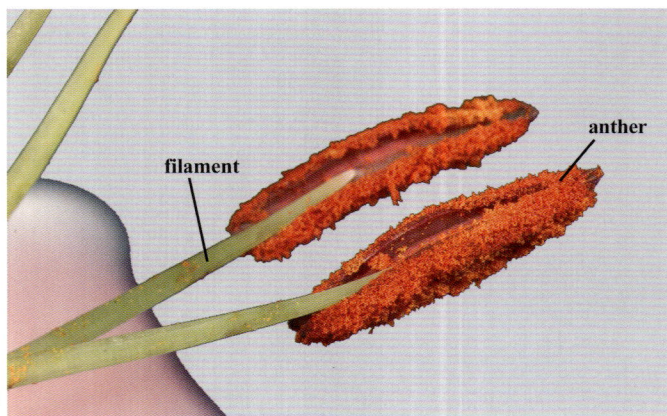

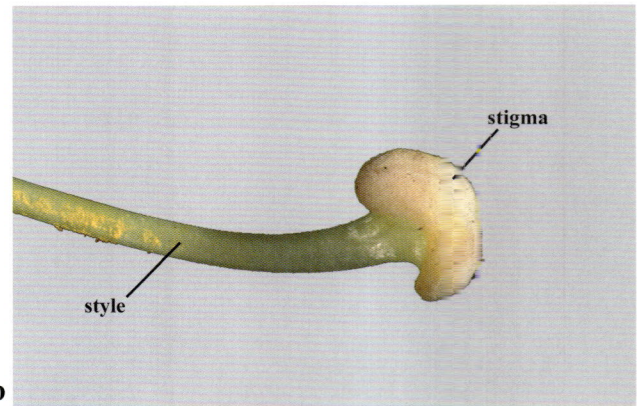

Fig. 966   Reproductive structures of lily. a) CU of anthers. b) CU of stigma.   *Lilium* sp.

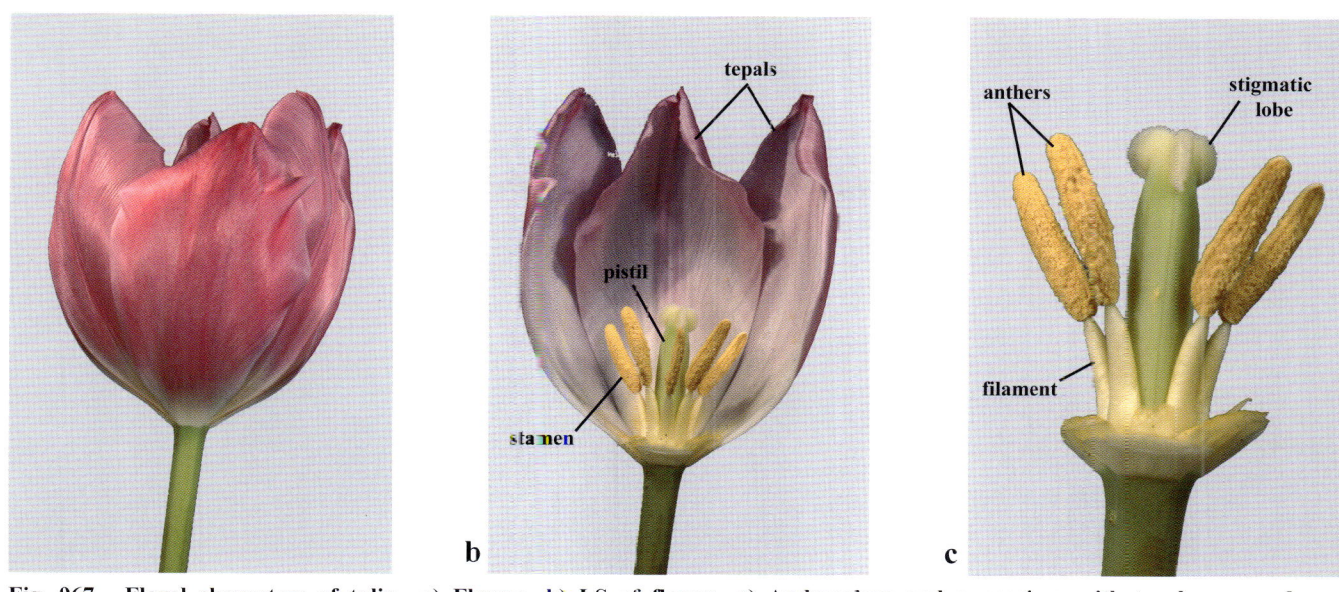

Fig. 967   Floral characters of tulip. a) Flower. b) LS of flower. c) Androecium and gynoecium with tepals removed. *Tulipa* sp.

Fig. 968   Floral characters of Easter lily. a) Flower. b) Gynoecium and part of androecium with tepals removed. *Lilium* sp.

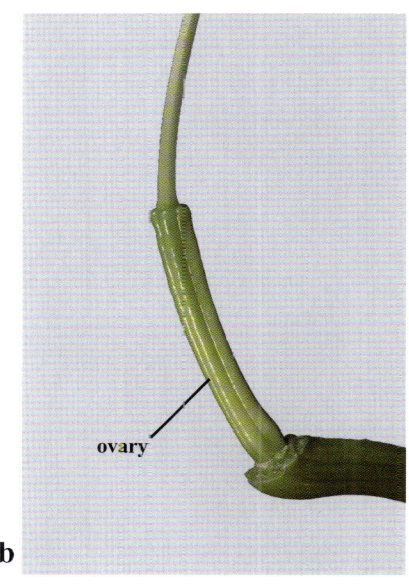

Fig. 969   Floral characters of Easter lily. a) LS of flower. b) CU of ovary.   *Lilium* sp.

Trilliaceae - *Trillium Family*

Habit: Herbs. Grow from rhizomes.

Leaves: Whorled, simple, entire. Palmate to reticulate venation.

Inflorescences: Terminal, reduced to a single flower.

Flowers: Showy, radial, bisexual.

Calyx: 3-4 sepals, distinct.

Corolla: 3-4 petals, distinct.

Stamens: 6-8.

Carpels: 3-10.

Ovary: Superior.

Placentation: Axile.

Fruit: Fleshy capsule or berry.

Seeds: Arillate, not black (lacking phytomelan).

Genera: *Trillium.*

Note: This family is traditionally included in the Liliaceae.

Fig. 970   Habit of trillium.
Trillium maculatum

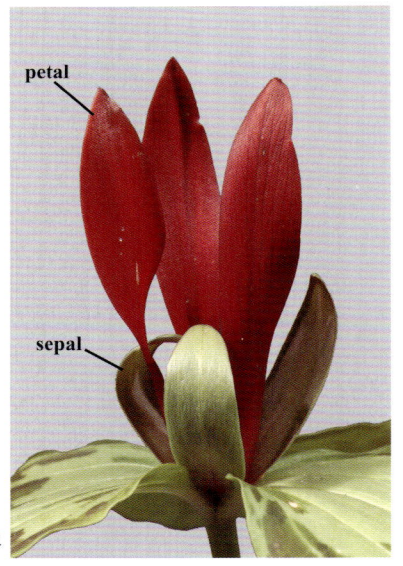
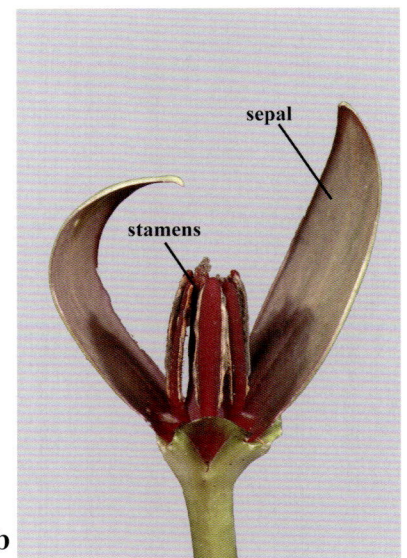

a   b   c

Fig. 971   **Floral characters of trillium. a) Flower. b) Flower with petals and one sepal removed. c) Pistil with stamens and sepals removed.**   *Trillium maculatum*

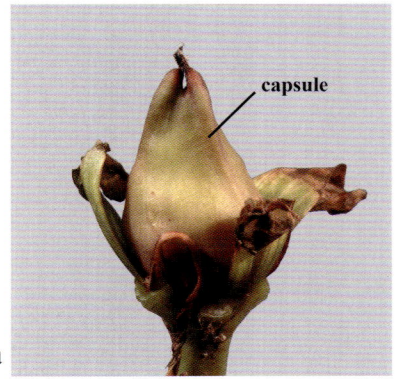

a   b   c

Fig. 972   **Fruit characters of trillium. a) Capsule. b) LS of capsule showing seeds. c) Three seeds with arils.**
Trillium maculatum

Asparagaceae - *Asparagus Family*

Habit: Herbs growing from rhizomes, shrubs, vines.
Leaves: Reduced to small, scale-like structures, sometimes with a basal spine. Photosynthesis through branchlets called **phylloclades**.
Inflorescences: Racemes, umbels, or solitary flower.
Flowers: Inconspicuous, bisexual or unisexual.
Perianth: Tepals not spotted.
Stamens: Usually 6. Adnate to tepals at base.
Carpels: 3, fused.
Ovary: Superior.
Placentation: Axile.
Fruit: Berry.
Seeds: Black (with phytomelan).
Other: Nectar from septa of ovaries.
Genera: *Asparagus*.
Note: This family is traditionally included in the Liliaceae.

Fig. 973   Flowering branch of asparagus fern. *Asparagus aethiopicus*

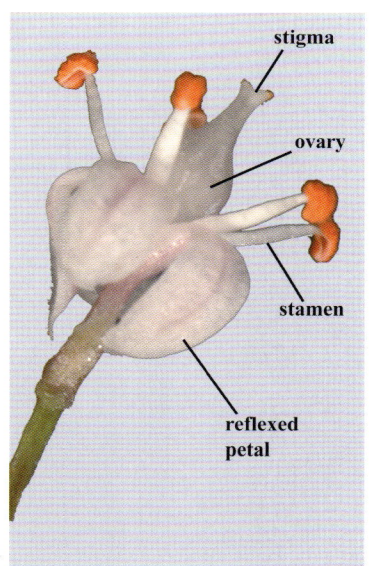

Fig. 974   Plant characters of asparagus fern. a) Flowers. b) Stem with axillary branchlets. c) Immature berries and phylloclades.   *Asparagus aethiopicus*

Fig. 975   Edible stalk of asparagus (young shoot with scale leaves).   *Asparagus officinalis*

Convallariaceae - *Lily of the Valley Family*

Habit: Herbs to trees. Grow from rhizomes.
Leaves: Alternate, simple, entire. Arranged spirally
or in a basal rosette. Parallel venation.
Inflorescences: Raceme, panicle or single flower.
Flowers: Small, radial, bisexual.
Perianth: 6 tepals, distinct or connate, petaloid, not spotted.
Stamens: 6.
Carpels: 3.
Ovary: Superior.
Placentation: Axile.
Fruit: Berry.
Seeds: Several, not black (lack phytomelan).
Other: Genus *Ruscus* bears flowers from flat
photosynthetic, leaf-like stems (**phylloclades**).
Genera: *Convallaria, Maianthemum, Polygonatum, Liriope, Ruscus.*
Note: Traditionally placed in the Liliaceae, but also recognized as
the Ruscaceae by some botanists.

Fig. 976   Inflorescence of border-
grass.   *Liriope spicata*

Fig. 977   Floral and fruit characters of border-grass. a) Flower.
b) Immature berries, one in LS.   *Liriope spicata*

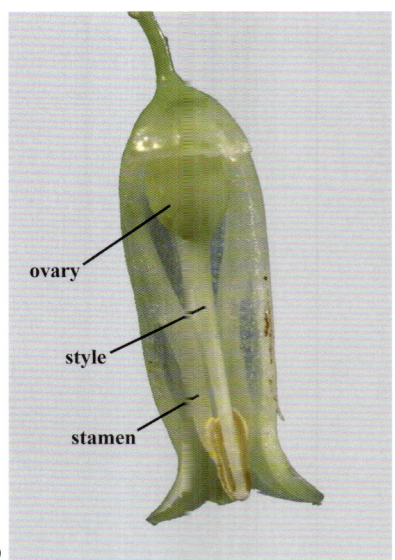

Fig. 978   Bud / flower of Solomon's
seal.   *Polygonatum biflorum*

Fig. 979   Floral characters of Solomon's seal. a) Flowering branch. b) LS of flower.
Polygonatum biflorum

Aloaceae - *Aloe Family*

Habit: Herbs, shrubs, trees.
Leaves: Succulent. Alternate, simple, entire to spiny or serrate. In basal or terminal rosettes. Filled with gelatinous material. Parallel venation. Sheathing bases.
Inflorescences: Terminal spike, raceme, panicle.
Flowers: Showy, radial or bilateral, bisexual. Often tubular.
Perianth: 6 tepals, petaloid, not spotted. Often fleshy.
Stamens: 6, distinct, adnate to perianth.
Carpels: 3.
Ovary: Superior.
Placentation: Axile.
Fruit: Loculicidal capsule or berry.
Seed: Black (with phytomelan). Numerous, flat or winged.
Other: Nectaries on septa of ovary.
Genera: *Aloe, Haworthia.*

Fig. 980   Habit of soap aloe.
Aloe saponaria

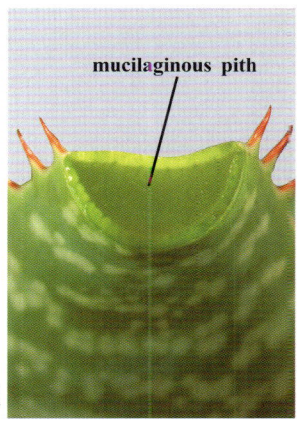

a   b   c   d

Fig. 981   **Floral and foliage characters of soap aloe. a) Inflorescence. b) Flower. c) LS of flower. d) XS of leaf.**
Aloe sopanaria

a   b   c   d

Fig. 982   **Fruit characters of soap aloe. a) Developing infructescence. b) Immature capsule. c) Mature capsule. d) XS of capsule.**   *Aloe saponaria*

Agavaceae - *Agave Family*

Habit: Herbs, shrubs, and trees. Grow from rhizomes. Dry habitat.
Often large with sturdy trunk or stem.
Leaves: Succulent. Alternate, simple, entire to spinose-serrate.
Sharp spine at tip. Arranged in rosettes at base or at tip of branches.
Extremely fibrous. Much longer (10x-20x) than wide.
Inflorescences: Terminal, determinate, paniculate. Often very large.
Flowers: Showy, radial to slightly bilateral, bisexual.
Perianth: 6 tepals, distinct to connate, petaloid, not spotted.
Stamens: 6.
Carpels: 3, fused.
Ovary: Superior or inferior.
Placentation: Axile.
Fruit: Usually a capsule, sometimes fleshy and berry-like.
Seed: Flat, contains phytomelan.
Genera: *Agave, Camassia, Chlorogalum, Furcraea, Hosta,
Manfreda, Yucca.*

Fig. 983   Habit of century plant.
Agave sp.

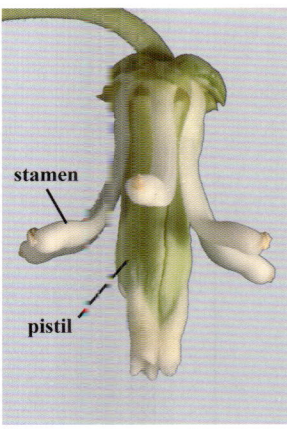

a   b   c   d

tepal

stamen

pistil

Fig. 984   Floral characters of Spanish bayonet. a) Flowering plant. b) Flower. c) LS of flower. d) Androecium and
gynoecium with tepals removed.
Yucca aloifolia

a   b   c

Fig. 985   Floral characters of mother-in-law's tongue.
a) Flowering plant. b) Flower. c) Foliage.
Sanseviera spp.

Fig. 986   Fruit of dracaena
'palm'.
Dracaena sp.

Alliaceae - *Onion Family*

Habit: Herbs. Grow from bulbs.

Leaves: Alternate, simple, entire. Usually basal. Parallel venation.
Bases sheathing. Strap-like.

Inflorescences: Umbel at end of long scape, subtended by bracts.

Flowers: Showy, radial or bilateral, bisexual.

Perianth: 6 tepals, distinct to connate, petaloid, not spotted.

Stamens: 6, sometimes adnate to tepals.

Carpels: 3.

Ovary: Superior.

Placentation: Axile.

Fruit: Loculicidal capsule.

Seed: Black (with phytomelan).

Other: Leaves and stem emit onion-like (sulfuric) odor when crushed.
Stem with more or less clear latex. Nectaries on septa of ovaries.

Genera: *Allium, Northoscordum, Tulbaghia.*

Note: Some botanists treat the Alliaceae as part of the Liliaceae

Fig. 987   Sprouting onion bulb.
Allium cepa

Fig. 988   Floral and fruit characters of onion. a) Inflorescence. b) Flower and immature capsules. c) Mature dehisced
and intact capsules. d) Black seeds with phytomelan.   *Allium cepa*

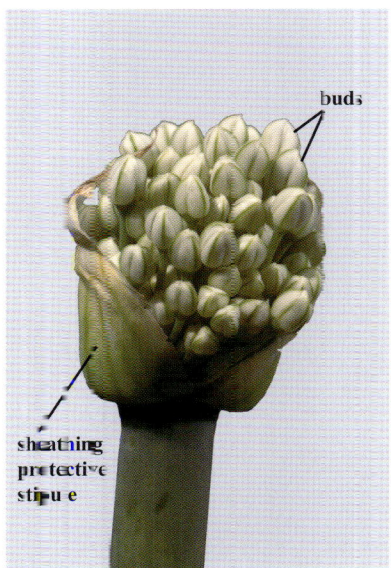

Fig. 989   Foliage and floral characters of onion. a) Bases of strap-like leaves
at stem. b) Inflorescence prior to bloom.   *Allium cepa*

Fig. 990   Inflorescence with bulbils
of wild onion.   *Allium canadense*

Amaryllidaceae - *Amaryllis Family*

Habit: Herbs. Grow from bulbs.

Leaves: Alternate, simple, entire, flat. Arranged basally.
Parallel venation. Bases sheathing. Long and strap-like.

Inflorescences: Umbel or single flower at the end of a long scape.

Flowers: Showy, radial to bilateral, bisexual.

Perianth: 6 tepals, distinct to connate, petaloid, not spotted.

Stamens: 6.

Carpels: 3.

Ovary: Inferior.

Placentation: Axile.

Fruit: Loculicidal capsule, rarely a berry.

Seed: Black or blue, phytomelan usually present.

Other: Nectaries usually on septa of ovary.

Genera: *Amaryllis, Crinum, Hippeastrum, Hymenocallis,
Narcissus, Zephyranthes.*

Fig. 991   Habit of amaryllis.
Amaryllis sp.

a

b

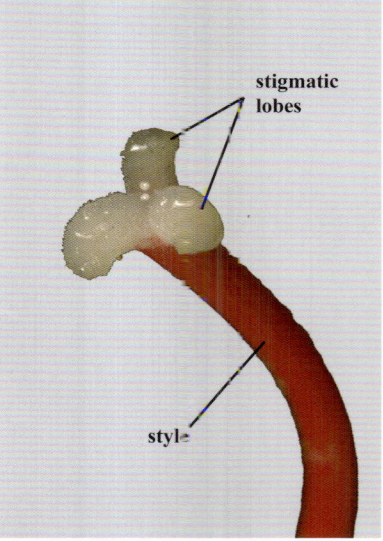

c

Fig. 992   Foliage and floral characters of amaryllis. a) Strap-like leaves and base of scape. b) Flower. c) Stigma.
Amaryllis sp.

a

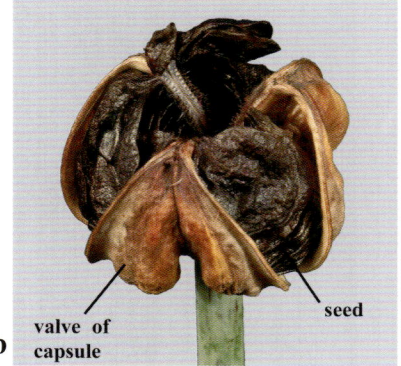

b

c

Fig. 993   Fruit characters of amaryllis. a) Developing capsule. b) Mature capsule dehiscing. c) Black seed indicating
phytomelan is present.   *Amaryllis* sp.

Fig. 994   Plant characters of daffodil. a) Sprouting corm. b) Inflorescence. *Narcissus* sp.

Fig. 995   Inflorescence and leaves of narcissus.   *Narcissus* sp.

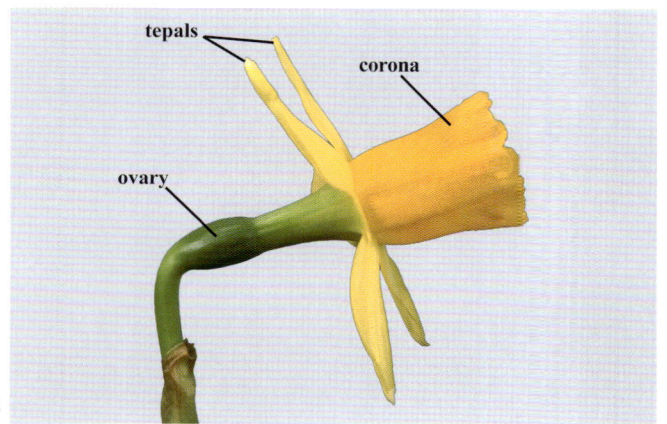

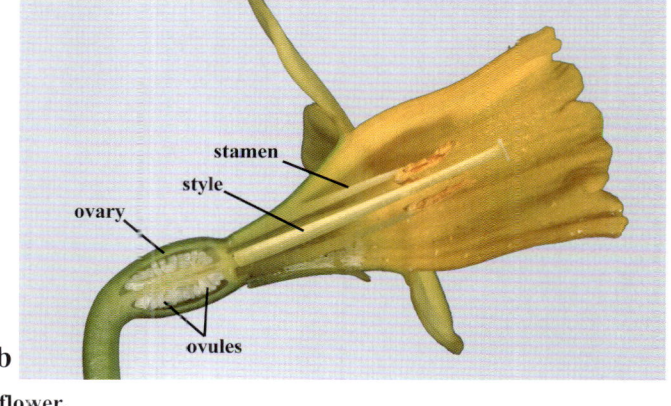

Fig. 996   Floral characters of daffodil. a) Flower. b) LS of flower. *Narcissus* sp.

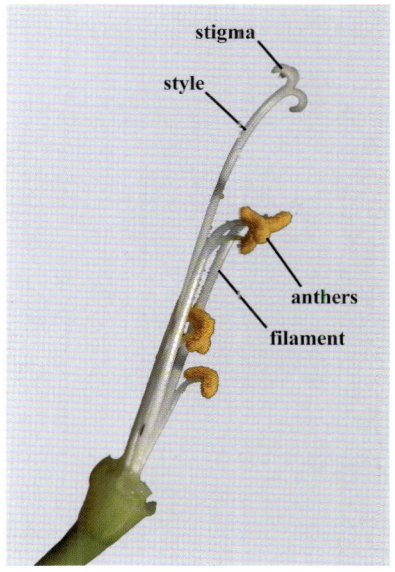

Fig. 997   Floral characters of rain-lily. a) Flower. b) LS of flower. c) Androecium and gynoecium with tepals removed. *Habranthus* sp.

Iridaceae - *Iris Family*

Habit: Herbs. Grow from rhizomes, bulbs, or corms.
Leaves: Alternate, simple, entire. 2-ranked. Parallel venation. Bases sheathing. Usually linear, narrow, flattened, equitant.
Inflorescences: Cyme, raceme, panicle, or reduced to a single flower at the end of a scape.
Flowers: Showy, radial to bilateral, bisexual.
Perianth: 6 tepals, sometimes differentiated, distinct or connate, petaloid, sometimes spotted.
Stamens: 3.
Carpels: 3.
Ovary: Inferior.
Placentation: Axile.
Fruit: Loculicidal capsule.
Seeds: Brown (lack phytomelan). Sometimes arillate.
Other: Nectaries on septa of ovaries.
Genera: *Calydorea, Crocus, Freesia, Gladiolus, Iris, Sisyrinchium, Trimezia.*

Fig. 998　　Flower of iris. *Iris* sp.

a　　　　　　　　　b　　　　　　　　　c

Fig. 999　　Floral and fruit characters of iris. a) Flower. b) Flower partially dissected. c) Immature capsules. *Iris* sp.

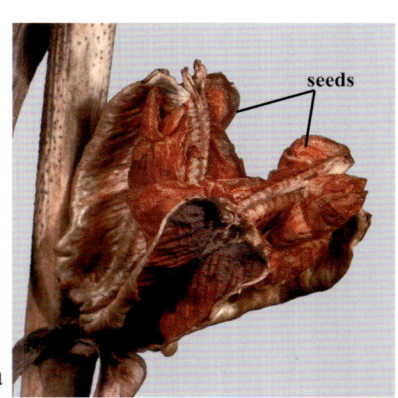

a　　　　　　　　　b　　　　　　　　　c

Fig. 1000　　Plant characters of iris. a) Mature capsule dehiscing. b) Seed. c) Sprouting rhizome. *Iris* sp.

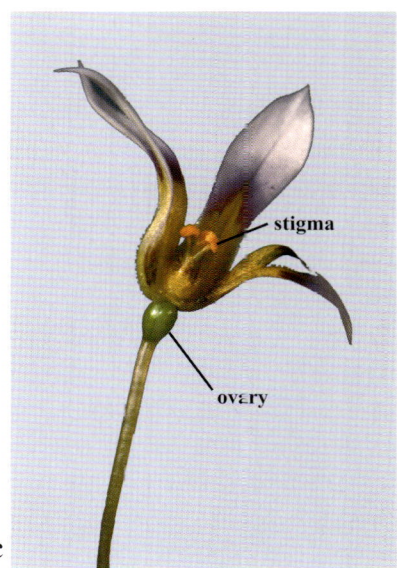

Fig. 1001   Floral characters of blue-eyed grass. a) Bud. b) Flower. c) Flower partially dissected. *Sisyrinchium angustifolium*

Fig. 1002   Floral characters of gladiolus. a) Inflorescence. b) Flower. c) LS of flower. *Gladiolus* sp.

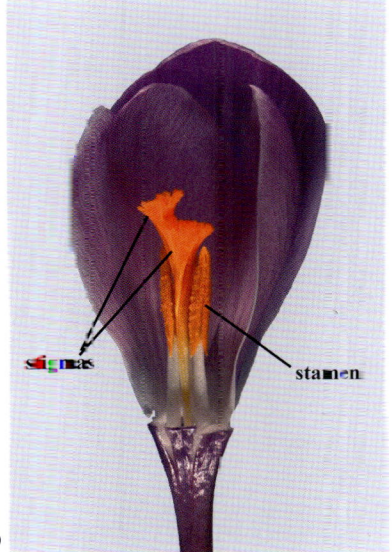

Fig. 1003   Plant characters of crocus. a) Habit. b) Androecium and gynoecium with most tepals removed. c) Corm. *Crocus* sp.

Orchidaceae - *Orchid Family*

Habit: Herbs, sometimes vines. Epiphytic or terrestrial. Grow from rhizomes, corms, and tubers. Stems usually swollen at base into **pseudobulbs** (water storage organs). Sometimes parasites.

Leaves: Alternate, simple, entire. Parallel venation. Bases sheathing. Usually leathery or succulent.

Inflorescences: Raceme, panicle, spike, or single flower.

Flowers: Showy, bilateral, bisexual. Often twist 180° during development (**resupinate**).

Perianth: 6 tepals, distinct. Central tepal larger and modified into a lip (**labellum**).

Reproductive Structures: Stamens (1 or 2), style, and stigma adnate to one another to form a **column**. Pollen clumped into distinct masses called **pollinia**.

Ovary: Inferior.

Placentation: Parietal or axillary.

Fruit: Capsule.

Seeds: Minute, lacking phytomelan.

Other: Often with aerial roots which may appear white due to a velamen layer. Extremely large family with more than 20,000 species.

Genera: *Aplectrum, Cattleya, Corallorhiza, Cymbidium, Cypripedium, Dendrobium, Epidendrum, Habenaria, Oncidium, Paphiopedilum, Phalaenopsis, Spiranthes, Tipularia, Vanda, Vanilla.*

Note: Probably the largest angiosperm family.

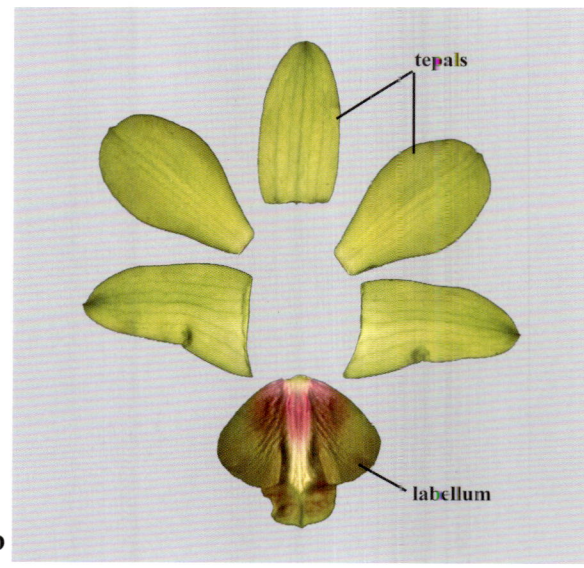

Fig. 1004   Floral characters of orchid. a) Flower intact. b) Flower dissected and expanded. *Dendrobium* sp.

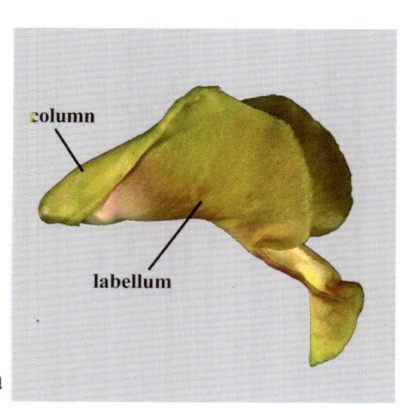
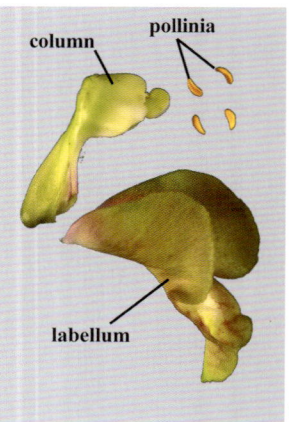
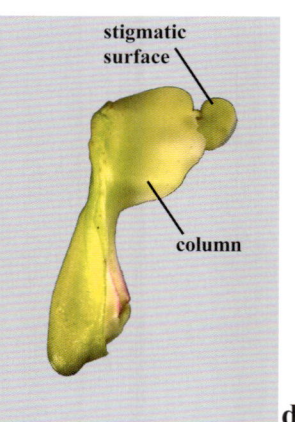
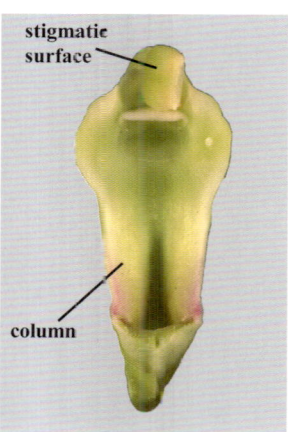

Fig. 1005   Characters of column. a) Labellum intact. b) Labellum separated. c) Lateral view. d) Ventral view. *Dendrobium* sp.

Fig. 1006   Flower of spotted oncidium. a) Front view. b) Lateral view.
Oncidium maculatum

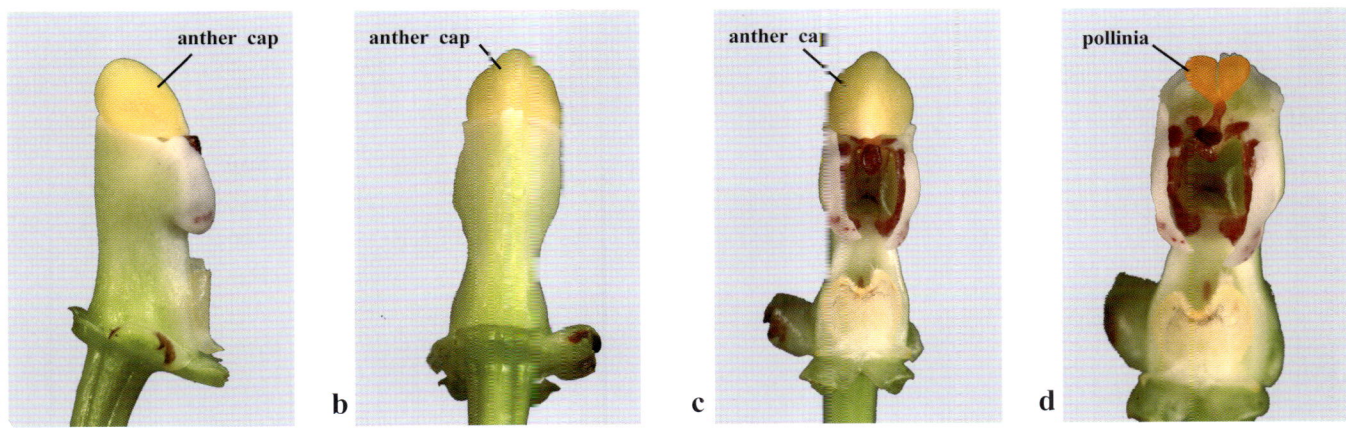

Fig. 1007   Column characters of spotted oncidium. a) Lateral view. b) Dorsal view. c) Front view.
d) Front view with anther cap removed revealing yellow pollinia.   *Oncidium maculatum*

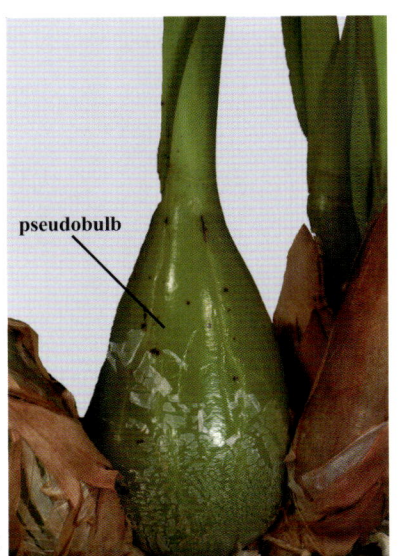

Fig. 1008   Orchid characters. a) Pseudobulbs at base of stem. b) Stems with whitish leaf sheaths. c) Immature fruits
with persistent perianth showing distinctively inferior ovary position.

Arecaceae - *Palm Family*

Note: Two correct Latin names for this family: Arecaceae and Palmae.

Habit: Trees or shrubs with unbranched trunks.

Leaves: Alternate, large and leathery, often spirally arranged and clustered into a crown at the apex. Simple, often highly dissected and appearing compound. Long petioles with sheathing, persistent bases often present. Blade of leaf is folded (**plicate**), attaching to the petiole in a zizag manner. A woody ligule or **hastula** is sometimes present at the petiole-blade junction. Palmate or pinnate venation.

Inflorescences: Large and many branched panicle to compound spike. Subtended by a large woody bract in some species.

Flowers: Numerous, small, sessile. Radial, unisexual or bisexual.

Calyx: 3 sepals.

Corolla: 3 petals.

Stamens: 3 or 6, sometimes numerous.

Carpels: 3.

Ovary: Superior.

Placentation: Axile.

Fruit: Fleshy drupe.

Seeds: 1 per fruit.

Other: Protective fibers with woven appearance often cover leaf bases on trunk.

Genera: *Areca, Chamaedorea, Coccothrinax, Cocos, Phoenix, Pritchardia, Rhapidophyllum, Roystonea, Sabal, Serenoa, Washingtonia.*

a   b   c   d

Fig. 1009   Floral and fruit characters of jelly palm. a) Portion of inflorescence with flower buds. b) Flowers. c) Developing drupes. d) CU of immature drupe.   *Butia capitata*

  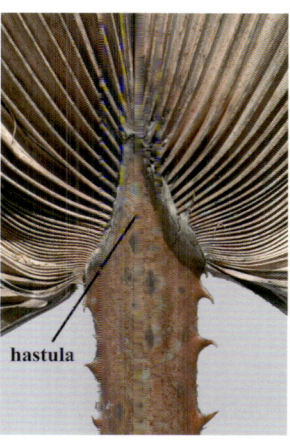

Fig. 1010   Mature drupe of jelly palm. *Butia capitata*

Fig. 1011   Inflorescence of pygmy date palm showing large woody bract.   *Phoenix roebelinii*

Fig. 1012   Hastula of Washington fan palm. *Washingtonia robusta*

Palm Terminology

hastula - Woody flap (ligule) where blade of leaf joins the petiole.
costa - An extension of the petiole that continues through the center of the leaf forming a midrib.
costapalmate - Palmate leaf where the lobes or segments originate from a costa.
palmate - Leaf whose segments originate at a common point like the fingers on the palm of your hand.
pinnate - Leaf whose segments originate from an elongate rachis like the pinnae of a feather.
plicate - Referring to the folds and folded segments making up the blade of most palms.
induplicate - Folded upward, or a segment that is V-shaped in cross section.
reduplicate - Folded downward, or a segment that is A-shaped in cross section.

Fig. 1013   PINNATE LEAF
Bamboo palm.
Chamaedorea microspadix

Fig. 1014   PALMATE LEAF
European fan palm.
Chamaerops humilis

Fig. 1015   COSTAPALMATE LEAF
Cabbage palm.
Sabal palmetto

Fig. 1016   Leaflet attachment in palms.   a) & c) Induplicate attachment on date palm leaf.   *Phoenix* sp.
b) & d) Reduplicate attachment on jelly palm leaf.   *Butia capitata*

Bromeliaceae - *Bromeliad Family*

Habit: Epiphytic or terrestrial herbs.
Leaves: Alternate, simple, entire to serrate or spiny margins.
Typically arranged in a basal rosette that may form a central 'cup'
or 'tank'. Parallel venation. Leathery and strap-like.
Inflorescences: Various, containing many flowers in
the axils of brightly colored bracts.
Flowers: Showy to inconspicuous, radial, bisexual.
Calyx: 3 sepals.
Corolla: 3 petals.
Stamens: 6.
Carpels: 3.
Ovary: Superior to inferior.
Placentation: Axile.
Fruit: Septicidal capsule, berry, or multiple of berries.
Seeds: Often with a tuft of hair or winged.
Other: Water absorbing peltate hairs on leaves.
Genera: *Aechmea, Ananas, Catopsis, Guzmania, Tillandsia, Vrisea.*

Fig. 1017   Epiphytic bromeliad on
tree trunk.
Aechmea sp.

a

b

c

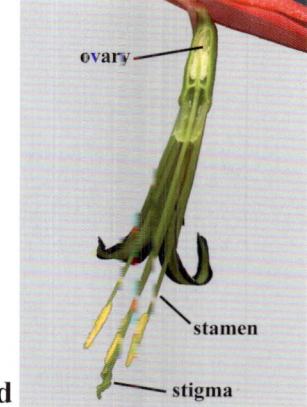

d

Fig. 1018   Floral characters of bromeliad. a) Habit of flowering plant. b) Inflorescence. c) Flower. d) LS of flower.
Bilbergia nutans

Fig. 1019   Young plant of ball moss
growing as epiphyte on tree branch.
Tillandsia recurvata

Fig. 1020   Spanish moss in flower.
Tillandsia usneoides

Fig. 1021   Blue flower subtended
by pink bracts. *Tillandsia cyanea*

Fig. 1022   Floral and fruit characters of pineapple. a) Inflorescence. b) CU of flowers. c) Immature multiple fruit.
Ananas comosus

Fig. 1023   Floral and fruit characters of a terrestrial bromeliad. a) Partial inflorescence. b) Nearly mature capsule.
c) Mature capsule after dehiscence. d) Seeds.
Dyckia sp.

Fig. 1024   Partial infructescence
with developing berries.
Aechmea angustifolia

Fig. 1025   XS of ovary.
Billbergia nutans

Fig. 1026   Tank bromeliad with
water trapped in leaf axils.
Neoregelia sp.

Pontederiaceae - *Water Hyacinth Family*

Habit: Aquatic herbs. Floating, emergent, or freestanding in wet habitats. Grow from rhizomes or stolons.
Leaves: Opposite or whorled, simple, entire. Two-ranked or in a basal rosette. Ovate to squarish to lanceolate. Bases sheathing. Parallel to palmate venation. Some floating species with petioles swollen and spongy.
Inflorescence: Raceme, spike, or a single flower. Subtended by a spathe-like bract.
Flowers: Showy, bisexual, radial to bilateral. Often heterostylous.
Perianth: 6 tepals, petaloid, one of which is differentiated.
Stamens: 6, adnate to perianth tube.
Carpels: 3.
Ovary: Superior.
Placentation: Axile.
Fruit: Loculicidal capsule or nutlet. Enclosed by persistent portion of perianth tube.
Seeds: Small, oval, longitudinally ribbed.
Genera: *Eichhornia, Heteranthera, Pontederia.*

Fig. 1027   Stand of pickerel weed in flower.
Pontederia cordata

Fig. 1028   Floral and fruit characters of pickerel weed. a) Inflorescence. b) Infructescence. c) Leaf and flower bud.
Pontederia cordata

nutlet

Fig. 1029   Floral and fruit characters of pickerel weed. a) Flower. b) Fruit in-situ. c) Fruit with persistent perianth (left) and nutlet after perianth has been removed (right) .   *Pontederia cordata*

Fig. 1030   Plant characters of water hyacinth. a) Floating mat in flower. b) Inflorescence.
Eichhornia crassipes

Fig. 1031   Plant characters of water hyacinth. a) Base of two plants showing connecting stolon and roots. b) Flower.
Eichhornia crassipes

Fig. 1032   Plant characters of water hyacinth. a) Habit showing inflated petioles used for buoyancy. b) Leaf showing squarish blade and expanded petiole. c) XS of petiole.   *Eichhornia crassipes*

Commelinaceae - *Spiderwort Family*

Habit: Succulent herbs with stems swollen at nodes.
Leaves: Alternate, simple, entire. Flat or folded.
Bases sheathing. Parallel venation.
Inflorescences: Cyme, or single flower.
Subtended by a spathe-like or leafy bract.
Flowers: Showy, radial or bilateral, bisexual.
Calyx: 3 sepals.
Corolla: 3 petals, typcially clawed, ephemeral.
One petal is sometimes reduced.
Stamens: 6, or 3 stamens and 3 staminodes.
Brightly colored moniliform hairs often present.
Carpels: 3.
Ovary: Superior.
Placentation: Axile.
Fruit: Loculicidal capsule.
Seeds: With a conspicuous disk-like cap.
Genera: *Callisia, Commelina, Murdannia, Tradescantia.*

Fig. 1033   Flower of dayflower.
Commelina erecta

a

b

c

Fig. 1034   Plant characters of small-leaf spiderwort. a) Habit. b) Inflorescence. c) Flower.
Tradescantia fluminensis

a

b

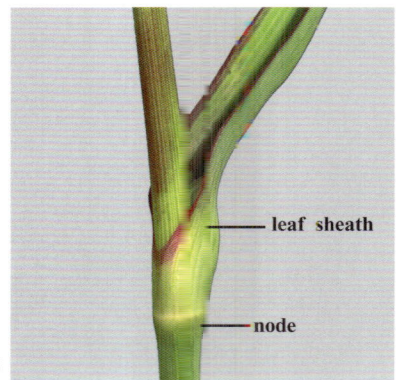

c

leaf sheath

node

Fig. 1035   Plant characters of spiderwort. a) Inflorescence. b) CU of stamens and moniliform hairs. c) Stem showing
sheathing leaf base and swollen node.   *Tradescantia ohiensis*

Typhaceae - *Cattail Family*

Habit: Aquatic, emergent herbs. Grow from rhizomes. Occurs in large colonies or stands.

Leaves: Alternate, simple, entire. Long and strap-like, basal, 2-ranked. Thick and spongy. Bases sheathing. Parallel venation.

Inflorescences: Dense clusters of flowers in form of cylinder or sphere. Staminate flowers above, carpellate below.

Flowers: Minute, radial, unisexual.

Perianth: 1-6 tepals, bract-like or bristles.

Stamens: 1-8.

Carpels: 3.

Ovary: Superior.

Placentation: Apical.

Fruit: Drupe or achene-like follicle.

Seeds: 1 seed per flower.

Other: Stems and leaves mucilaginous.

Genera: *Sparganium* and *Typha*.

Note: *Sparganium* has been traditionally recognized as the family Sparganiaceae.

Fig. 1036 Stand of cattails with one infructescence showing. *Typha domingensis*

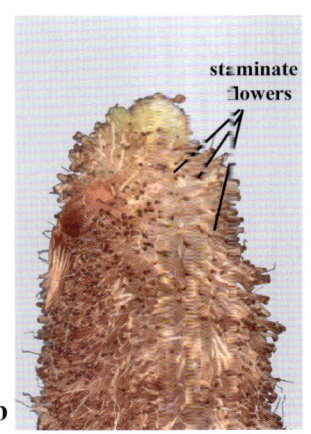
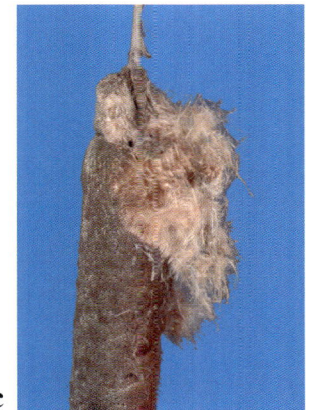
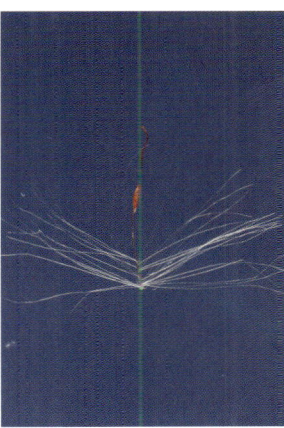

a b c d

Fig. 1037 Floral and fruit characters of cattail. a) Inflorescence. b) Staminate portion of inflorescence. c) Infructescence breaking apart. d) Tiny achene-like follicle with attached hairs for aerial dispersal. *Typha domingensis*

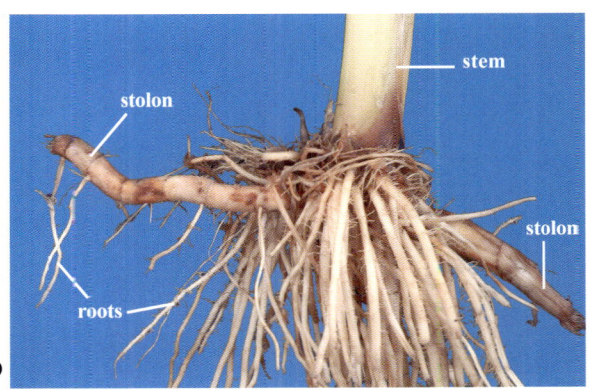

a b

Fig. 1038 Plant characters of cattail. a) XS of leaves. b) Basal portion of stem showing roots and stolon. *Typha domingensis*

Eriocaulaceae - *Pipewort Family*

Habit: Herbs with shortened stems. Grow in damp to wet habitats.

Leaves: Alternate, simple, entire. Narrow and grass-like. Typically in basal rosette. Parallel venation.

Inflorescences: Formed into a head with an involucre of papery bracts. Terminal on scapes basally sheathed by bracts.

Flowers: Inconspicuous, radial to bilateral, unisexual.

Perianth: Fringed with hairs.

Calyx: 2-3 sepals.

Corolla: 2-3 petals, sometimes reduced, often with nectaries.

Stamens: 2-6, frequently unequal, adnate to petals.

Carpels: 2-3.

Ovary: Superior, on a stalk.

Placentation: Axile.

Fruit: Loculicidal capsule.

Genera: *Eriocaulon, Lachnocaulon, Syngonanthus.*

Fig. 1040   Inflorescence of hatpins consisting of a head of many flowers.
Syngonanthus flavidulus

Fig. 1041   LS of inflorescence of hatpins.
Syngonanthus flavidulus

Fig. 1039   Habit of hatpins plant.
Syngonanthus flavidulus

Fig. 1042   Habit of bogbutton plant.
Lachnocaulon anceps

Xyridaceae - *Yellow-Eyed Grass Family*

Habit: Herbs with short bulbous stems or rhizomes. Grow in damp to wet habitats.

Leaves: Alternate, simple, entire. Narrow and grass-like. Flat to cylindrical, 2-ranked, equitant. Sheathing bases.

Inflorescences: Formed into a cone-like head of persistent bracts. Terminal on a long scape.

Flowers: Slightly bilateral, bisexual. Sessile. Each borne in the axil of a stiff bract.

Calyx: 3 sepals, unequal with one ephemeral.

Corolla: 3 petals, clawed, typically yellow, ephemeral.

Stamens: 3, adnate to petals. 3 plumose staminodes covered with moniliform hairs.

Carpels: 3.

Ovary: Superior.

Placentation: Various.

Fruit: Loculicidal capsule with persistent corolla tube.

Seeds: Minute, longitudinally ridged.

Genus: *Xyris*.

Fig. 1044   Inflorescence of yellow-eyed grass in lateral view.   *Xyris* sp.

Fig. 1045   Inflorescence of yellow-eyed grass in dorsal view.   *Xyris* sp.

Fig. 1043   Habit of yellow-eyed grass. *Xyris* sp.

Fig. 1046   Cone-like infructescence of yellow-eyed grass.   *Xyris* sp.

Juncaceae - *Rush Family*

Habit: Grass-like herb. Grows from rhizome. Damp to wet habitats.
Stems round in x.s., solid between nodes.
Leaves: Alternate, simple, entire. 3-ranked. Basal, linear and grass-
like, flat to cylindrical. Parallel venation. Open basal sheath.
Inflorescences: Greatly branched cyme or panicle,
or condensed into a head.
Flowers: Inconspicuous, radial, unisexual and bisexual.
Perianth: 6 tepals, frequently membranous.
Stamens: 6.
Carpels: 3.
Ovary: Superior.
Placentation: Axile or parietal.
Fruit: Loculicidal capsule containing many tiny seeds.
Seeds: Small, 3-numerous.
Genera: *Juncus, Luzula.*

**Fig. 1047   Habit of rush growing
at water's edge.**
Juncus effusus

a

b

c

Fig. 1048   Floral and fruit characters of rush. a) Inflorescence. b) Flower buds. c) Capsules on portion of infructescence.
Juncus effusus

a

b

Fig. 1049   Stem characters of rush. a) XS of stem showing round shape. b) LS of stem showing solid nature.
Juncus effusus

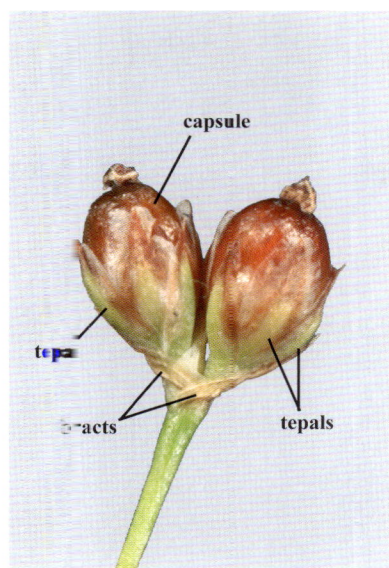

Fig. 1050   Fruit characters of rush. a) Infructescence. b) Two immature capsules. c) Mature capsule. (Note line of dehiscence.)   *Juncus elliottii*

Fig. 1051   Partial infructescence with one open capsule showing many small seeds within.   *Juncus elliottii*

Fig. 1052   Open sheath of leaf base of rush.   *Juncus effusus*

Fig. 1053   Fruit characters of rush. a) Immature (left) and mature (right) infructescences. b) CU of infructescence.   *Juncus megacephalus*

Fig. 1054   Capsules of shore rush. *Juncus marginatus*

Cyperaceae - *Sedge Family*

Habit: Grass-like herbs. Grow from rhizome. Damp to wet habitat. Stems usually 3-sided, triangular in x.s.

Leaves: Alternate, simple, entire. 3-ranked. Basal, linear and grass-like, flat. Parallel venation. Closed basal sheath.

Inflorescences: Multiple spikelets, each with many or a single flower, subtended by bracts.

Flowers: Inconspicuous, unisexual and bisexual. Each subtended by 1 papery bract (2 in *Carex*).

Perianth: Absent or replaced by several hairs, bristles, or scales.

Stamens: 1-3, distinct.

Carpels: 2-3.

Ovary: Superior.

Placentation: Basal.

Fruit: Achene or nutlet. Frequently with reduced perianth structures. Each fruit with a single seed.

Other: Long feathery stigmas.

Genera: *Carex, Cyperus, Eleocharis, Rhynchospora, Scirpus, Scleria.*

Fig. 1055   Habit a****l inflorescence of sedge.
Carex lupuliformis

Fig. 1056   Floral characters of sedge. a) Carpellate inflorescence. b) Carpellate flowers.
Carex lupuliformis

Fig. 1057   White-top sedge showing 3-ranked habit.
Rhynchospora colorata

Fig. 1058   XS of stem showing triagular shape.

a   b   c

Fig. 1059   Plant characters of sedge. a) Habit. b) Inflorescence. c) Flowers. *Cyperus esculentus*

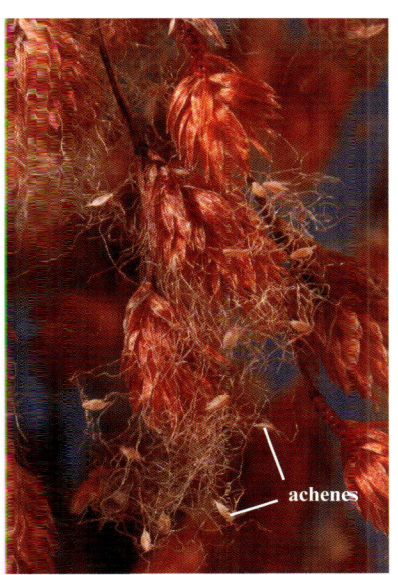

a   b   c

Fig. 1060   Plant characters of woolgrass. a) Spikelet. b) Infructescence releasing achenes.   *Scirpus cyperinus*

Fig. 1061   Node showing closed leaf sheath.   *Carex* sp.

a   b

Fig. 1062   Plant characters of sedge. a) Habit showing 3-ranked leaves. b) Inflorescence.   *Cyperus* sp.

Fig. 1063   Fruit capsule of the netted nutrush.   *Scleria reticularis*

Poaceae - *Grass Family*

Note: Two correct Latin names for family: Poaceae or Graminae.

Habit: Herbs, some woody and tree-like (growing over 20 meters tall). Grow from rhizomes or stolons. Stems round or elliptical, circular or oval in cross-section and hollow between nodes.

Leaves: Alternate, simple, entire. 2-ranked. Basal, linear, grass-like, flat. Parallel venation. Open basal sheath with membranous flap or fringe of hairs (**ligule**). Sheath may envelope and conceal stem.

Inflorescences: Multiple spikelets, subtended by bracts (**glumes**).

Flowers: Inconspicuous, unisexual and bisexual. In theory, each subtended by two bracts (**lemma** and **palea**), but many modifications exist. Flowers are very reduced and termed **florets**.

Perianth: Reduced to 1-3 minute, translucent scales (**lodicules**).

Stamens: 1-3, distinct. Anthers sagittate at both ends.

Carpels: 2-3. Stigmas elongate and feathery.

Ovary: Superior.

Placentation: Parietal.

Fruit: Caryopsis (grain)(seed coat fused to fruit wall). Each fruit with a single seed.

Other: Stigmas stick out.

Genera: *Andropogon, Bambusa, Bromus, Cynodon, Digitaria, Eleusine, Eragrostis, Festuca, Hordeum, Muhlenbergia, Oryza, Panicum, Paspalum, Phragmites, Poa, Saccharum, Secale, Setaria, Sorghum, Triticum, Zea.*

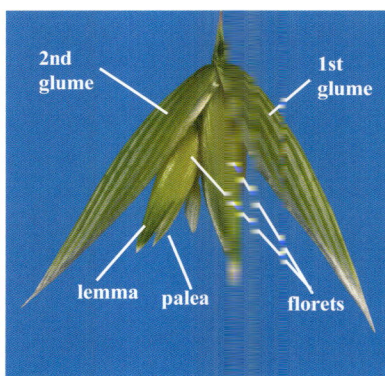

Fig. 1064   Floral characters of oats. a) Inflorescence. b) Spikelet. c) Spikelet with florets extended. *Avena sativa*

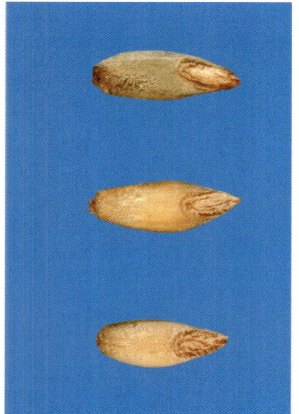
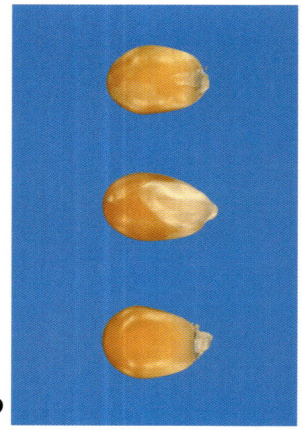

Fig. 1065 Caryopses or fruits of various grain grasses. a) Rye *Secale cereale*. **b) Corn** *Zea mays*. **c) Wheat** *Triticum aestivum*. **d) Rice** *Oryza sativa*

Fig. 1066   Floral characters of Johnson grass.
a) Inflorescence. b) Flowers with stamens extended.
Sorghum halapense

Fig. 1067   Floral characters of sandspur.
a) Inflorescence. b) CU of flower.
Cenchrus sp.

Fig. 1068   Floral characters of rye. a) Inflorescence.
b) Flowers with stamens extended.
Secale cereale

Fig. 1069   Grass leaf pulled away from stem to show
open leaf sheath and ligule.

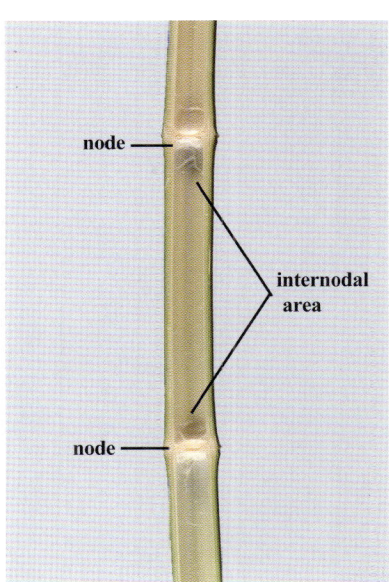

Fig. 1070   Grass leaf showing
parallel venation.

Fig. 1071   Grass leaf in-situ
showing how base may appear
closed even though it is open.

Fig. 1072   LS of bamboo stem
showing hollow areas between
nodes.   *Bambusa* sp.

Zingiberaceae - *Ginger Family*

Habit: Small to large herbs. Grow from rhizomes.

Leaves: Alternate, simple, entire. 2-ranked. Large, oblong to broadly elliptic. Petiole with air canals. Open basal sheath with ligule. Pinnate venation.

Inflorescences: Cyme.

Flowers: Showy, bilateral, bisexual. Ephemeral. Subtended by bracts.

Calyx: 3 sepals, connate.

Corolla: 3 petals, connate, one often differentiated.

Stamens: 1, style enclosed by filament and anther. 4 staminodes, 2 of which form petaloid lip (**labellum**).

Carpels: 3.

Ovary: Inferior.

Placentation: Axile.

Fruit: Fleshy loculicidal capsule or berry.

Seeds: Arillate.

Other: Pseudostem formed by overlapping leaf bases. Aromatic (ethereal oils present).

Genera: *Alpinia, Curcuma, Hedychium, Zingiber.*

Fig. 1073   Habit of butterfly ginger. *Hedychium coronarium*

   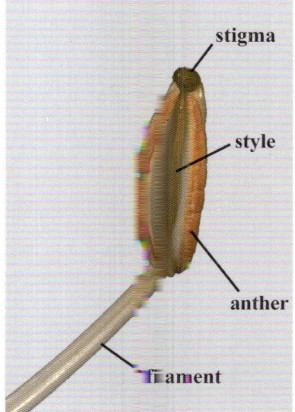

a   b   c   d

Fig. 1074   **Plant characters of butterfly ginger. a) Portion of stem showing ligulate leaf bases. b) Inflorescence with buds. c) Inflorescence with flower. d) CU of anther grasping style.** *Hedychium coronarium*

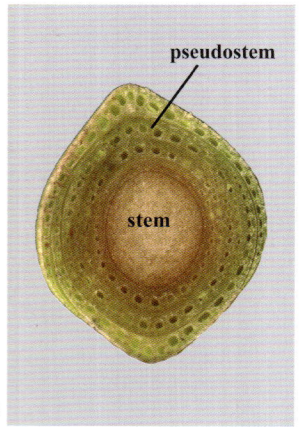   

a   b   c   d

Fig. 1075   **Plant characters of butterfly ginger. a) XS of stem and surrounding leaf sheaths that form a pseudostem. b) Developing infructescence. c) Dehiscing capsule. d) LS of capsule.** *Hedychium coronarium*

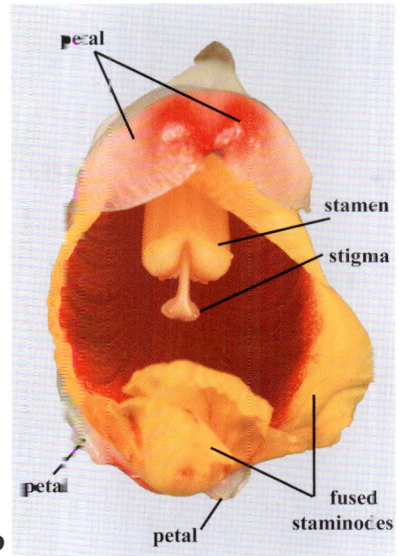

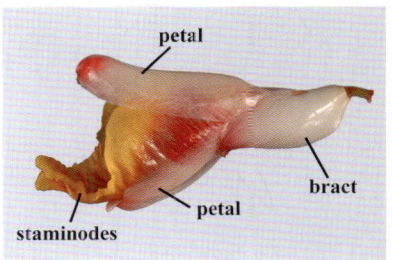

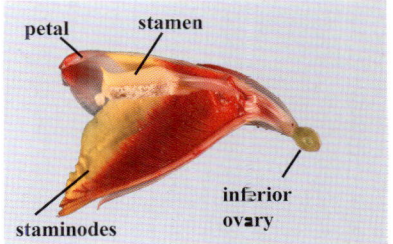

Fig. 1076   Floral characters of shell ginger. a) Inflorescence. b) Flower in front view. c) Flower in lateral view.
d) LS of flower.   *Alpinia speciosa*

Fig. 1077   Inflorescence of torch
ginger.   *Hedychium coccineum*

Fig. 1078   Inflorescence of hidden
lily.   *Curcuma* sp.

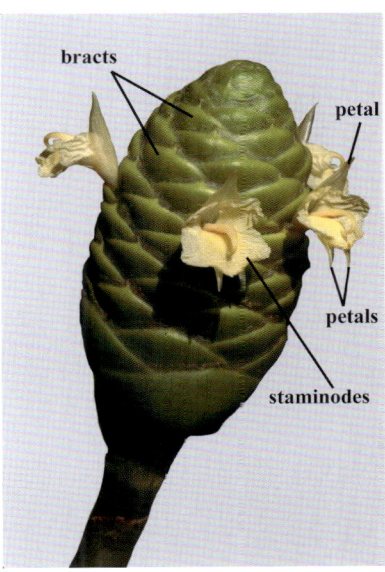

Fig. 1079   Inflorescence of pine
cone lily.   *Zingiber zerumbet*

Fig. 1080   Plant characters of peacock ginger. a) Habit and inflorescence. b) Flower. (Note the staminodes are the most
visible structures. The three petals and small sepals are hidden in this photo.)   *Kaempferia pulchra*

Costaceae - *Costus Family*

Habit: Small to medium herbs. Grow from rhizomes.
Leaves: Spirally arranged, simple, entire. Large, oblong to
broadly elliptic. Closed basal sheath with short ligule. Pinnate venation.
Inflorescence: Spike or head. Bracts spirally arranged
into cone-like structure.
Flowers: Showy, bilateral, bisexual. Ephemeral.
Subtended by bracts.
Calyx: 3 sepals, connate.
Corolla: 3 petals, one differentiated.
Stamens: 1, style enclosed by anther.
5 staminodes forming showy petaloid lip (**labellum**).
Carpels: 3.
Ovary: Inferior.
Placentation: Axile.
Fruit: Loculicidal capsule with persistent calyx.
Seeds: Arillate.
Other: Stem frequently spirally twisted.
Genera: *Costus, Dimerocostus, Monocostus.*

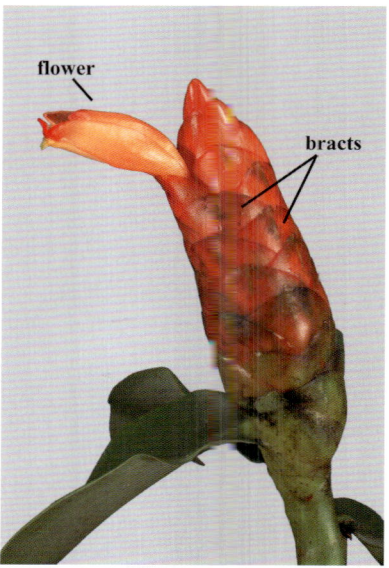

Fig. 1081 Inflorescence of candlestick ginger. *Costus scaber*

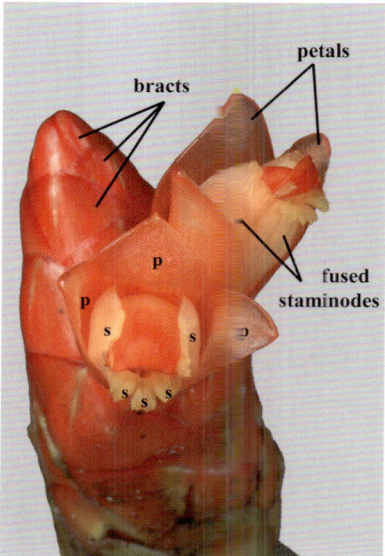

Fig. 1082 Plant characters of candlestick ginger. a) Habit. b) Inflorescence borne on cone-like structure formed of bracts. (s = one of five fused staminodes, p = one of three petals) *Costus scaber*

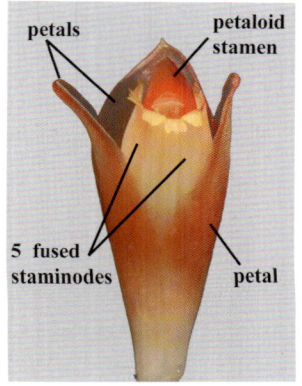

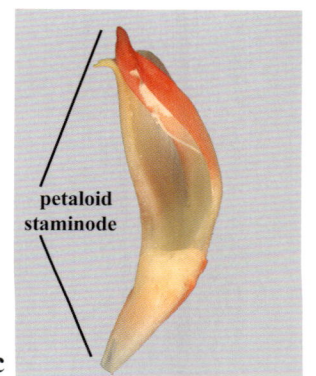
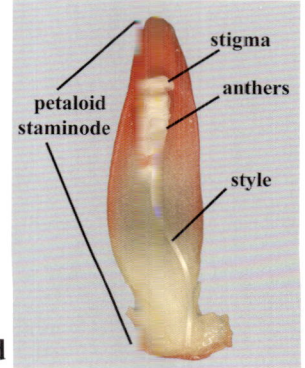

Fig. 1083 Floral characters of candlestick ginger. a) Ventral view of flower. b) Lateral view of flower with petals removed. c) Lateral view of petaloid staminode. d) Ventral view of petaloid staminode. *Costus scaber*

Marantaceae - *Arrowroot Family*

Habit: Erect herbs. Grow from rhizomes.
Leaves: Alternate, simple, entire. Oblong to broadly elliptic.
Open basal sheath with no ligule. Long petiole with upper pulvinus.
Pinnate venation. Blade folds upward at night (prayer plants).
Inflorescences: Cymes, often subtended by bracts.
Flowers: Showy, asymmetrical, bisexual. Each subtended
by 1-2 bracts. Arranged in mirror-image pairs.
Calyx: 3 sepals, distinct.
Corolla: 3 petals, connate. 1 enlarged.
Stamens: 1. 3-4 petaloid staminodes with 1 forming hood over style.
Stamen and staminodes connate and adnate to corolla.
Carpels: 3.
Ovary: Inferior.
Placentation: Axile.
Fruit: Loculicidal capsule or berry.
Seeds: Typically arillate.
Genera: *Calathea, Maranta, Thalia*

Fig. 1084   Leaf blade of alligator
flag.   *Thalia geniculata*

pulvinus

a　　　　　　　　　　　b　　　　　　　　　　　c

Fig. 1085   Plant characters of alligator flag   a) Pulvinus of leaf. b) Inflorescence. c) Flower.
Thalia geniculata

a

Fig. 1086   Plant characters of prayer plant   a) Foliage. b) Leaf. c) Flower.
Maranta leuconeura

Cannaceae - *Canna Family*

Habit: Erect herbs. Grow from rhizomes.
Leaves: Spirally arranged, simple, entire. Large, oblong to broadly elliptic. Open basal sheath with no ligule. Pinnate venation.
Inflorescences: Cymes. Main axis triangular in x.s.
Flowers: Large, showy, asymmetrical, bisexual. Each subtended by a bract. Ephemeral.
Calyx: 3 sepals, sometimes unequal.
Corolla: 3 petals, connate.
Stamens: 1. 3-4 petaloid staminodes with 1 differentiated. Stamen and staminodes connate and adnate to corolla.
Carpels: 3. Style petaloid.
Ovary: Inferior, surface papillate.
Placentation: Axile.
Fruit: Capsule with bumpy surface. Splits irregularly as fruit wall deteriorates.
Seeds: Black, spherical. Borne in clusters of hairs.
Genera: *Canna.* (A monogeneric family.)

Fig. 1087   Flowering plant in stand of canna.   *Canna* sp.

a

b

Fig. 1088   Floral characters of canna. a) Flowering plant with developing fruits. b) Asymmetrical flower. *Canna* sp.

a

b

c

d

Fig. 1089   Fruit characters of canna. a) Immature capsule. b) LS of immature capsule. c) Mature capsule. d) Mature capsule with portion of capsule removed to show large black seeds.   *Canna* sp.

Musaceae - *Banana Family*

Habit: Herbs, large to tree-like. Grow from corm.
Leaves: Spirally arranged, alternate, simple, entire. Long, oval to broadly elliptic. Clustered in whorl at top of plant. Long petiole with open basal sheath. Pinnate venation. Blade often tears into parallel strips attached to midvein.
Inflorescences: Terminal racemes, pendulous or erect. Spirally arranged with conspicuous bracts.
Flowers: Bilateral, unisexual.
Calyx: 3 sepals, petaloid.
Corolla: 3 petals, differentiated. 2 adnate to sepals.
Stamens: 5-6, distinct, adnate to corolla.
Carpels: 3.
Ovary: Inferior.
Placentation: Axile.
Fruit: Fleshy capsule or berry.
Seeds: Few to many.
Other: Overlapping leaf bases form pseudostem. Stem contains mucilage and sometimes milky sap.
Genera: *Ensete, Musa.*

Fig. 1090   Habit of banana. *Musa* sp.

Fig. 1091   Fruit and floral characters of banana. a) Pendent flower stalk with large purple bracts and green immature fruits. b) Flowers positioned atop bud-shaped bracts. c) Erect flower stalk.   *Musa* sp.

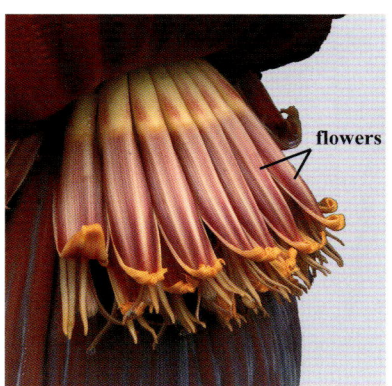

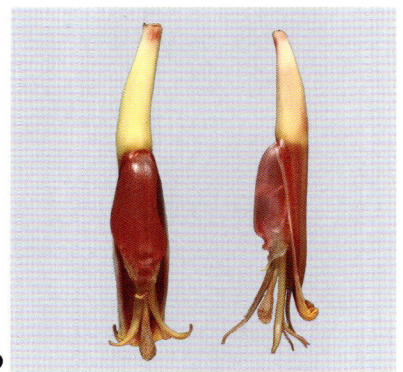

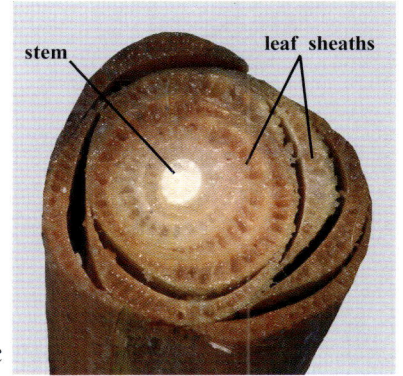

Fig. 1092   Floral and stem characters of banana. a) Cluster of flowers in-situ on stalk. b) Flowers in ventral and lateral views. c) XS of stem and surrounding overlapping leaf sheaths forming pseudostem.   *Musa* sp.

Heliconiaceae - *Heliconia Family*

Habit: Medium to large erect herbs.
Leaves: Alternate, simple, entire. 2-ranked. Large, oblong to elliptic.
Long petiole with open basal sheath. Pinnate venation.
Blade often tears into parallel strips attached to midvein.
Inflorescences: Erect or pendent. Showy 2-ranked bracts of red-
yellow-orange are most obvious part of inflorescences.
Flowers: Inconspicuous, zygomorphic, bisexual.
Subtended by boat-shaped bracts.
Calyx: 6 sepals. Fused, but with 1 more free than others and distinctive.
Corolla: 6 tepals.
Stamens: 5, 1 staminode.
Carpels: 3.
Ovary: Inferior.
Placentation: Axile.
Fruit: Drupe-like schizocarp. Frequently blue.
Seeds: Without aril.
Other: Overlapping leaf bases form pseudostem.
Genera: *Heliconia.*

Fig. 1093   Inflorescence of
heliconia.
Heliconia sp.

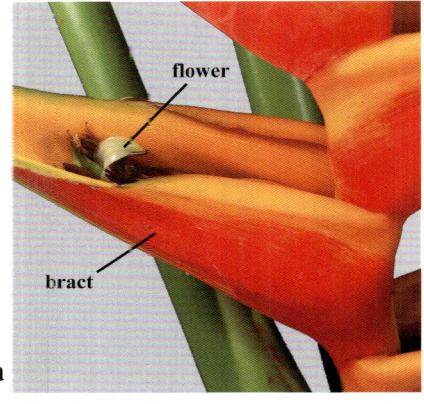
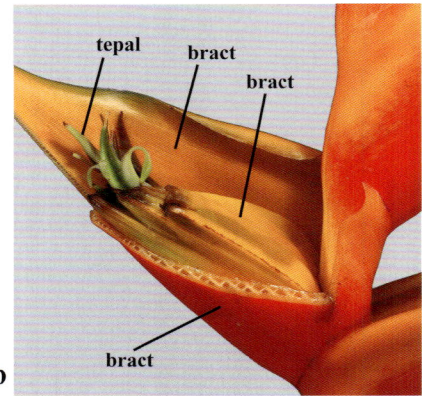
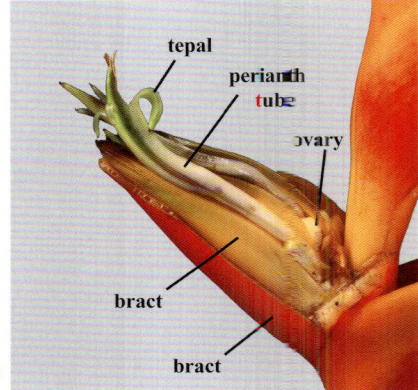

Fig. 1094   Floral characters of heliconia. a) Bract with small portion of flower showing. b) Portion of bract removed to
show flower within. c) Flower with subtending bract removed and one tepal removed.
Heliconia sp.

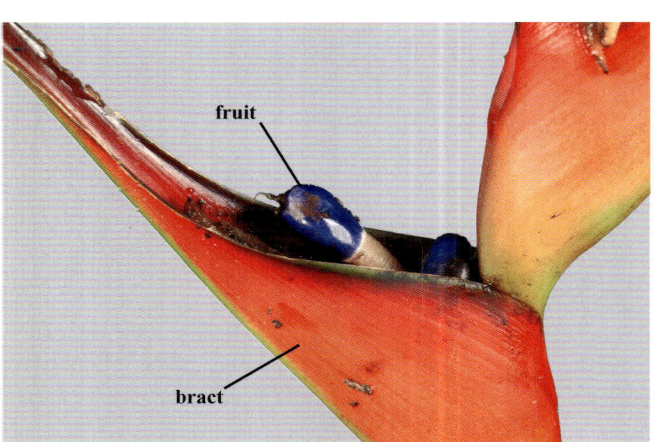

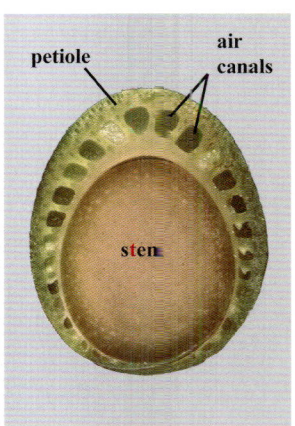

Fig. 1095   Plant characters of heliconia. a) Fruit in sheathing bract. b) Inflorescence. c) XS of stem and a sheathing
petiole showing air canals.
Heliconia spp.

Strelitziaceae - *Bird-of-Paradise Family*

Habit: Herbs, medium to tree-like.
Leaves: Alternate, simple, entire. 2-ranked. Oval to lanceolate. Long petiole with open basal sheath. Pinnate venation. Blade often tears into parallel strips attached to midvein.
Inflorescences: 1-several cymes, each subtended by a boat-like bract.
Flowers: Showy, bilateral, bisexual. Well-developed hypanthium.
Calyx: 3 sepals, petaloid.
Corolla: 3 petals, differentiated.
Stamens: 5, 1 or no staminodes.
Carpels: 3.
Ovary: Inferior.
Placentation: Axile.
Fruit: Loculicidal capsule.
Seeds: Bear brightly colored aril.
Genera: *Strelitzia, Ravenala.*

Photo by F.W. Howard

Fig. 1096 Traveler's palm habit. *Ravenala madagascariensis*

a

b

Fig. 1097 **Floral characters of bird-of-paradise. a) Inflorescence. b) Flower.** *Strelitzia reginae*

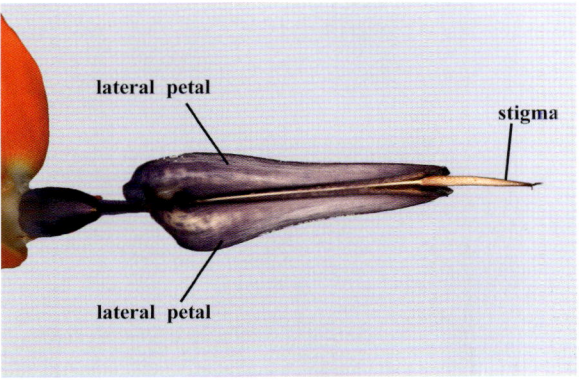

Photo by F.W. Howard

Fig. 1098 **Plant characters of bird-of-paradise. a) Fused lateral petals (blue) subtending androecium, style, and stigma. b) Habit.** *Strelitzia reginae*

Glossary

abaxial - Facing away from the stem axis, such as the underside of a leaf. Equivalent of dorsal.

acaulescent - Without a stem, or appearing that way such as when the stem is underground.

accessory fruit - Fruit or cluster of fruits that is made up predominantly or entirely of tissue not derived from the ovary.

accrescent - Increase in size with age following flowering.

achene - Typically small, dry indehiscent fruit with a thin wall surrounding a single seed.

acicular - Pointed or needle-shaped.

acorn - Nut partially to entirely enclosed by a scaly cupule. Fruit of trees or shrubs in the genus *Quercus*, and their relatives.

actinomorphic - Flower or structure that is **radially symmetrical** or divisible into equal halves by two or more planes. Equivalent of **regular** when applied to flowers.

acuminate apex - Leaf that tapers to a narrowly elongate tip with the sides slightly concave.

acute apex or **acute base -** Terminating in a point less than 90 degrees with slightly convex tapering sides.

adaxial - Facing towards the stem axis, such as the upper side of a leaf. Equivalent of ventral.

adnate - Fusion of dissimilar parts or structures to one another, such as stamens to petals.

adventitious root - Root that does not originate from root tissue, such as from a leaf or stem.

aerial root - Root growing above ground or water, often used in lieu of adventitious root.

aggregate fruit - Compound fruit resulting from several to many carpels of a single flower.

alternate leaf arrangement - One leaf per node. Single leaves at various heights along the stem, typically alternating from one side of the stem to the other. (If internodes are reduced, may appear whorled or fascicled.

ament - Inflorescence of small, inconspicuous commonly wind-pollinated flowers. Equivalent of catkin.

androecium - Collective term for all the stamens of a flower. All the male reproductive structures.

androgynophore - Stalk that bears all the stamens (androecium) and all the carpels (gynoecium).

annual - Plant that develops from seed to maturity within one year.

annulus - Row or cluster of thick-walled cells on a fern sporangium that function in spore release.

anther - Apical portion of the stamen that contains the pollen sacs and produces the pollen.

antheridium - Male reproductive structure of ferns that is a tiny capsule containing sperm. Found at the base and wings of a fern prothallus.

anthesis - Period during which a flower opens.

apical - Towards the tip or apex of a structure.

apical meristem - Tissue at the tip of a stem or a root from which new cells are created.

apical placentation - Condition where the ovules are attached at the tip or apex of an ovary.

apocarpous - Gynoecium with carpels separate or unfused.

apopetalous - Flower with separate or unfused petals.

aposepalous - Flower with separate or unfused sepals.

apotepalous - Flower with separate or unfused tepals.

appressed - Condition where structures are pressed closely against one another.

archegonium - Female reproductive structure of ferns that is tiny and flask-shaped, and that produces and encloses an egg. Found on the central area of the fern prothallus below the notch.

areole (= areola) - Small areas on the stem of Cactaceae that contains spines or hairs or glochids. Also used for any small well-defined area on a surface, especially on cacti.

aril - Outgrowth of the funiculus, associated with the seed. Typically brightly-colored and protein- or lipid-rich, functioning to attract seed dispersers.

asymmetrical - Not symmetrical. Lacking a plane of symmetry, such as a leaf with an oblique base where the opposite sides do not allign.

attenuate apex - Leaf tip, perianth, or other structure that gradually tapers to a narrow apex.

awn - Bristle-like or horn-like structure, generally used to refer to pointed floral or fruit appendages.

axil - Angle between the upper surface of a leaf and the stem.

axile placentation - Ovules are attached to the central axis or inner angle formed by septa of an ovary with two or more locules.

axillary branch - Branch originating from an axillary bud, often synonomous with lateral branch.

axillary bud - Bud originating in the leaf axil.

banner petal - Largest and uppermost petal on the papilionaceous flowers of the Subfamilies Faboideae and Caesalpinoideae of the Fabaceae. Also called the **flag petal** or **standard petal**.

bark - All tissues outside the vascular cambium in a woody stem.

basal placentation - Ovules are attached at the base of the ovary.

beak - Extended, pointed portion of a structure.

berry - Indehiscent fruit that is fleshy and typically contains several to many seeds.

betalains - Red, yellow, and orange pigments found in most plants in the Caryophyllales.

biennial - Plant that grows from seed to maturity in two years. Vegetative growth, often as a basal rosette, occurs during the first, while flowering and fruiting occurs during the second.

bifoliate - Compound leaf with two leaflets.

bilabiate - Having two lips. Typically applied to a calyx or corolla with dimorphic upper and lower portions.

bilateral symmetry - Flower or structure that is divisible into two equal halves by only a single plane. Equivalent of **irregular** or **zygomorphic** when applied to flowers.

bipinnate - Doubly pinnate. Primary leaf segments of a compound leaf are themselves divided into leaflets.

biseriate - Arranged in two whorls, such as the stamens or petals of a flower.

bisexual flower - Flower containing both a gynoecium and androecium. Equivalent with a **perfect flower**.

blade - Flat, expanded photosynthetic portion of a leaf. Equivalent to the **lamina**.

bract - Reduced or modified leaf associated with a flower or inflorescence.

bracteole - Small bract.

bractlet - Small bract.

bristle - Stiff hair.

bud - Very young undeveloped stem, leaf, or flower that is often covered by protective structures such as bud scales, stipules, or hairs.

bud scale - Small, modified leaf that covers and protects the bud.

bud scale scar - Scar on a twig or stem that indicates where a bud scale was attached.

bulb - Short underground stem surrounded by layers of thick, fleshy modified leaves (**scales**).

bulbil - Small bulb or vegetatively produced bulb-like structure that gives rise to a new plant.

buttress root - Large, flattened fin-like extension of the trunk base of some trees.

caducous - Falling off quickly, such as petals that only remain briefly after a flower opens.

calyx - Collective term for all the sepals, which form the outermost whorl of the perianth.

campanulate - Shaped like a bell.

capitate - Enlarged at the tip with a narrow neck-like base and a circular head-like tip, or compacted into a dense head such as the flowers in an inflorescence.

capitulum - Compact inflorescence with short axis and sessile flowers. Equivalent with **head**.

capsule - Dry dehiscent fruit produced from two or more carpels.

carnivorous - Flesh-eating, such as some plants that trap insects in the Droseraceae, Lentibulariaceae, and Sarraceniaceae.

carpel - Individual female reproductive structure composed of stigma, style, and ovary. Equivalent to the pistil if there is only one carpel, or if all carpels are separate and unfused.

carpellate flower - Flower with a functioning gynoecium but without a functioning androecium. Equivalent to a **female flower**.

caruncle - Outgrowth of integument (outer seed coat) at the point of seed attachment. Can be warty or ridge-like and is sometimes considered an aril.

caryopsis - Small, dry indehiscent fruit which has a thin coat fused to a single seed. Characteristic fruit of grasses. A grain.

cataphyll - Protective leaves varying from small and scale-like to large and pointed, such as those found on cycads around the terminal bud.

catkin - Erect or pendent spike of densely packed apetalous flowers. Typically wind-pollinated.

caudate - Bearing a tail-like structure.

caulescent - Having a stem.

cauliflory - Bearing flowers or inflorescences directly on woody branches or trunks.

cauline - Pertaining to the stem.

chaffy - Thin membranous tissues or structures that are dry and papery.

chartaceous - Having a papery texture.

chlorophyll - Green pigment of plant cells that is light sensitive and essential for photosynthesis.

chloroplast - Cellular organs that contain chlorophyll and serve as the site for photosynthesis.

circinate - Coiled at the apex with the tip in the center, such as in fern fronds and some inflorescences.

circumscissile capsule - Capsule that dehisces along a horizontal line that circumscribes the entire fruit, resulting in the top portion falling off like a lid.

clavate - Shaped like a club.

clawed - Having the form of a slender basal stalk and expanded apical limb, such as in petals or tepals.

cleistogamous - Flower that does not open and is self-pollinated or sets seed asexually.

collenchyma - Cells that form a supporting tissue in regions of primary growth.

colleter - Multicellular hair that produces a sticky secretion. Usually associated with the nodal region or calyx.

column - Structure formed by the fusion of the androecium and gynoecium in the Orchidaceae.

coma - Tuft of hairs. Usually associated with small fruits or seeds that are wind-dispersed.

complete flower - Flower with four whorls of floral parts: sepals, petals, stamens, carpels.

compound inflorescence - Inflorescence containing two or more orders of branches.

compound leaf - Leaf with a blade having two or more leaflets.

cone - Female reproductive structure of conifers consisting of ovule bearing scales clustered terminally on a stem.

conifer - Tree or shrub that bears cones.

connate - Fusion of similar organs or parts to one another, such as petals to petals.

connective - Portion of an anther that connects the two pollen sacs.

contorted - Twisted.

contractile root - Root that grows by alternately shrinking and expanding, resulting in anchoring the stem into the soil or supporting substrate.

convolute - Rolled with edges overlapping.

cordate - Heart-shaped. Usually a leaf base with the lobes rounded.

coriaceous - Thick and leathery.

corm - Short, vertical underground stem surrounded by thin papery leaves.

corolla - Collective term for all the petals of a flower.

corona - Outgrowth or appendage between the corolla and androecium.

cortex - Tissue of a root or stem located between the epidermis and the vascular tissue.

corymb - Inflorescence that is short, wide, and at one level or flat at the top.

cosmopolitan - Worldwide in distribution.

costapalmate - Palmate leaf with a petiole that extends through the blade forming a midrib.

cotyledon - Primary leaf in the plant embryo.

crenate - Margin characterized by rounded teeth or appearing scalloped.

compound fruit - Fruit with two or more closely associated, mature, unfused carpels. Either a multiple fruit or an aggregate fruit.

connivent - Term for unfused structures (especially floral) that grow closely together and appear fused.

cross-pollination - Transfer of pollen from the anther of one plant to the stigma of another plant.

cruciform - Shaped like a cross.

cucullate - Hooded or shaped like a hood.

cuneate - Triangular in shape but tapering to a point. Narrowly wedge-shaped.

cupule - Structure shaped like a cup and found at the base of certain fruits, such as on nuts in the Fagaceae and drupes in the Lauraceae.

cuticle - Thin, waxy layer on the outer wall of epidermal cells.

cyathium - Specialized inflorescence of some Euphorbiaceae where a single carpellate flower and multiple staminate flowers are clustered together and imitate a single flower.

cyme - Compound inflorescence consisting of clusters of three flowers. Determinate inflorescence with the central flower developing first.

deciduous - Shedding leaves at the end of the growing season, so that the plant is left leafless.

decurrent - Extending downward, such as projection of leaf base along stem.

decurrent base - Base of leaf that tapers to a slender point of attachment.

decussate - Opposite leaves where each pair is at right angles to the previous and following pair.

dehiscence - Act or process by which a structure splits or opens when mature, such as in fruit or anthers.

deltoid - Shaped like a triangle.

dentate - Toothed.

dentate margin - Toothed edge where the teeth are at right angles to the line of the margin.

denticidal capsule - Capsule that opens by means of apical teeth.

determinate inflorescence - Inflorescence in which the main axis terminates in a flower that opens first causing growth of the primary axis to stop.

diadelphous - Stamens arranged in two groups and united by their filaments.

didynamous - Four stamens in two pairs, one pair of which is longer than the other.

dimorphic - Having two different and distinct shapes or forms.

dioecious - Having staminate and carpellate flowers on different plants of the same species. Equivalent with **unisexual**.

disk floret or **disk flower -** Tubular flower with radial symmetry found in many species of the Asteraceae.

dissected - Divided deeply.

distichous - Structures placed on only two sides of an axis, such as leaves along a stem. Equivalent with **two-ranked**.

distinct - Separate or not fused to similar structures, such as stamens or petals.

distyly - Having two different style lengths on different plants of the same species

domatium - Plant structure such as a bubble-like outgrowth of tissue, tuft of hairs, or cavity serving as a home or protected location for small insects or arthropods that often live symbiotically with the plant.

dorsal - Facing away from the axis, such as the underside of a leaf. Equivalent with **abaxial**.

doubly serrate - Leaf margin bearing both large and small teeth.

drupe - Indehiscent fleshy fruit with one or more hard pits called stones or pyrenes in the center.

elater - Thickened band of moisture-sensitive tissue attached to the spores of horsetails (Equisetaceae) that uncoils as the spores dry and helps in their dispersal from the sporangium.

elliptic - Any plant structure that is widest at the middle or oval in shape.

emarginate apex - Leaf with a shallow notch at the tip or apex.

endocarp - Innermost layer of the mature ovary wall.

endodermis - Specialized layer of cells enclosing the vascular region of some roots and stems.

endosperm - Nutritive tissue of many angiosperm seeds.

ensiform - Shaped like a sword.

entire - Margin that is continuous and without any teeth or divisions.

ephemeral - Existing for a brief period, usually a day or less.

epicalyx - Whorl of bracts surrounding the true calyx.

epidermis - Outermost layer of cells of primary plant tissues (leaves, young stems, roots).

epigynous flower - Flower with an inferior ovary and the perianth and stamens arising from the floral cup or hypanthium (that is adnate to the ovary).

epipetalous stamens - Stamens that are adnate or attached to the petals.

epiphylly - Condition of a structure (ex. flower) being borne on a leaf or an organism (ex. liverwort) growing on a leaf of a different organism.

epiphyte - Plant that grows on another plant and uses it for support.

epitepalous stamens - Stamens that are adnate or attached to the tepals.

equitant - Flattened, overlapping leaves arranged in two ranks such as in the Iridaceae and Typhaceae.

ericoid leaf - Thick, needle-like leaf whose margins roll towards the middle of the underside of the leaf.

ethereal oils - Secondary plant compounds that are highly aromatic.

eusporangium - Sporangium without a stalk and with a wall several cell layers thick.

eustele - Stele with the vascular bundles distinct and arranged around a central pith.

evergreen - Plant or tree that is green and with leaves throughout the year.

exserted - One structure projecting beyond another, such as stamens projecting beyond petals.

exstipulate - Without stipules.

extrafloral nectary - Nectar-producing gland that is not physically associated with a flower.

extrastaminal nectar disk - Nectar disk located between the stamens and the perianth.

extrorse - Opening outwards, such as anthers that open towards the outer part of the flower.

false indusium - Recurved leaf margin that folds over sori developing at the leaf's edge. It is a part of the leaf and not a separate structure like a true indusium.

fascicle - Bundle or cluster of structures, such as flowers on an inflorescence or leaves on a short shoot.

fenestration - Window-like opening.

fertile - Bearing fruit or spores.

fertilization - Fusion of the sperm from a pollen grain with the egg in an ovule.

fibrous root system - Root system characterized by many branching roots of approximately the same size.

fiddlehead - Coiled immature leaf of a fern with the apex at the center of the roll. Fiddleheads have a **circinate** shape.

filament - Basal stalk of a stamen on top of which is the anther.

filiform - Long and slender in shape or thread-like.

fimbriate - Having a fringe.

flag petal - Largest and uppermost petal on the papilionaceous flowers of the Subfamilies Faboideae and Caesalpinoideae of the Fabaceae. Also called the **banner petal** or **standard petal**.

floral cup - Cup-like or tubular structure that bears the perianth and stamens, and is sometimes adnate to the ovary. Equivalent to the **hypanthium** and to the **floral tube**.

floral formula - Abbreviated and formulaic way of writing a description of a flower that includes the information regarding symmetry, the number of parts, and their position and insertion.

floral tube - Cup-like or tubular structure that bears the perianth and stamens, and is sometimes adnate to the ovary. Equivalent to the **hypanthium** and to the **floral cup**.

floret - Very small individual flower, such as those characteristic of the Asteraceae, Poaceae, and Cyperaceae.

flower - Reproductive structure of angiosperms composed of a modified shoot and modified leaves.

flower bud - Bud with very young, undeveloped flowers.

foliaceous - Shaped like a leaf.

follicle - Dry, dehiscent fruit that originates from a single carpel and splits along a single suture.

free - Separate and unfused.

free-central - Placentation where the ovules are attached to the central axis or column of an ovary with one locule.

frond - Leaf of a fern, consisting of a stipe and blade.

fruit - Mature ovary together with any adjacent parts that have become fused to it.

fugaceous - Falling off quickly.

funiculus - Stalk of an ovule where it attaches to the placenta. Also called a **funicle**.

funnelform - Shaped like a funnel.

fusiform - Thick in the middle and tapering at both ends. Shaped like a spindle.

gametangium - Structure that produces gametes.

gamete - Sex cell containing only half the regular number of chromosomes (haploid). Two sex cells fuse to make a zygote, which then has the full number of chromosomes (diploid).

gametophyte - Haploid phase of the plant reproductive cycle that produces the gametes.

geniculate - With a bent shape, like an elbow. Jointed.

glabrous - Smooth, without hairs.

glandular hair - Hair bearing a secretory gland.

glaucous - Waxy covering, usually whitish in color.

globose - Spherical, shaped like a globe.

glochid - Tiny barbed bristle that occurs in clusters on some species of the Cactaceae.

glume - One of two small papery bracts at the base of a grass spikelet, absent in some genera.

grain - Small, dry indehiscent fruit which has a thin coat fused to a single seed. Characteristic fruit of grasses. Equivalent with **caryopsis**.

ground meristem - Meristematic tissue that produces ground tissues. Equivalent with **primary meristem**.

ground tissue - All tissues other than vascular tissues, the epidermis, and the periderm.

gymnosperm - Seed plant in which the seeds are not enclosed by an ovary, such as the conifers.

gynoecium - Collective term for all the carpels of a flower. All the female reproductive structures.

gynophore - Stalk that bears the gynoecium.

gynostegium - Compound organ in the Asclepiadaceae formed by the fusion of the stigma and stamens.

gynostemium - Compound organ in some Aristolochiaceae formed by the fusion of the style and stamens. Also found in some orchids where the stamens fuse with the style and stigma to form a column.

habit - Overall or general appearance of a plant.

hair - Extension of the epidermis with varied forms and functions. Equivalent with **trichome**.

hastate - Shaped like an arrowhead.

hastula - Woody structure (ligule) located where the petiole joins the blade on many leaves of palms (Arecaceae).

haustorium - Penetrating root of a parasitic plant that invades the vascular tissue of the host plant. (Also called a **haustorial root**).

head - Compact inflorescence with short axis and sessile flowers. Equivalent with **capitulum**.

helicoid cyme - Coiled cyme with all lateral branches developing from the same side of the main axis.

hemiparasite - Plant that parasitizes a host but is also capable of photosynthesis. It does not rely on the host for all of its nutritional needs, only tapping into the xylem or water transport system.

herb - Plant that does not have persistent woody tissue above ground.

herbaceous - Without woody tissue, and usually dying at the end of the growing season.

hesperidium - Specialized berry found in citrus and its relatives. Consists of a leathery exterior or rind and a fleshy, often juicy interior that is divided into several to many sections.

heterophylly - Having leaves of more than one distinct shape on the same plant. Such leaves may differ due to age, due to environmental factors such as light intensity, or they may have a genetic basis.

heterosporous - Having two kinds of spores (microspores and megaspores).

hilum - Scar on a seed where the funiculus was attached.

hirsute - Hairy with coarse hairs that are rough in texture.

hispid - Hairy with stiff or rigid hairs.

homogamous flower - Flower that sheds pollen at the same time that the stigma of the same flower is receptive.

homogamous head - Flower head consisting of only one kind of floret, such as a head of disk florets in members of the Asteraceae.

homosporous - Having one kind of spore.

hood - Extension of the filaments shaped like a hood, such as in members of the Asclepiadaceae.

horn - Extension of the filaments shaped like a peg or horn, such as in members of the Asclepiadaceae.

husk - Exterior covering of some fruits and nuts.

hyaline - Transparent, thin, membranous material.

hypanthium - Cup-like or tubular structure that bears the perianth and stamens, and is sometimes adnate to the ovary. Equivalent to the **floral cup** or **floral tube**.

hypodermis - One to several differentiated layers of cells beneath the epidermis.

hypogynous flower - Flower with a superior ovary and the perianth and stamens arising from below the ovary, with no hypanthium.

imbricate - Overlapping, such as the arrangement found with some bracts, leaves, and petals. One unit overlapping the other like the shingles on a roof.

imperfect flower - Flower without either a functional androecium or a functional gynoecium. Equivalent with a **unisexual flower**.

included structures - Structures that are hidden or enclosed and therefore not readily visible.

incomplete flower - Flower missing one or more of the four whorls of floral parts: sepals, petals, stamens, carpels.

indehiscent - Structure or fruit that does not open.

indehiscent pod - Fruit that is typically dry and does not split open at maturity.

indeterminate inflorescence - Inflorescence in which the main axis produces only lateral flowers.

indumentum - Collective term for the hairs that cover a plant's surface.

induplicate - Folded down or inwards so that when viewed in XS it is V-shaped.

indusium - Membranous outgrowth of fern leaf that acts as a protective flap over the developing sorus.

inferior ovary - Ovary located in a position below the point where the perianth parts are attached to the receptacle or flower stalk, or below the basal point of a hypanthium or floral cup.

inflorescence - Arrangement of flowers on a floral axis, sometimes consisting of many flowers in a cluster or a solitary flower.

infructescence - Arrangement of fruits on a floral axis.

insertion of floral parts - Arrangement or pattern of floral parts as they are attached to the receptacle.

integument - Outer portion of an ovule that develops into the seed coat.

interfascicular region - Region between the vascular bundles of a stem.

internode - Portion of a stem between two successive or adjacent nodes.

interpetiolar stipules - Stipules located at a node between the bases of the petioles of opposite leaves.

intrastaminal nectar disk - Nectar disk located between the stamens and the gynoecium of a flower.

introrse - Opening inwards towards the center.

intruded placenta - Placenta intruding into the locule of an ovary.

involucre - Cluster of bracts surrounding a flower or inflorescence.

involute - Having margins rolled inwards towards the upper surface.

irregular symmetry - Flower or structure that is divisible into two equal halves by only a single plane. Equivalent of **bilateral symmetry** or **zygomorphic** when applied to flowers.

jaculator - Thickened funiculus that aids in dispersing the seed from the fruit, such as in many species of the Acanthaceae. Equivalent of **retinaculum**.

keel petals - Two fused petals typically lowermost in position forming part of the papilionaceous flowers found in species of the Subfamily Faboideae in the Fabaceae. Or the lowermost canoe-like petal in many Polygalaceae.

labellum - Lip. Often used to refer to a particular petal that is differentiated from all others, especially in orchids but also in other families.

laciniate - Composed of narrow, irregular lobes.

lacticifer - Cell containing latex.

lamina - Blade of a leaf.

lanceolate - Shaped like a lance. Longer than wide, being broadest at the base and tapering to a point.

lateral bud - Bud that is not located at the tip of a shoot.

lateral root - Root that originates from an older root. Equivalent with **branch root** or **secondary root**.

latex - Plant sap produced by specialized cells (lacticifers), often milky or colored but not necessarily.

latrorse anther - Anther with locules that open on the side.

leaf - Flat lateral appendage of a stem or branch that serves as the plant's primary photosynthetic organ.

leaf bud - Bud made up of young undeveloped leaves.

leaf primordium - Tissue forming a lateral outgrowth of the apical meristem that will become a leaf.

leaflet - One of the blades making up a compound leaf.

leaf scar - Scar on a twig after a leaf has fallen that indicates its point of attachment.

legume - Dry fruit originating from a single carpel that typically dehisces by means of two longitudinal sutures. Found in species in the Fabaceae, although there are many modifications.

lemma - Lowest of the two bracts subtending a grass floret. Found on species of the Family Poaceae.

lenticel - Outgrowth, usually corky, of a plant surface that functions in gas exchange.

lenticular - Shaped like a lens that is convex on both sides.

leptosporangium - Sporangium that is borne on a stalk, opens via an annulus, and is only one cell layer thick.

liana - Woody vine.

ligulate flower - Flat, strap-like fertile floret consisting of a petaloid limb with 5 apical teeth. Found in some species of the Asteraceae.

ligule - Outgrowth of tissue at the top of the leaf sheath in some plant families such as the Poaceae.

limb - Flattened expanded portion of a petal, or the expanded free portion of a fused corolla.

linear - Structure that is long and narrow.

lip - Differentiated petal of an orchid (= **labellum**), or one of two unequal portions of a calyx or corolla that is composed of fused petals or sepals

lobate base - Leaf base with distinct rounded lobes.

lobed margin - Leaf margin consisting of rounded projections.

locule - Cavity or chamber such as is found in an anther or ovary.

loculicidal capsule - Capsule that dehisces into the locule, forming segments or valves that have parts of two adjacent locules and that are divided longitudinally by a septum.

lodicules - Scale-like reduced perianth parts in the flowers of species of the Poaceae.

loment - Dry fruit produced from a single carpel that breaks along transverse lines into segments of one seed each. Only found in the Fabaceae.

long shoot - Stem or branch with long internodes on a plant where both long and short internodes occur.

lorate - Long and narrow or shaped like a strap.

mangrove - Tropical trees that grow in coastal areas that are flooded by tides.

megagametophyte - Female gametophyte, found within the ovule of seed plants and of heterosporous ferns and their allies.

megaphyll - Large leaf with several to many veins, as exemplified by ferns and seed plants.

megasporangium - Sporangium that produces megaspores.

megaspore - Large haploid spore that grows into a female gametophyte (megagametophyte)

megasporophyll - Leaf-like structure that produces megasporangia.

membranous - Consisting of a membrane. Thin and flexible in texture.

mericarp - One segment of a schizocarp fruit.

-merous - Suffix referring to the base number of parts, usually in reference to petals/sepals. (Ex. 3-merous signifies petals in groups of 3 so that the total numbers may be 3, 6, 9, 12, etc.)

mesocarp - Middle layer of the ovary wall.

microgametophyte - Male gametophyte. Mature pollen grain in seed plants.

microphyll - Small leaf with one vein and no leaf gap found in members of the Lycopsids.

microsporangium - Sporangium that produces microspores.

microspore - Spore that develops into a male gametophyte.

microsporophyll - Foliaceous plant structure that bears microsporangia.

midvein - Large central vein of a leaf.

monadelphous - Stamens of a flower united together by their filaments forming a tube.

moniliform - Shaped like a string of beads, such as with some hairs.

monoecious - Having staminate and carpellate flowers on the same plant.

mucilage - Substance that is slimy or sticky. Mucous-like.

mucronate apex - Apex that ends in a short point.

multiple fruit - Compound fruit formed from multiple, closely packed flowers where the developing structures coalesce into a single unit.

multiseriate - Having many layers.

muricate - With short hard projections, thus often rough to the touch.

mustard oils - Aromatic compounds found in mustards and related plants.

mycorrhiza - Symbiotic relationship between certain fungi and the roots of vascular plants.

naked bud - Bud that is not covered or protected by scales or stipules.

nectar - Sugary liquid produced by plants to attract pollinators or other beneficial animals.

nectar disk - Nectar-producing structure that is disk-like in shape.

nectar gland - Gland that produces nectar. Equivalent with **nectary**.

nectar spur - Hollow projection of a petal or sepal that contains nectar.

nectary - Structure or gland that produces nectar. Equivalent with **nectar gland**.

node - Point along the stem from which the leaves and buds arise.

nut - Indehiscent dry fruit consisting of a single seed surrounded by a hard, smooth wall.

nutlet - Same as nut, only smaller in size. Determination between nut and nutlet is relative.

oblique base - Leaf base that is asymmetrical.

oblong - Leaf where the margins are more or less parallel-sided.

obovate - Leaf that is broader at the apex than at the base.

obtuse - Leaf apex/base that is blunt.

ocrea - Tubal sheath formed by the fusion of two stipules at the nodal areas in most Polygonaceae.

odd pinnate - Pinnately compound leaf with an odd number of leaflets, as seen by a single terminal leaflet.

operculate - Having or opening by means of a cap or lid.

operculum - Cap or lid.

opposite - Leaf arrangement where two leaves arise from the stem at opposite sides of the same node.

ovary - Enlarged basal portion of a carpel, or of a gynoecium with fused carpels, that contains the ovules and which at maturity becomes a fruit.

ovate - Leaf where the base of the blade is broader than the apex.

ovulate scale - Scale-like shoot of some conifers to which the ovule is attached.

ovule - Organ that develops into a seed after fertilization within the locule of the ovary. It is attached to the ovarian wall at the placenta by a stalk (funiculus).

palea - Upper bract of the two that typically subtend a floret in members of the Poaceae.

palisade mesophyll - Columnar layers of chloroplast-containing cells located between the layers of the epidermis.

palmate - Arising from a common point and diverging.

palmate leaf - Simple leaf with lobes that diverge outward from the base.

palmate venation - Pattern of venation where three or more main veins diverge outward from the area of the leaf base. The veins may remain divergent or loop back and converge near the apex.

palmately compound leaf - Compound leaf with three or more leaflets attached to a common central point.

panicle - Inflorescence with compound branching.

papilionaceous - 'Butterfly' type flower found in members of the Subfamily Faboideae (formerly the Papilionoideae) in the Fabaceae. Flower with a corolla consisting of a banner petal, two wing petals, and two fused petals forming a keel. Superficially similar flowers are found in some Polygalaceae.

papillate - Surface covered with minute, nipple-like bumps.

pappus - Reduced calyx of members of the Asteraceae that varies in form including hairs, scales, and bristles.

parallel venation - Pattern of venation where the veins are generally parallel to the leaf margins, such as in many of the monocots.

parenchyma cells - Common thin-walled cells that vary in size and shape.

parietal - Placentation characterized by the ovules being attached to the walls of the ovary.

pedicel - Stalk of a flower or fruit.

peduncle - Stalk of a flower or fruit cluster. Can be difficult to distinguish from a pedicel if only one flower is present.

pellucid dots - Clear dots caused by cavities in the tissue found in various plant structures

peltate leaf - Leaf where the petiole attaches near the center of the blade rather than at the margin.

pendulous - Hanging downward or drooping.

pepo - Specialized berry with a leathery to hard exterior or rind and a fleshy interior. Found in some species in the Cucurbitaceae.

perennial - Plant that lives three years or more, tending to flower and fruit repeatedly.

perfect flower - Flower with both a functional androecium and gynoecium. Equivalent with a **bisexual flower**.

perianth - Collective term for the calyx and corolla. All the sepals and all the petals collectively.

pericarp - Wall of a fruit (the mature ovary).

perigynous flower - Flower with a superior ovary and a hypanthium (perianth parts and stamens attached to the margin of a floral cup that is separate from and surrounds the ovary.) Sometimes the hypanthium is reduced to a small rim.

persistent - Structure that does not fall off but remains attached to the plant.

petal - Single unit of the corolla, usually colorful and serving to attract pollinators.

petaloid - Petal-like in appearance.

petiole - Stalk of a leaf.

petiolule - Stalk of a leaflet.

phloem - Vascular tissue that transports sugars.

photosynthesis - Reaction occurring in chloroplasts that converts light energy from the sun into chemical energy. Production in green plants of carbohydrates and oxygen from carbon dioxide and water in the presence of light.

phyllary - One of many bracts that make up an involucre found on the flower heads in the Asteraceae.

phylloclade - Stem that is somewhat leaf-like (flat and expanded) in appearance and in function.

phylogeny - Evolutionary history of a group of organisms.

phytomelan - Black, crust-like substance that covers seeds in many of the families of the Asparagales.

pilose - Covering of long, soft hairs.

pinna - Leaflet of a pinnately compound leaf or frond. Sometimes it is further divided.

pinnate venation - Pattern of venation characterized by a primary midvein with smaller secondary veins arising from both sides like the pinnae of a feather.

pinnately compound leaf - Compound leaf with more than three leaflets, arising from both sides of the rachis like the pinnae of a feather.

pinnule - Ultimate leaflet or segment of a compound leaf (usually used with bi- or tripinnate leaf).

pistil - Ovule producing organ of a flower consisting of a stigma, style, and one or more carpels.

pit - Inner hard portion of a drupe containing the seed. Equivalent with **pyrene** and **stone**.

pith - Soft tissue at the center of a stem.

placenta - Part of the ovary to which the ovules are attached.

placentation - Manner of attachment or arrangement of ovules in an ovary.

plicate - Folded in a fan-like manner.

plinervy venation - Subset of palmate venation with 3 strong primary veins at base, often with pinnate venation above middle of leaf.

plumose - Fuzzy or feather-like.

plunger pollination - Mechanism by which pollen is secondarily offered to pollinators on a modified style that passes through the anthers within the flower.

pneumatophore - Specialized roots found on mangroves and other swamp species that grow up out of the water and aid in gas exchange.

pollen - Microspores containing the male gametophyte, produced in the anthers. Ultimately responsible for transferring sperm to egg.

pollen sacs - Chambers that contain the pollen in the anthers.

pollen tube - Tube that grows from the stigma to the ovary after a pollen grain has germinated. Functions to carry sperm to the ovules.

pollination - Transfer of pollen from an anther to a stigma in angiosperms, or of pollen from a male strobilus to an ovule on a female cone in gymnosperms.

pollinium - Cluster or mass of pollen grains that is transported as a single unit during pollination in the Orchidaceae and Apocynaceae.

polygamous - Having both perfect (bisexual) and imperfect (unisexual) flowers on the same plant.

pome - Fleshy indehiscent fruit where the outer portion is soft and formed from the expansion of the floral structures that surround the ovary and enlarge as the fruit matures. The central portion contains the seeds and is surrounded by a cartilaginous or papery structure that represents the ovary wall. Only found in the Subfamily Maloideae of the Rosaceae.

pore - Small circular or elliptical opening.

poricidal capsule - Capsule that releases seeds by means of pores or flaps.

prickle - Sharp or pointed outgrowth of the epidermis that can occur anywhere on a plant.

primary growth - Growth resulting from the apical meristems of shoots and roots.

primordium - Cell or organ in its earliest stage of development and differentiation.

procambium - Primary meristematic tissue that gives rise to primary vascular tissue.

prop root - Adventitious root growing from the trunk that functions to support the plant.

protandry - Shedding of pollen by anthers before the stigma of the same flower is receptive.

prothallus - Photosynthetic gametophyte of ferns.

protogyny - Maturation and receptiveness of the stigma before the anthers of the same flower shed pollen.

pseudanthium - Cluster of reduced flowers that collectively imitate a single flower.

pseudobulb - Expanded and bulb-like stem of many of the Orchidaceae.

pseudostem - Tightly overlapping leaf bases surrounding the actual stem in many of the Zingiberales.

pubescent - Hairy.

pulvinus - Thickened portion of a petiole or petiolule that functions in movement of the leaf or leaflet.

pyrene - Inner hard portion of a drupe containing the seed. Equivalent with **pit** and **stone**.

pyxis - Capsule that dehisces along a horizontal line that circumscribes the entire fruit, resulting in the top portion falling off like a lid. Equivalent with **circumscissile capsule**.

raceme - Indeterminate inflorescence of stalked flowers from a single axis.

rachis - In ferns, the axis of a leaf or frond that bears the pinnae. In compound leaves, the extension of the petiole that bears the pinnae.

radial symmetry - Form of symmetry that permits a flower or structure to be divided into equal halves by two or more planes. Equivalent with **actinomorphic** and with **regular** when applied to flowers.

ray floret - Bilateral, sterile or carpellate flower with corolla flat and strap-like with 2-3 apical lobes or teeth. No stamens are present.

receptacle - Apex of the flower axis that bears the parts of the flower.

recurved - Curved backwards or downwards.

reduplicate - Creased or folded downwards so that the segment appears A-shaped in cross section with the underside or abaxial surface within, as with the segments of some palm leaves.

reflexed - Bent backward or downward, away from the main axis.

regular symmetry - Form of symmetry that permits a flower or srtucture to be divided into equal halves by two or more planes. Equivalent with **actinomorphic** or **radial symmetry**.

reniform - Shaped like a kidney.

replum - Thickened outer rim that forms a persistent part of the silique fruits of the Brassicaceae and of some Papaveraceae fruits.

resupinate - Twisted 180 degrees during development resulting in the structure being upside-down.

reticulate venation - Pattern of venation that is net-like in appearance. Typically tertiary veins that branch off the secondary veins but can include higher order veins.

retinaculum - Thickened funiculus that aids in dispersing the seed from the fruit, such as in species of the Acanthaceae. Equivalent with **jaculator**.

retuse apex - Rounded or blunt apex with a small notch.

revolute - Margin of a leaf or other structure that is rolled abaxially (or towards the underside).

rhizome - Horizontal stem, usually underground but also occurring on the soil surface.

root - Part of the plant axis that is usually descending and below ground, lacking nodes and leaves, and serving to conduct water and other materials from the environment into the plant while anchoring it.

root nodule - Small nodule containing bacteria that take nitrogen from the environment and convert it to a form usable to the plant. Found on the roots of legume plants (Fabaceae) and in a few other families.

rosette - Circular or spiral cluster of leaves arising very close to one another on a stem, often basal or at ground level.

rotate flower - Corolla with petals united at their bases and the flattened limbs at right angles to the corolla tube.

rotate venation - Pattern of venation where the main veins project from a central point like the spokes of a wheel. Only found on peltate leaves.

rounded apex / base - Apex or base that has a round shape without points or notches.

ruminate endosperm - Endosperm with a wrinkled pattern of irregular ridges.

runner - Typically slender and elongated horizontal stem that grows on or near the surface of the ground, producing new plants at the nodes and the tip.

saccate - Shaped like a pouch or bag.

sagittate - Triangular with basal lobes. Shaped like an arrowhead.

salverform flower - Corolla with petals united at their bases forming a narrow tube and the limbs flattened and extending outward.

samara - Dry indehiscent winged fruit that contains a single seed.

scabrous - Rough texture, like a cat's tongue.

scale - Broad term for small leaves, bracts, and other structures.

scape - Bare, upright stem that bears a flower or inflorescence at its tip.

schizocarp - Typically dry but sometimes fleshy fruit derived from two or more carpels that split into two or more mericarps, each of which contains one seed.

sclerenchyma cells - Cells with diverse size and shape that have a thickened secondary cell wall.

scorpioid cyme - Coiled cyme where lateral flowers or branches develop on opposite sides of the main axis.

secondary growth - Growth resulting from lateral meristems that results in stem diameter increasing.

secondary vein - Vein that diverges or branches off from a primary vein.

seed - Final form of an ovule after fertilization.

self-fertilization - When the sperm and egg that unite come from the same plant.

self-pollination - When both the pollen and the stigma it lands on are from the same plant.

sensu lato - To interpret widely or in the broad sense. Abbreviated **s.l.**

sensu stricto - To interpret narrowly or in the strict sense. Abbreviated **s.s.**

sepal - One of the outer perianth parts that collectively form the calyx, and that serve to protect the other flower parts while they are in the bud.

septicidal capsule - Capsule that splits along the septa to release the seeds, resulting in segments or valves that represent a single entire locule without a septum dividing it internally.

septifragal capsule - Capsule with longitudinal valves that split away from the septa.

septum - Partition or dividing wall, especially between adjacent locules.

sericeous - Silky.

serrate - Margin with teeth that point forwards (towards the apex).

sessile - Direct attachment of one structure to another without a stalk.

shoot - Stem or branch.

short shoot - Stem or branch with short internodes on a plant where both long and short internodes occur. (Clusters of leaves on a short shoot may appear as a fascicle or whorl.)

shrub - Woody perennial plant of moderate height with multiple stems or trunks.

silicle - Fruit produced from a gynoecium with two carpels, with the seeds attached to a persistent thickened rim (replum) and membrane. Found only in the Brassicaceae. Equivalent with **silique**, only shorter.

silique - Fruit produced from a gynoecium with two carpels, with the seeds attached to a persistent thickened rim (replum) and membrane. Found only in the Brassicaceae. Equivalent with a **silicle**, only longer.

simple fruit - Fruit that develops from one carpel, or from several fused carpels of one flower.

simple inflorescence - Inflorescence with only one axis.

simple leaf - Leaf with only a single blade and not divided into leaflets.

sorus - Small cluster of sporangia appearing as brown to yellow spots or lines on the underside of a fern leaf.

spadix - Thickened spike axis of certain inflorescences, such as those found in the Araceae.

spathe - Large foliaceous bract that subtends or surrounds the inflorescence, such as found in the Araceae.

spike - Inflorescence with sessile flowers or flower clusters arranged on a central axis.

spikelet - Small spike and characteristic inflorescence of the Poaceae and Cyperaceae.

spine - Sharp modified leaf or stipule that arises at a node.

sporangiophore - Stalk that bears sporangia.

sporangium - Hollow typically thin-walled structure that produces spores.

spore - Reproductive cell produced in a sporangium that is capable of developing into an adult by itself.

sporocarp - Reproductive structure that is a modified leaf found in ferns of the Marsileaceae.

sporophyll - Modified leaf that bears sporangia.

sporophyte - Diploid phase of the plant reproductive cycle that produces spores.

spur - Outgrowth of a stamen.

stamen - Organ that produces pollen, consisting of pollen sacs at the end of a stalk or filament. Male reproductive organ.

staminate flower - Flower with a functional androecium but lacking a functional gynoecium. Equivalent to a male flower.

staminode - Sterile stamen, sometimes petaloid in shape, sometimes glandular.

standard petal - Largest and uppermost petal on the papilionaceous flower of the Subfamilies Faboideae and Caesalpinoideae in members of the Fabaceae. Equivalent with the **banner petal** or the **flag petal**.

stele - Central cylinder containing the vascular tissue of roots and stems.

stem - Portion of plant axis with leaves and nodes, typically occurring above ground.

sterile stem - Stem that is not bearing fruit or reproductive structures.

stigma - Apical part of the pistil that receives the pollen.

stigmatic crest - Pollen-receptive area on stigmas that are horn-like or with extended shapes.

stipe - Supporting stalk or the petiole of a fern frond.

stipule - Often leaf-like structure that occurs in pairs or surrounds the base of the petiole on the stem.

stipule scar - Scar on the stem indicating where a stipule was attached.

stolon - Typically slender and elongated horizontal stem that grows on or near the surface of the ground, producing new plants at the nodes and tips. Equivalent with **runner**.

stoma - Tiny pore in the epidermis of leaves and stems through which gases pass. Surrounded by two guard cells that control its opening and closure.

stomatal complex - Collective term for the stoma and its associated cells, such as the guard cells.

storage root - Large swollen root that serves as a storehouse for plant carbohydrates.

strigose - Bearing stiff, unidirectional hairs.

strobilus - Apical cluster of reproductive leaves shaped like a cone.

style - Elongated portion of the pistil that connects the stigma with the ovary.

substomatal chamber - Air space below a stoma.

succulent - Plant with thick, fleshy and relatively soft leaves and/or stems modified for saving water.

suffrutescent - Plant that is woody at the base but with herbaceous shoots. Intermediate between an herb and a woody plant.

superior ovary - Ovary located in a position above where the perianth parts are attached to the receptacle or flower stalk, or above the basal point of a hypanthium or floral cup.

superposed bud - Bud located adjacent to an axillary bud, can be above or below.

syconium - Multiple and accessory fruit of figs (genus *Ficus* in the Moraceae) consisting of a hollow pear-shaped structure with the flowers and fruits on the inner surface.

sympetalous flower - Flower with the petals fused, at least at the base.

synandrous - Stamens fused or connate.

synangium - Compound structure of fused sporangia found in some ferns and fern allies.

syncarp - Fruit formed from the fusion of the carpels of one or more flowers and their associated parts resulting in a single structure.

syncarpous - Carpels fused or connate.

syngenesious - Connate stamens attached at the anthers and forming a tube or cylinder.

synsepalous flower - Flower with the sepals fused, at least at the base.

syntepalous flower - Flower with the tepals fused, at least at the base.

taproot - Root or root system characterized by one large, downward-growing root that is significantly larger than the lateral roots growing in association with it.

tendril - Modified stem, leaf, or inflorescence that has elongated and functions in supporting the plant. Typically coiled or twining or with adhesive sucker disks.

tepal - One of the outer floral parts when the perianth is not divided into petals and sepals.

terete - Circular or oval in cross section, cylindrical.

terminal - Located at the tip or apical portion of a structure.

terminal bud - Bud occurring at the tip of a stem.

tertiary vein - Vein that diverges or branches off from a secondary vein, often forming a reticulate pattern with other tertiary and quaternary veins.

tetradynamous stamens - Androecium of six stamens with the two outer ones shorter than the four inner ones.

thallus - Undifferentiated plant body that is often flat and leaf-like.

thorn - Sharp modified branch originating at nodes from the leaf axil.

three-ranked - Structures placed on three sides of an axis, such as leaves along a stem. Equivalent with **tristichous**.

tomentose - Densely covered with soft hairs.

toothed - Having a serrate or dentate margin. Bearing teeth.

translator arm - Slender structure that connects the pollinia of two adjacent anthers, as in species of the Apocynaceae.

tree - Perennial woody plant, typically medium to large, with a single trunk that is usually without branches near the base.

trichome - Hair.

trifoliate - Compound leaf with three leaflets.

trigger hair - Hair found on carnviorous plants which causes an insect-trapping mechanism to be activated when it is touched.

tripinnate - Pinnately compound leaf with compound leaflets.

tristichous - Structures placed on three sides of an axis, such as leaves along a stem. Equivalent with **three-ranked**.

trumpet-shaped - Shaped like a tube that flares out at the end like a trumpet.

truncate apex / base - Apex or base that ends abruptly as if it were cut off squarely or straight across.

trunk - Portion of the tree stem between the ground and the branches. Large vertical stem.

tuber - Thick, fleshy underground part of a stem that functions in water and food storage.

tubular - Shaped like a tube.

twining - Climbing without tendrils by using the stem to spiral around a support plant or object.

two-ranked - Structures placed on two sides of an axis, such as leaves along a stem. Equivalent with **distichous**.

umbel - Inflorescence with multiple branches that originate at the same point and terminate in a flower or secondary umbel. In the latter case, the inflorescence is a **compound umbel**.

unifoliate - Compound leaf with one leaflet that often can be differentiated from a simple leaf by a joint or pulvinus where the blade meets the petiole.

unisexual flower - Flower lacking either a functional gynoecium or a functional androecium.

urceolate - Shaped like an urn.

utricle - Indehiscent dry fruit, typically small, with a thin wall that is loose from the single seed.

valvate - Meeting cleanly at the edges without overlapping.

valve - Segment of a fruit that separates from the other valves at dehiscence to release the seeds.

vascular - Referring to conductive structures or tissues such as xylem, phloem, and vascular cambium.

vascular bundle - Collective term for tissue containing xylem and phloem, usually surrounded by a sheath of parenchyma cells.

vascular cambium - Cylinder of dividing meristematic cells that produce secondary xylem and secondary phloem, increasing the girth of a stem. Equivalent with **vascular cylinder**.

vein - Vascular bundle forming part of the conductive system of plants.

velamen - Absorbent epidermis of the aerial roots of some species of Orchidaceae and Araceae.

venation - Arrangement of veins in a leaf.

ventral - Facing towards the axis. Equivalent with **adaxial**.

vestiture - Covering of a surface, such as with hairs.

villose - Dense covering of long, fine hairs. Also spelled **villous**.

vine - Herbaceous trailing or climbing plant. When woody, a vine is called a **liana**.

viscin - Plant substance that is elastic or sticky or both.

viviparous - Germination of fruit while still attached to the parent plant, as in some mangroves.

xerophyte - Plant adapted to arid or dry habitats and conditions.

xylem - Vascular tissue that conducts most of the water and minerals through a plant.

whorl - Group of structures attached at the same point or level on a central axis or stem.

whorled leaves - Three or more leaves at a single node.

wing - Membranous extension of a structure.

wing petal - One of a pair of lateral petals in the papilionaceous flowers of the subfamilies Faboideae and Caesalpinoideae in the Fabaceae. Also used to refer to the lateral sepals of some Polygalaceae.

woody - Stiff or hard, usually due to the presence of secondary xylem.

zygomorphic symmetry - Flower or structure that is divisible into two equal halves by only a single plane. Equivalent with **bilateral symmetry** and **irregular**, when applied to flowers.

Botany Textbooks and Reference Books

Photographic Atlas Of Botany And Guide To Plant Identification. 2004. Castner, J.L. Feline Press. Gainesville, FL.

Medicinal And Useful Plants Of The Upper Amazon. 1998. Castner, J.L., Timme, S.L., and J.A. Duke. Feline Press. Gainesville, FL.

Plant Systematics - A Phylogenetic Approach. 1999. Judd, W.S., Campbell, C.S., Kellogg, E.A., and P.F. Stevens. Sinauer Associates. Sunderland, MA.

Guide To Flowering Plant Families. 1994. Zomlefer, W.B. University of North Carolina Press. Chapel Hill, N.C.

Biology Of Plants (6/e). 1999. Raven, P.H., Evert, R.F., and S.E. Eichhorn. Worth Publishers. New York, N.Y.

Contemporary Plant Systematics (3/e). 2000. Woodland, D.W. Andrews University Press. Berrien Springs, MI.

Vascular Plant Taxonomy (4/e). 1975. Walters, D.R. and D.J. Keil. Kendall-Hunt Publishing Co. Dubuque, IA.

Common Families Of Flowering Plants. 1997. Hickey, M. and C. King. Cambridge University Press. New York, N.Y.

Plant Identification Terminology - An Illustrated Glossary (2/e) 2000. Harris, J.G. and M.W. Harris. Spring Lake Publishing. Spring Lake, UT.

Flowering Plants Of The World. 1978. Heywood, V.H. (Editor). Mayflower Books. New York, N.Y.

How To Know The Ferns And Fern Allies. 1979. Mickel, J.T. William C. Brown Co. Dubuque, IA.

A Field Manual Of The Ferns & Fern-Allies Of The United States & Canada. 1985. Lellinger, D.B. Smithsonian Institution Press. Washington, D.C.

FAMILY INDEX